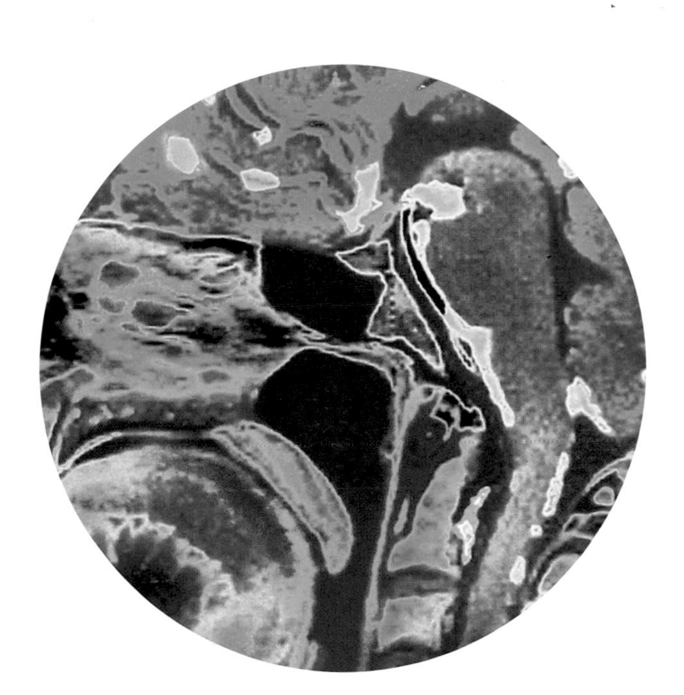

ENCYCLOPEDIA OF THE HUMAN BODY

LONDON, NEW YORK,
MUNICH, MELBOURNE, and DELHI

for PAGE*One*:
Cairn House, Elgiva Lane,
Chesham, Bucks HP5 2JD

Creative director Bob Gordon
Managing editor Judith Hannam
Editors Naomi Mackay, Michael Spilling, Charlotte Stock
Art editor Robert Law
Designers Annie Hiew, Tim Stansfield
Special photography Andy Crawford
Digital artwork Anthony Duke, Mark Tattam, Peter Bull
Picture research John Farndon
With special thanks to Tania Fuller, Joe Gordon, Alex Jones,
Sophie Pateman, Adrian Phillips, and Henry Spilberg

for Dorling Kindersley:

Senior editors Selina Wood, Shaila Awan
Art editors Polly Appleton, Steven Laurie
Project editor Lucy Hurst
Managing art editor Clare Shedden
Managing editors Marie Greenwood, Andrew Macintyre
Digital artwork Robin Hunter
DTP design Siu Yin Ho
Production Kate Oliver
Picture research Sean Hunter
DK pictures Rose Horridge

Editorial consultant and author Richard Walker

Contributing authors David Burnie, Daniel Carter,
Phil Gates, Penny Preston, Frances Williams

First published in Great Britain in 2002
by Dorling Kindersley Limited
80 Strand, London
WC2R 0RL

A Penguin Company

2 4 6 8 10 9 7 5 3

A CIP catalogue record for this book
is available from the British Library.

ISBN 1-4053-05266

Colour reproduction by Colourscan, Singapore
Printed in Slovakia by Neografia

See our complete catalogue at

www.dk.com

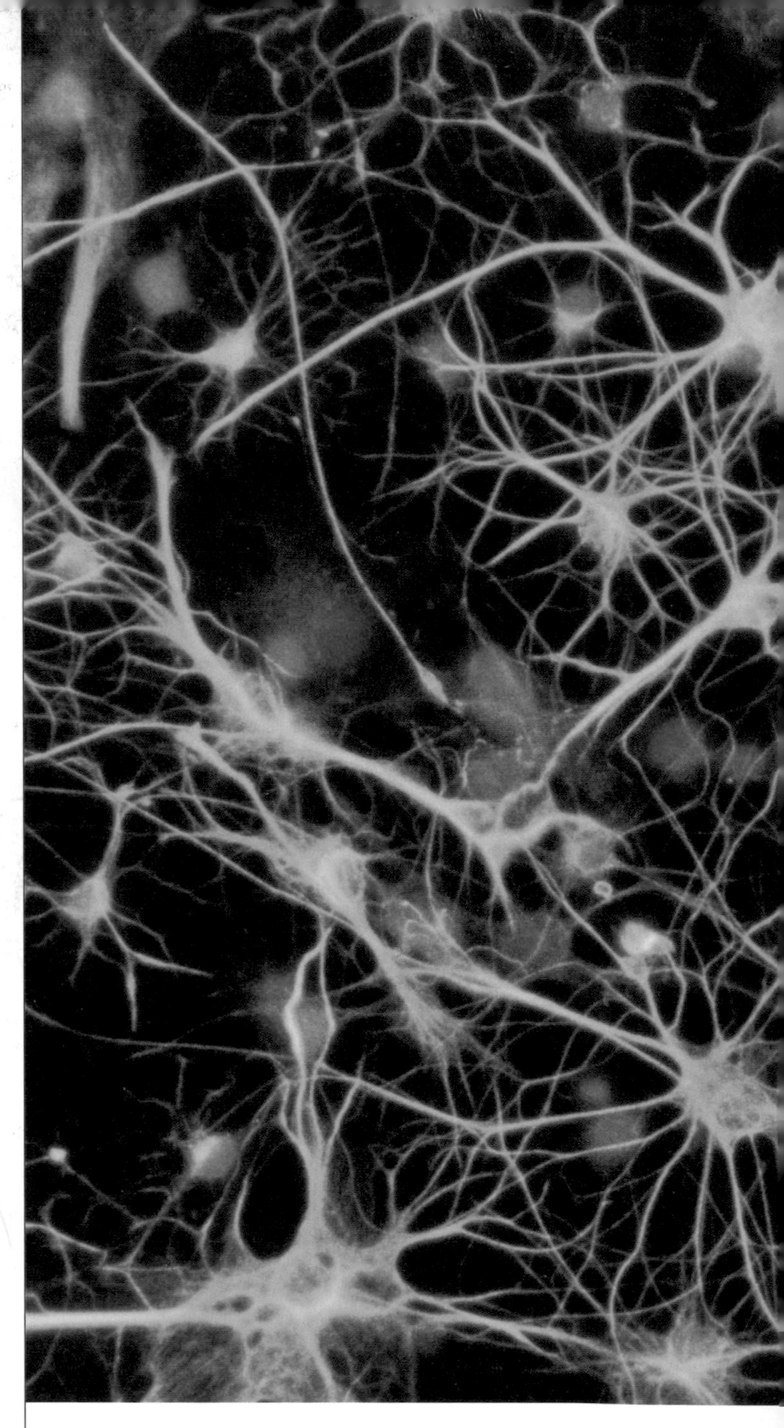

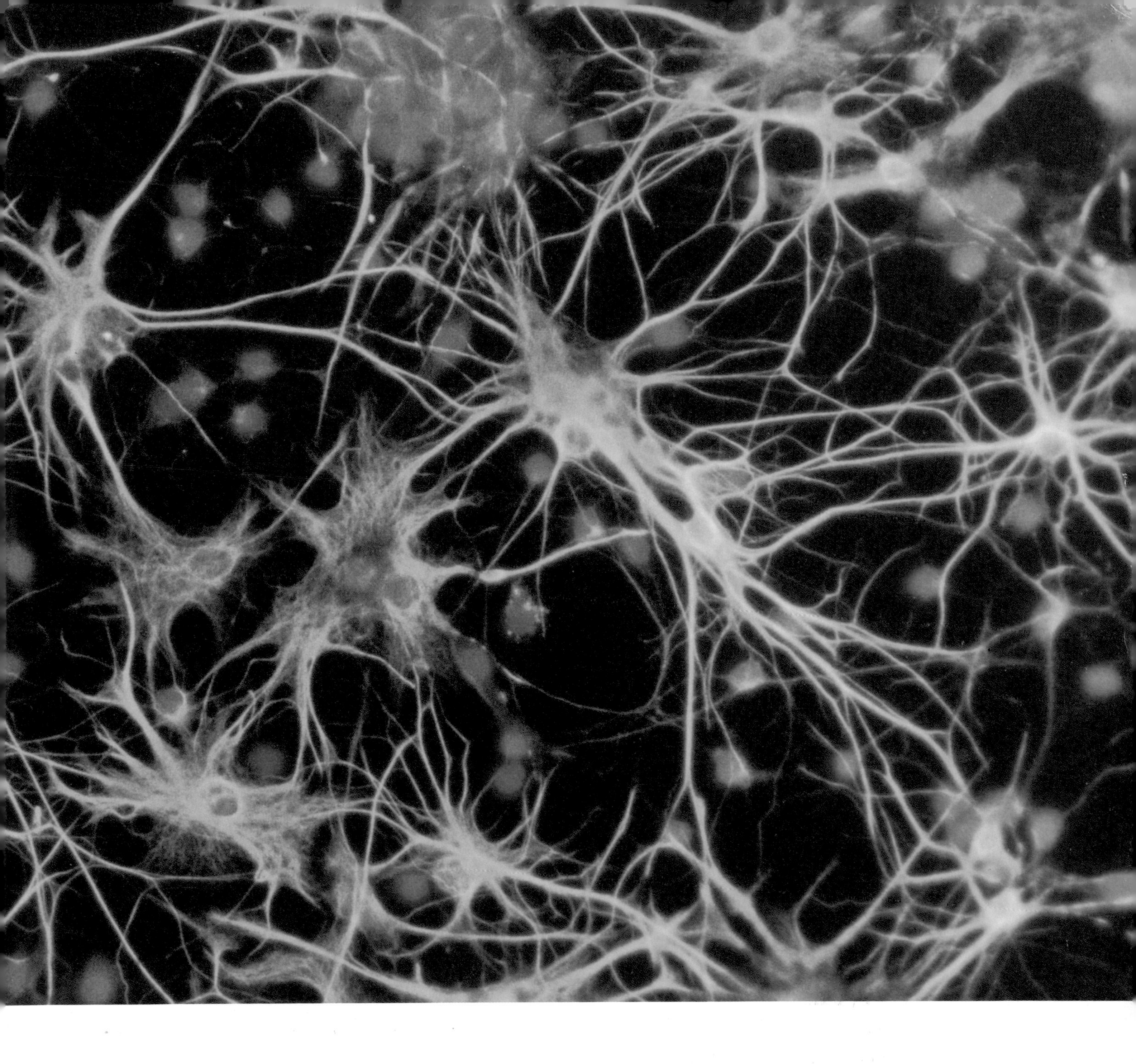

ENCYCLOPEDIA OF THE HUMAN BODY

Richard Walker

Contents

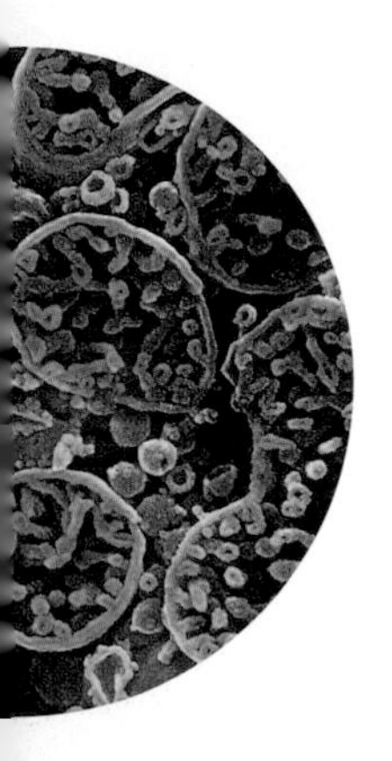

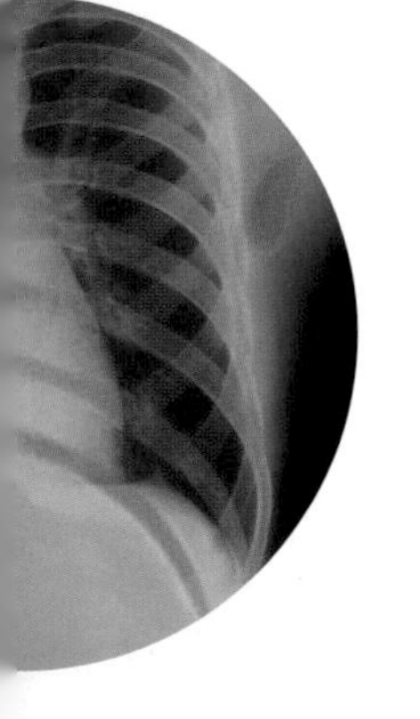

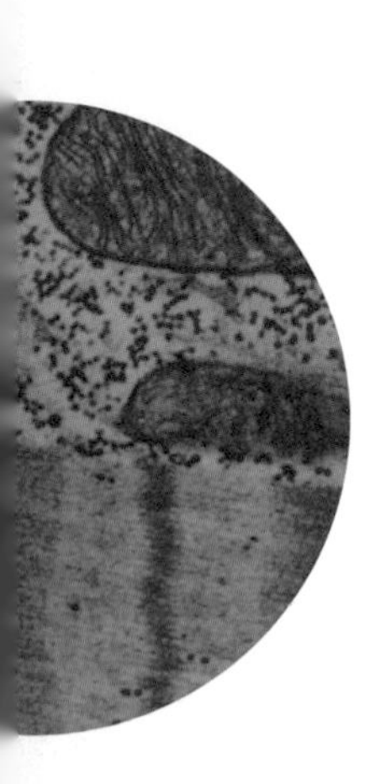

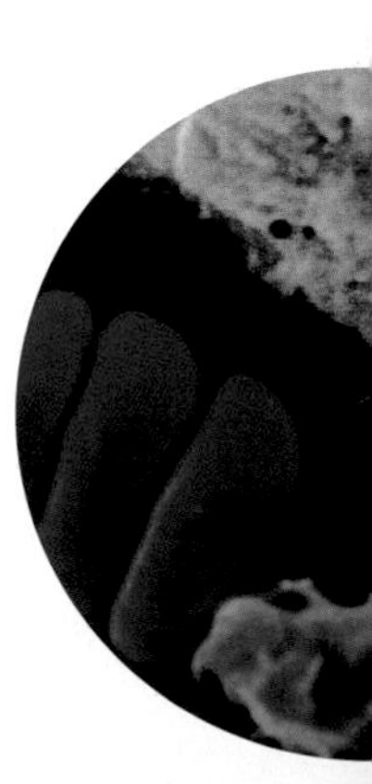

Contents

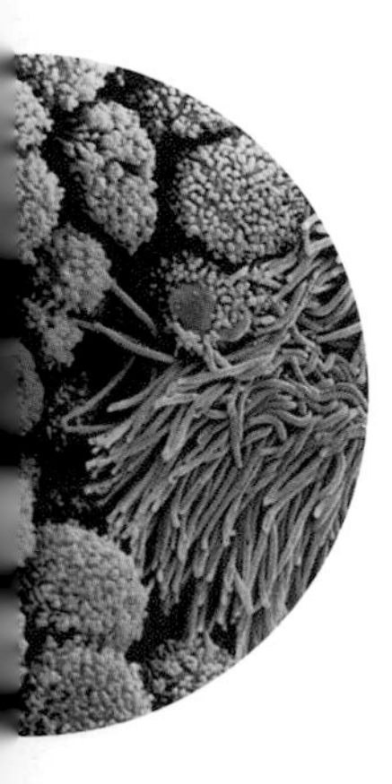

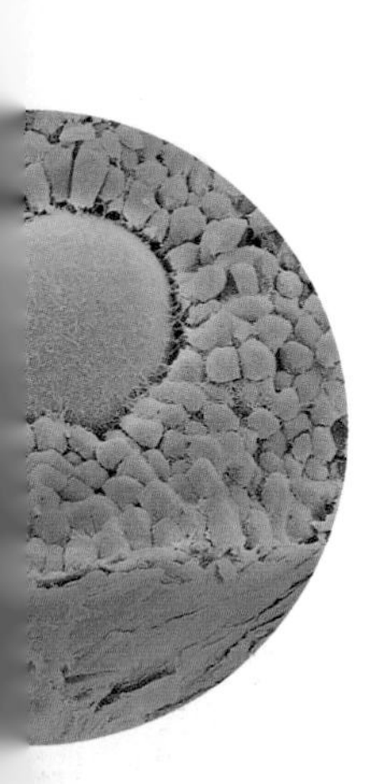

THE BODY THROUGH TIME 252–279

REFERENCE SECTION 280–304

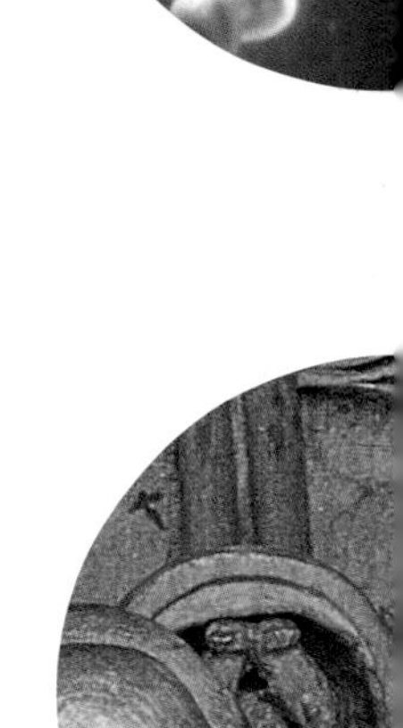

The following abbreviations have been used for technical imaging techniques:

CT = Computed Tomography
LM = Light micrograph
NMR = Nuclear Magnetic Resonance Imaging
MRI = Magnetic Resonance Imaging
PET = Positron Emission Tomography
SEM = Scanning Electron Micrograph
TEM = Transmission Electron Micrograph

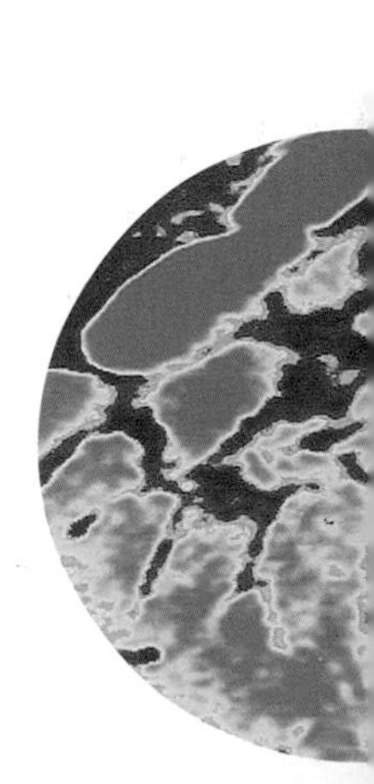

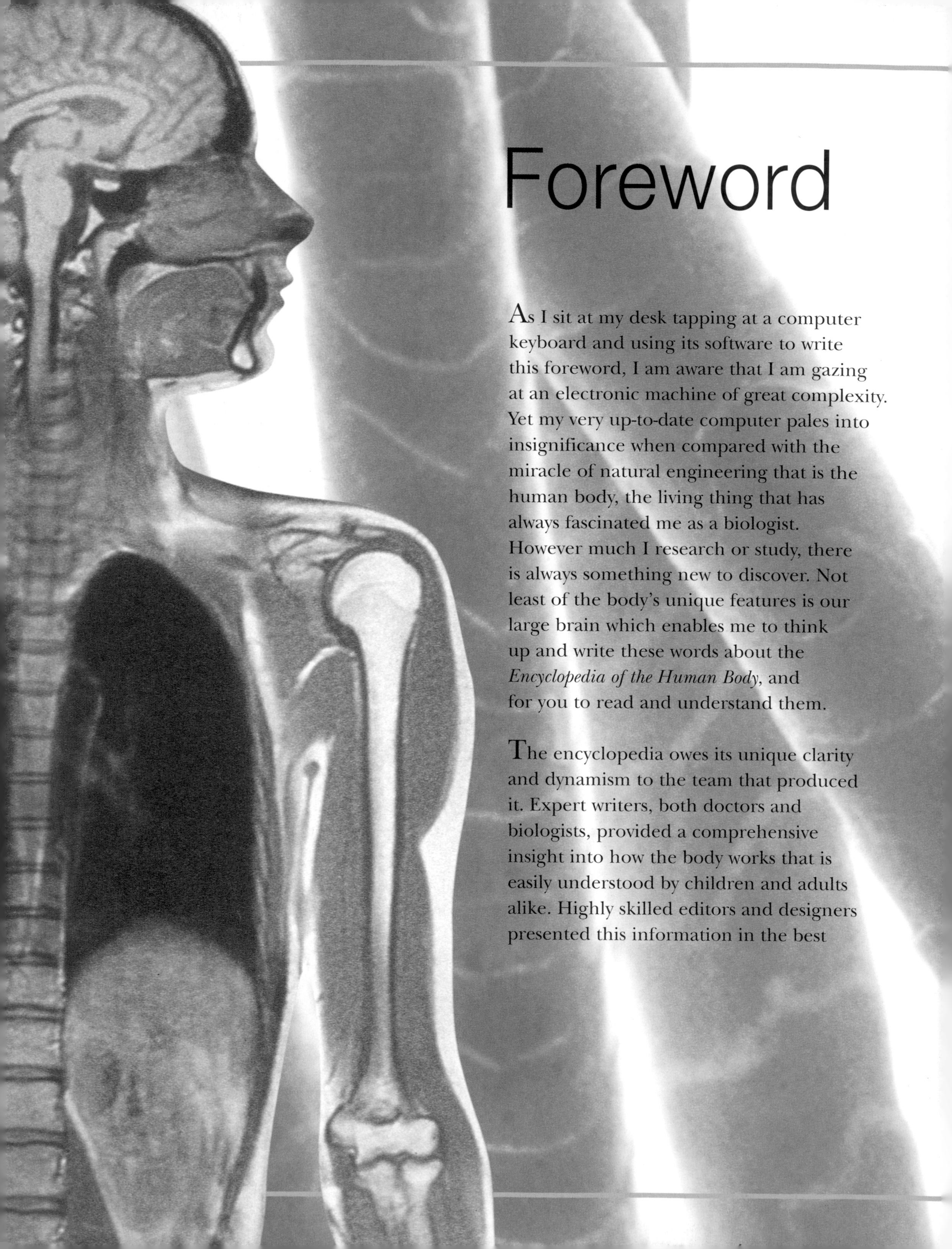

Foreword

As I sit at my desk tapping at a computer keyboard and using its software to write this foreword, I am aware that I am gazing at an electronic machine of great complexity. Yet my very up-to-date computer pales into insignificance when compared with the miracle of natural engineering that is the human body, the living thing that has always fascinated me as a biologist. However much I research or study, there is always something new to discover. Not least of the body's unique features is our large brain which enables me to think up and write these words about the *Encyclopedia of the Human Body*, and for you to read and understand them.

The encyclopedia owes its unique clarity and dynamism to the team that produced it. Expert writers, both doctors and biologists, provided a comprehensive insight into how the body works that is easily understood by children and adults alike. Highly skilled editors and designers presented this information in the best possible way, complementing the text with an incredible array of images that guide the reader on her or his journey through the body. These images include photographs, micrographs, scans, drawings, and diagrams that use colour and detail to, for example, reveal the slimy lining of the intestines, expose the hordes of bacteria on the skin, explain why we do, or do not, have blue eyes, and describe how a baby is born from first contraction to first yell.

To make all this fascinating information accessible for the reader, we have organized the encyclopedia into seven sections. The first five deal with how the body is constructed, how it is supported and moved, how it is controlled, how it maintains itself, and how it reproduces and changes during life. Throughout these sections there are selected pages that focus on specific issues, such as "Understanding the mind" or "Art and anatomy", adding extra breadth to the topics covered in each section.

The final, seventh section contains a detailed timeline and glossary, but it is the sixth section that adds a whole new dimension to this encyclopedia. It explores how understanding of the human body and the development of medicine have gone hand in hand from the earliest times to the modern age. It explains how we know and understand so many of the things we take for granted today – such as how blood circulates round the body or why people get ill – that were a mystery to our ancestors and took all their intelligence and creativity to resolve.

One day, somewhere in the future, a descendant of my desk computer may be smart enough to mimic the thoughts and actions of a human being, but, as you will see when you explore the body's remarkable structure and incredible workings in the *Encyclopedia of the Human Body*, it could never replace the real thing.

Richard Walker

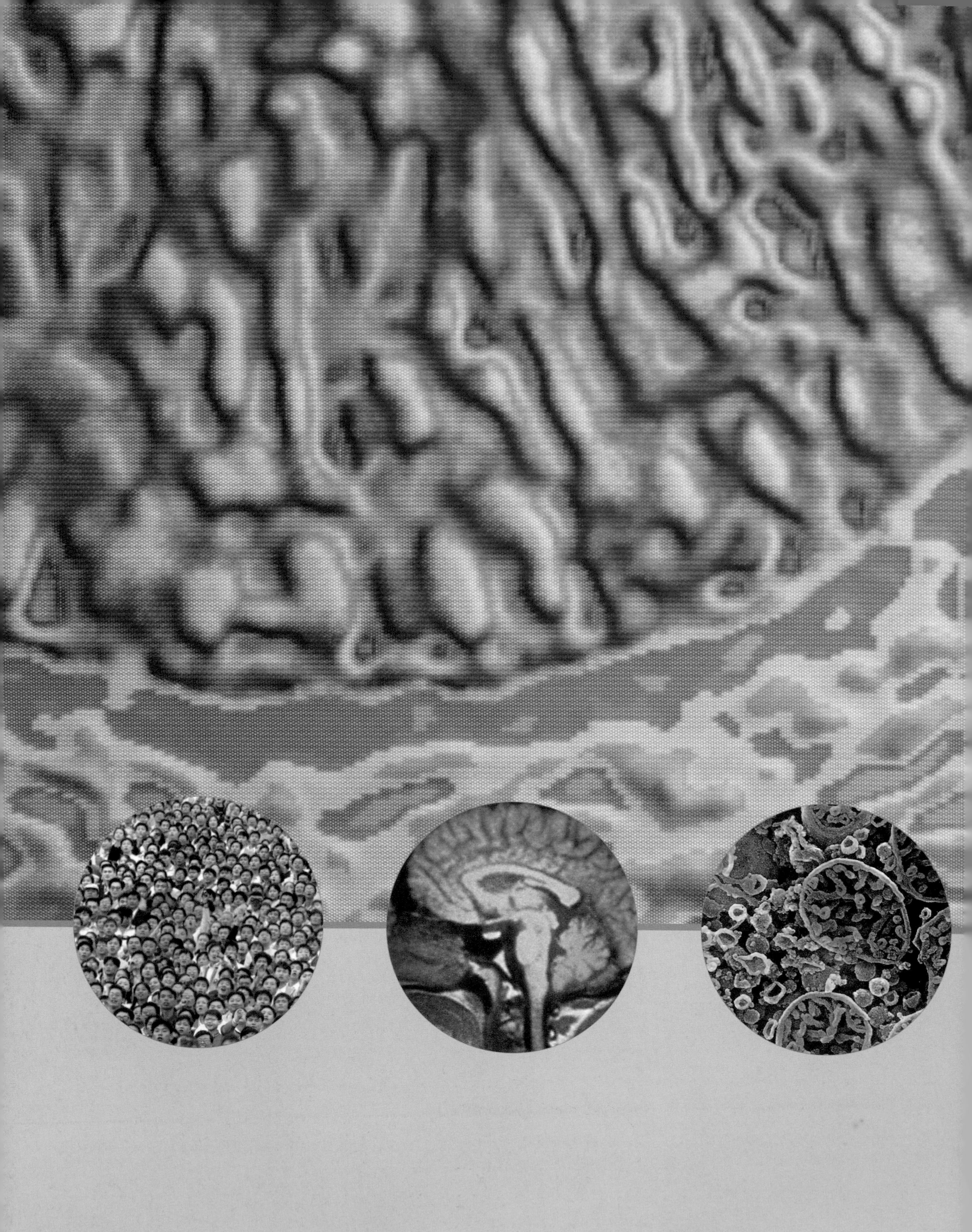

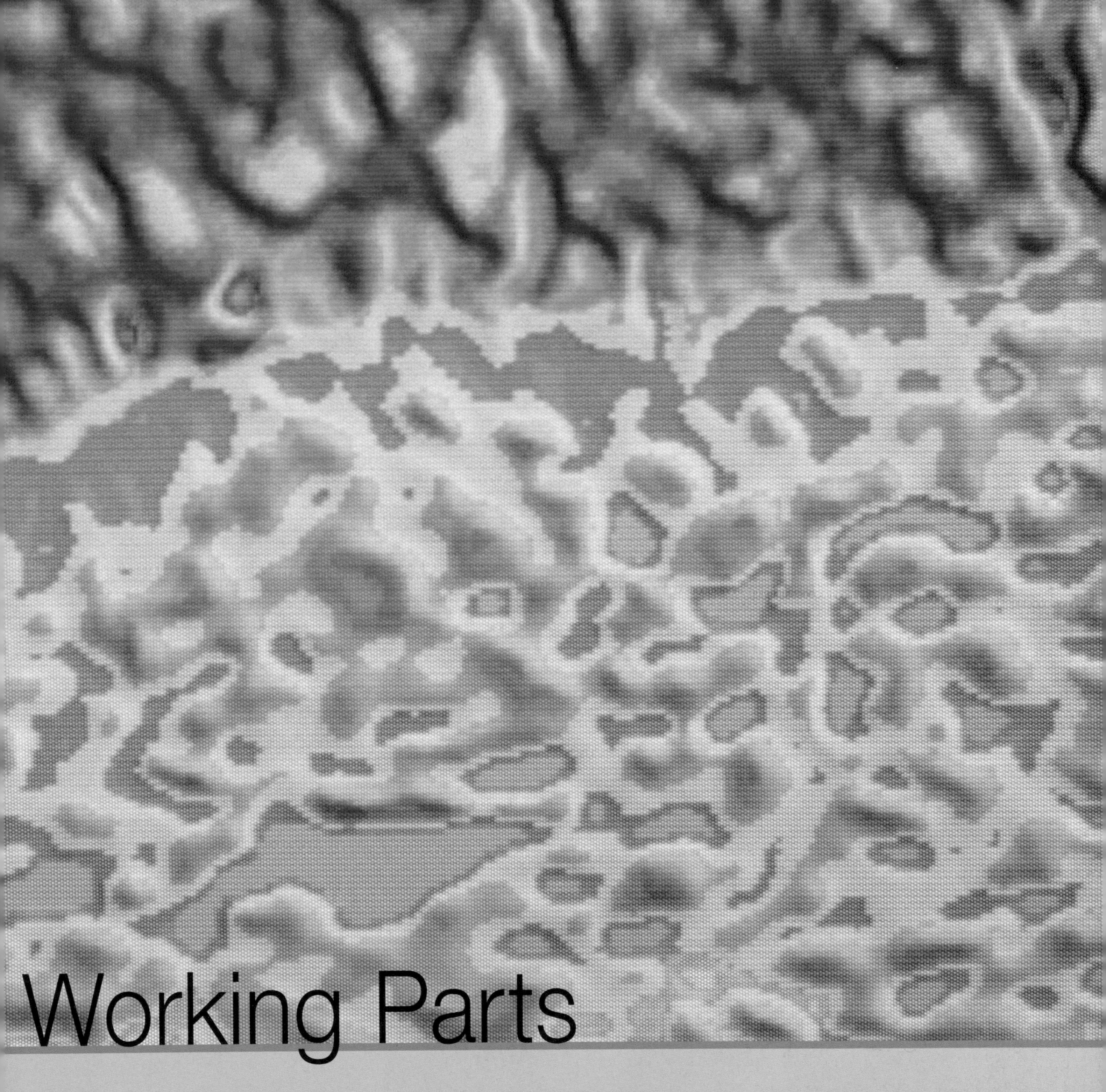

Working Parts

DESPITE THE INCREDIBLE VARIATION in body shapes and sizes, all humans share the same working parts. At the microscopic level, the human body is built from huge numbers of cells. Sharing the same basic structure, these tiny chemical plants are grouped together to make the organs that pump blood, digest food, take in air, and perform all other life-maintaining functions. The structure and workings of the living body can be explored, without cutting it open, using a range of different imaging techniques.

BEING HUMAN

HOMO SAPIENS – Latin for "wise man" – is the name used by scientists to identify you, the reader, and the other six billion humans on planet Earth as a species (type) of animal. It may come as a surprise to some people to realize that humans are actually animals, and just one of the 1.5 million named species that make up the animal kingdom. At the same time, humans stand apart from other animals because their brain power is so superior, even when compared to their close relatives, the apes. Their unique intelligence, communication skills, curiosity, and ability to problem-solve have enabled humans to colonize every continent, triumphing over environments and climates that have defeated other animal species, and to develop a complex understanding of themselves and the world around them.

SOCIAL ANIMALS

Humans are social animals that typically – although not always – live in family units led by a male and female partner, with one or more dependent children. Young humans usually remain with, and are nurtured by, their parents for about 18 years, until they are sufficiently mature and experienced to exist on their own. Several family units together form a larger social grouping or community, which can range in size from a village to a city. In modern industrial and agricultural societies, communities link together to form bigger units, the largest of which – nation states such as the United States or India – contain hundreds of millions of individuals.

Forward-facing eyes *are characteristic of chimpanzees, humans, and all other primates*

Chimpanzee (*Pan troglodytes*)

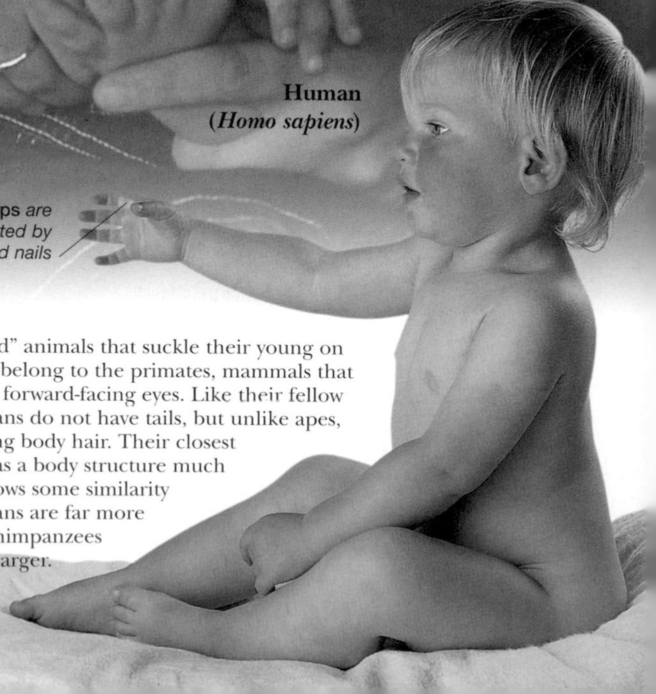

Human (*Homo sapiens*)

Fingertips *are protected by hard nails*

CLOSEST RELATIVES

Humans are mammals – hairy, "warm-blooded" animals that suckle their young on milk. Along with apes and monkeys, humans belong to the primates, mammals that have five fingers and toes tipped by nails and forward-facing eyes. Like their fellow apes, such as chimpanzees and gorillas, humans do not have tails, but unlike apes, they walk upright and lack long body hair. Their closest relative, the chimpanzee, has a body structure much like that of humans, and shows some similarity in behaviour. However, humans are far more intelligent and skilful than chimpanzees because their brain is much larger.

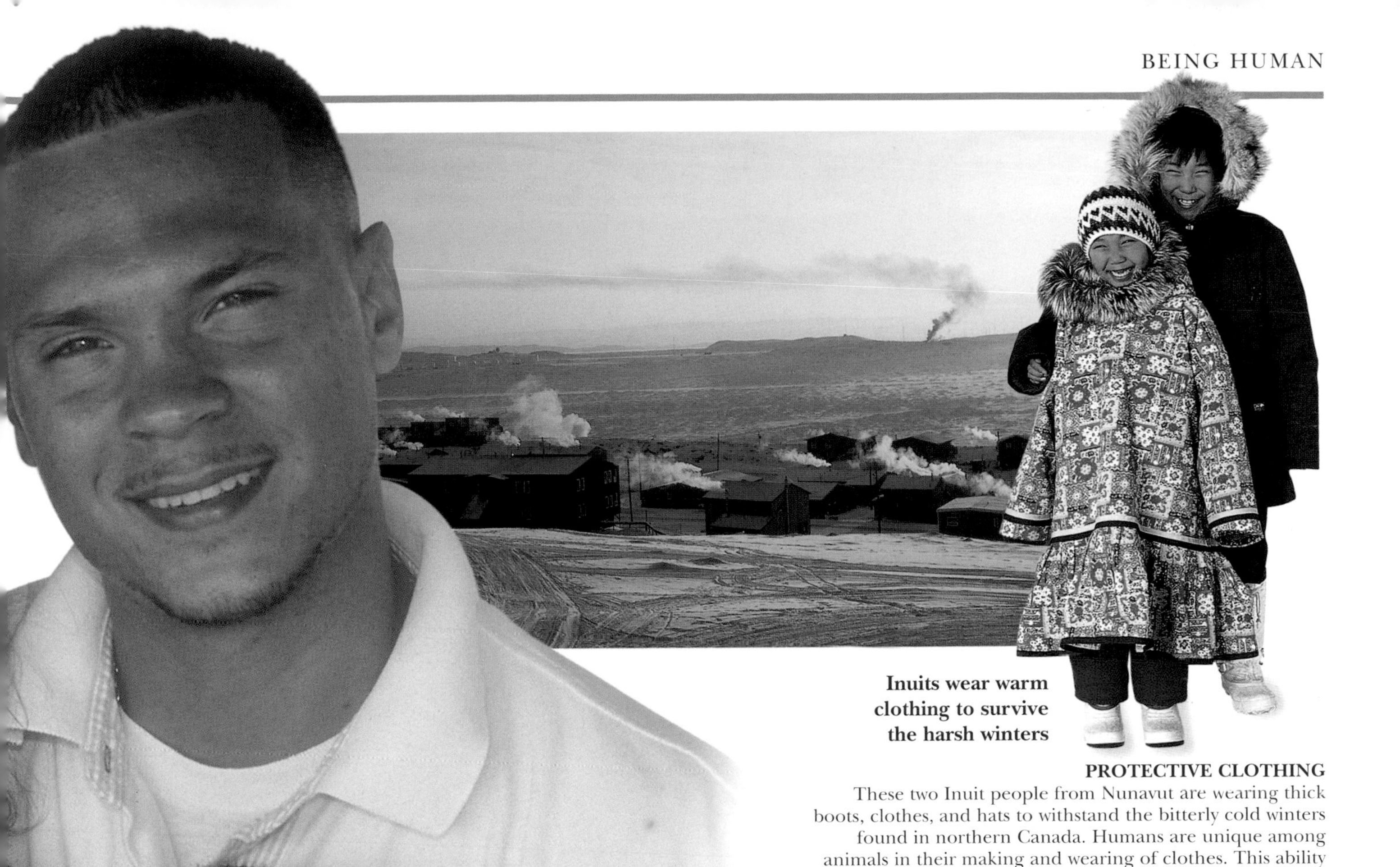

Inuits wear warm clothing to survive the harsh winters

PROTECTIVE CLOTHING

These two Inuit people from Nunavut are wearing thick boots, clothes, and hats to withstand the bitterly cold winters found in northern Canada. Humans are unique among animals in their making and wearing of clothes. This ability allowed humans to leave the confines of tropical Africa, where they first evolved, and move to cooler climates. While clothes have traditionally been made from animal and plant products, such as fur, wool, and cotton, today they are also made from synthetic materials such as nylon. Clothes do more than provide protection from the weather, they also send out messages about the wearer's status in society, their lifestyle, culture, and religious beliefs.

COMMUNICATION

The ability to communicate is essential for humans, as it is for all social animals, to ensure that they live together successfully. Like their ape relatives, humans use body language, gestures, and facial expressions to convey feelings and intentions to each other. But humans also have an additional, unique means of communication – language. By speaking and writing words they can share ideas, plans, decisions, memories, and wisdom. The existence of language means that information can be passed on from one generation to another, providing society with an ever-growing knowledge base.

HUMAN SPECIES FACT FILE

Class	*Mammalia (mammals)*	Body weight (average)	*Males 75 kg (165 lb); females 52 kg (115 lb)*
Order	*Primates*	Activity	*Ground-living, diurnal (active during the day)*
Species	*Homo sapiens*	Reproduction	*Normally 1 young per litter*
Distribution	*Worldwide*	Maximum lifespan	*90–100 years*
Habitat	*Most land habitats, living in houses and other shelters*	Conservation status	*Not endangered; population increasing worldwide*
Food	*Animals, plants, and their products*		

Keeping in touch by mobile phone

NATURAL VARIATION

IMAGINE STANDING IN a busy city watching the stream of humanity flowing past. One obvious thing an observer would notice is the sheer range of body variation. Everyone, unless they have an identical twin, has a unique combination of features – including their height, weight, shape, hair colour and texture, skin tone, eye colour, and the sound of their voice. However, these are but variations on a central theme that remains constant. Every human body is constructed to a fixed pattern and works in the same way, with minor differences between males and females. Externally, the basic design of a human is an upright body supported by two legs and feet, with two arms and hands that carry and hold, and a flattened face.

BODY REGIONS

Both female and male humans share the same body regions, although their overall shapes and reproductive organs differ. The main axis of the body is made up of the head and trunk. The head contains and protects the brain and sense organs, and is linked by the neck to the trunk, the centre of the body. The thorax, or chest, forms the upper part of the trunk and contains the lungs and heart. The lower part, the abdomen, contains the digestive, reproductive, and urinary organs. Attached to the trunk are the limbs – the arms and legs.

Head *contains the brain, which coordinates the body's movements and produces thoughts*

Neck *holds the head upright and connects it to the trunk*

Thoracic cavity, *within the thorax, contains the lungs and heart*

Abdominal cavity, *within the abdomen, contains most digestive organs*

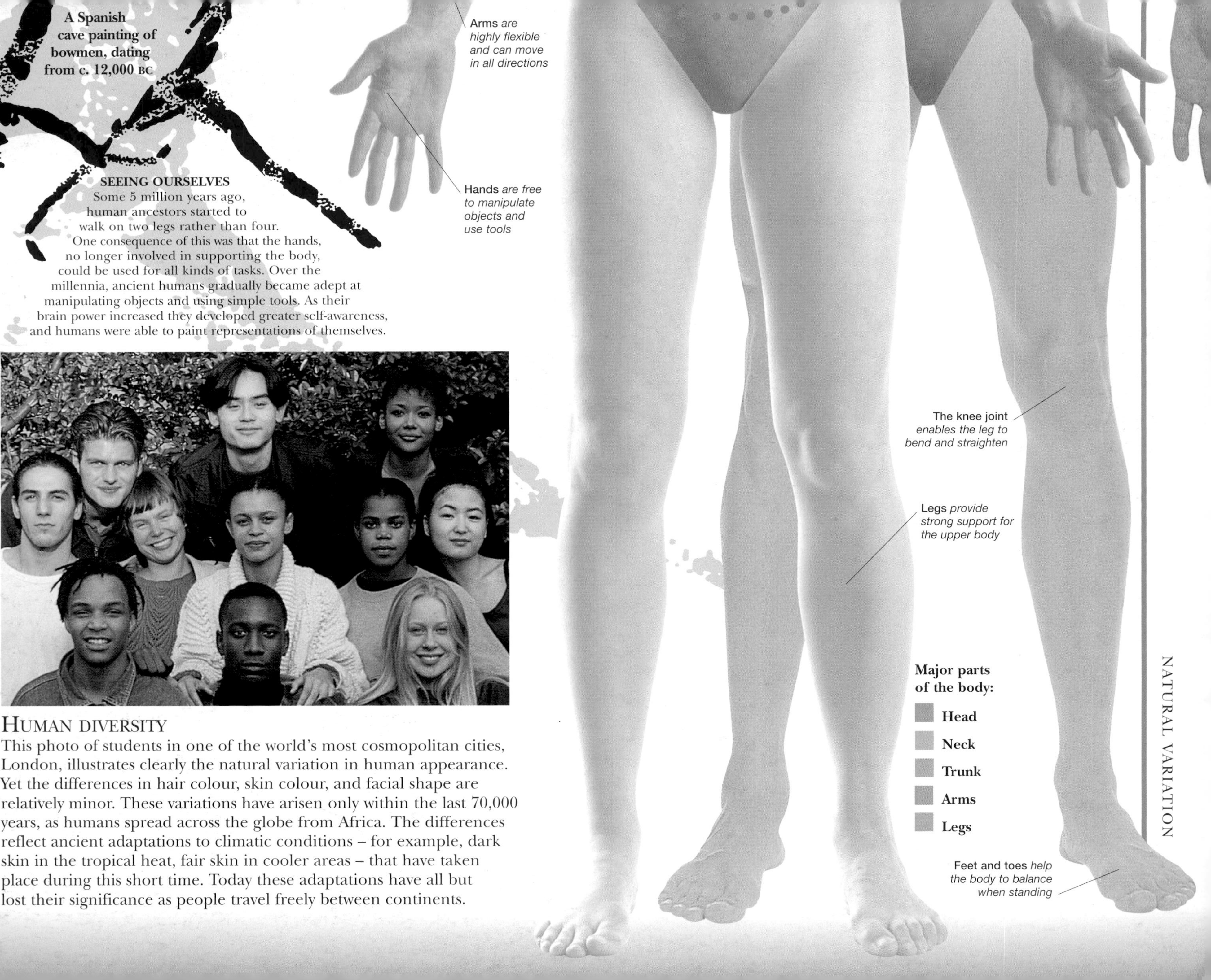
Arms are highly flexible and can move in all directions
Hands are free to manipulate objects and use tools
The knee joint enables the leg to bend and straighten
Legs provide strong support for the upper body
Major parts of the body:
Head
Neck
Trunk
Arms
Legs
Feet and toes help the body to balance when standing

A Spanish cave painting of bowmen, dating from c. 12,000 BC

SEEING OURSELVES

Some 5 million years ago, human ancestors started to walk on two legs rather than four. One consequence of this was that the hands, no longer involved in supporting the body, could be used for all kinds of tasks. Over the millennia, ancient humans gradually became adept at manipulating objects and using simple tools. As their brain power increased they developed greater self-awareness, and humans were able to paint representations of themselves.

HUMAN DIVERSITY

This photo of students in one of the world's most cosmopolitan cities, London, illustrates clearly the natural variation in human appearance. Yet the differences in hair colour, skin colour, and facial shape are relatively minor. These variations have arisen only within the last 70,000 years, as humans spread across the globe from Africa. The differences reflect ancient adaptations to climatic conditions – for example, dark skin in the tropical heat, fair skin in cooler areas – that have taken place during this short time. Today these adaptations have all but lost their significance as people travel freely between continents.

UNDER THE MICROSCOPE

THE END OF THE 16th century brought with it an invention that would open up a new world for scientists and doctors. In 1590, Dutch instrument maker Zacharias Janssen (1580–1638) made a magnifying device – later to be called the microscope – that, for the first time, made visible objects that were too small to be seen with the naked eye. Initially Janssen's invention made little impact, but in time it would be used to reveal the existence of cells and other previously unseen features of the living world. In the 20th century, the invention of the electron microscope took research into the microworld a step further.

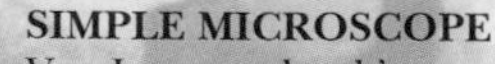

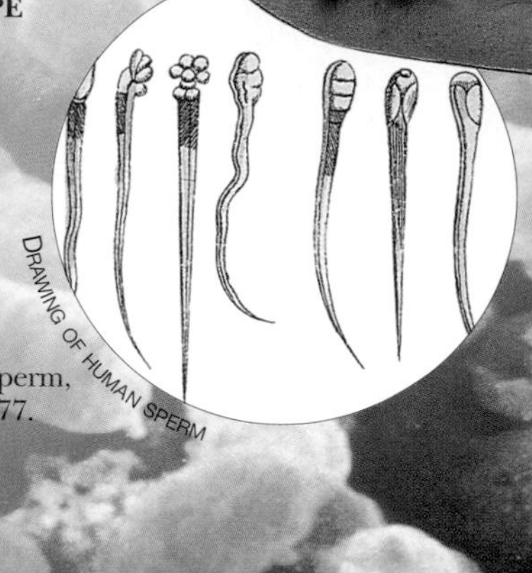

SIMPLE MICROSCOPE
Van Leeuwenhoek's microscope (right) was held upright with the eye close to the lens. The object to be examined was placed in front of the lens on a pin which was brought into focus by a series of screws. His drawing shows human sperm, discovered by him in 1677.

THE DRAPER'S LENS

Following on from Janssen's invention, Dutch cloth merchant Antoni van Leeuwenhoek (1632–1723) made a simple microscope by clamping a tiny convex lens between two brass plates. Although this device sounds primitive, it had a magnifying power of between 70 and 250 times. Van Leeuwenhoek's observations revealed the existence of single-celled organisms (now called protists), as well as some types of body cell. In 1683, he noted minute organisms in his own tooth scrapings, the first bacteria to be seen by the human eye.

HOOKE'S DRAWING OF A CORK SECTION

B

A

SEM OF LIGHT-SENSITIVE CELLS IN THE RETINA OF THE EYE

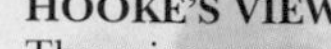

HOOKE'S VIEW
The microscope made by Robert Hooke (left) consisted of a pasteboard barrel, with an eyepiece lens at the top end and an objective lens at the bottom. Here, the specimen is lit by an oil lamp focused through a water-filled sphere. Hooke's sketch of a section through cork shows the chambers he called "cells".

Compound microscope

Despite van Leeuwenhoek's success, the future of microscopy lay in compound microscopes that used two or more lenses to produce their magnifying effects. British physicist Robert Hooke (1635–1703) made his own compound microscope and, in 1665, presented his observations in a book, illustrated with his own lavish drawings, called *Micrographia.* Among these observations was a section through cork, a dead plant material, showing tiny boxes that he called "cells". This term would later come into common use for a different purpose.

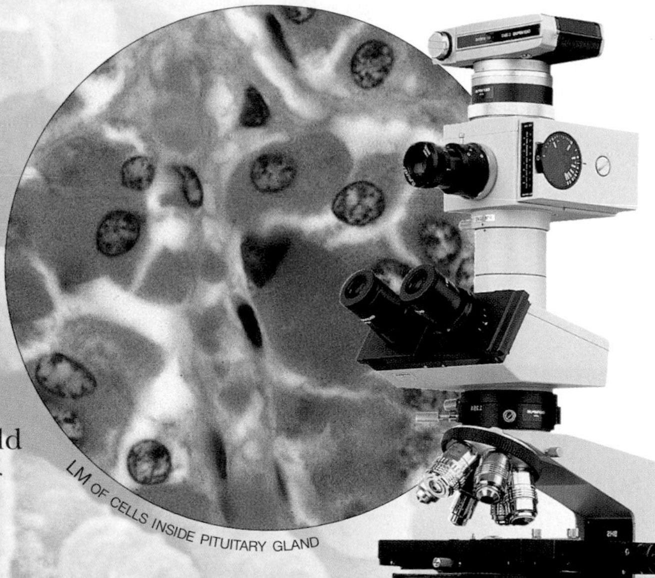

LM OF CELLS INSIDE PITUITARY GLAND

Seeing cells

Hooke's microscope and similar instruments of the time were hindered by poor-quality lenses. This defect prevented any great advances being made in microscopy until 1830, when a solution was provided in the form of achromatic lenses, which removed blurring and colour fringes. In the decades that followed, the improved compound microscope was used to discover, identify, and describe human cells and tissues.

THE MODERN VIEW
Used in science laboratories and hospitals worldwide, a modern compound microscope can provide a view of a section through the pituitary gland, as shown here. The otherwise transparent cells have been stained so the viewer can see their component parts.

Greater magnification

Light, or optical, microscopes have always been restricted by an upper limit of magnification of about 2,000 times. German scientist Ernst Ruska (1906–88) devised an alternative microscope in 1930 that could magnify objects 200,000 times or more. Rather than using light for illumination, it sent a beam of electrons through the thinly sectioned object. This beam – focused not by glass lenses but by electromagnets – was projected onto a phosphor screen, producing a visual image of the object. Called a transmission electron microscope, it revealed among other things the detailed internal structure of cells. The scanning electron microscope, invented in the 1960s, scans the object with an electron beam to produce a 3-D image.

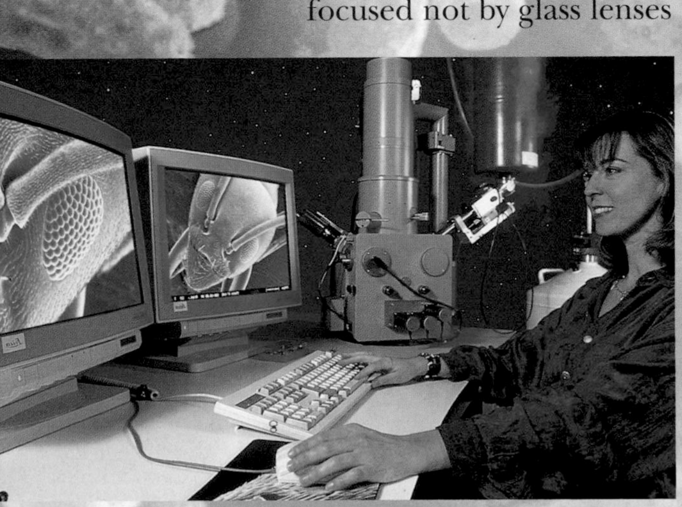

SCANNING ELECTRON MICROSCOPE
A scientist uses a scanning electron microscope to get a 3-D scanning electron micrograph (SEM) of an insect's head.

Cells

LIKE EVERY OTHER living thing on planet Earth, human beings are constructed from tiny, living units called cells. Individually, these cells are transparent, and so minute that they are only visible with a microscope. But the vast numbers of cells present in the body – about 100,000,000,000,000 or 100 trillion – collectively form a recognizable human being. These cells are not all the same; just as within any society certain people play specific roles, so do different types of cell inside the body. This organization is dynamic, not static. Every day the body produces billions of new cells to replace those that have become diseased, damaged, or just worn out. In younger humans, these new cells also enable the body to grow.

Nucleus

Squamous (flat) epithelial cell

LM OF EPITHELIAL CELLS

EPITHELIAL CELLS
The light micrograph above shows epithelial cells from the lining of the cheek. These flat cells fit together like paving stones to protect the inside of the mouth. Other epithelial cells – which may be flat, cube-shaped, or column-shaped – line and protect the body's tubes and cavities, such as blood vessels and the intestines. They also cover the body, forming the epidermis, the upper layer of the skin.

Osteocyte *sits within its own cavity in bone*

Tiny "threads" *link osteocyte to other bone cells*

SEM OF AN OSTEOCYTE

BONE CELLS
Osteocytes are bone cells that live isolated existences within their own tiny cavity, or lacuna, surrounded by the bony matrix that gives the body's bones their strength and hardness. Despite their isolation, these spider-like cells communicate with their neighbours through tiny "threads" that run along minute channels between lacunae. The role of these long-lived cells is day-to-day maintenance of the bony matrix.

CELL VARIETY

The trillions of cells that make up a person are all derived from a single fertilized egg cell. As this single cell divides repeatedly to make a human being, so groups of cells differentiate. This means that they become specialized, each group taking on its own appearance and role. By the time a person becomes an adult, they will have about 200 different types of cell, each adapted to a particular task. Four of these types – epithelial, bone, sperm, and blood cells – are described here as an example of how body cells differ, and how their appearance is related to their function.

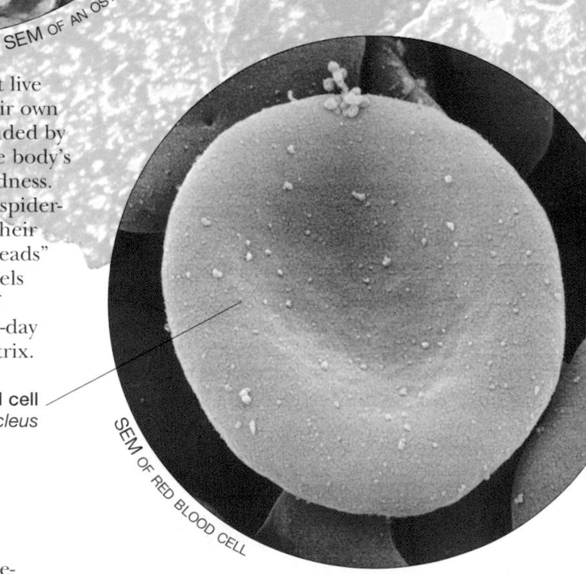

Red blood cell *has no nucleus*

Tail

Head

SEM OF SPERM

SPERMATOZOA
This scanning electron micrograph shows tadpole-shaped spermatozoa, or sperm, produced by a man's testes. The head of each sperm contains one part of the set of instructions needed to make a new human being; an ovum, or egg cell, produced by a woman, contains the other part. During sexual reproduction, the sperm's tail beats from side to side, pushing the streamlined cell towards an ovum. If the two fuse, a new human life is started.

RED BLOOD CELLS
As their name suggests, red blood cells, or erythrocytes, are found in the blood. Unlike other body cells, these dimpled, doughnut-shaped cells lack a nucleus. Instead they are packed with haemoglobin, a substance that both gives them their colour and carries oxygen. During a four-month lifespan, each red blood cell makes millions of circuits, picking up oxygen in the lungs and delivering it to all parts of the body.

Cell division

Cells reproduce by cell division. In this micrograph, a cell's nucleus (red) has just divided into two, with the cytoplasm (blue) about to follow. Short-lived cells, such as skin cells, which are constantly worn away, are continually replaced by cell division. At the other extreme, nerve cells do not divide again once they have been formed. The majority of new body cells are produced by a type of cell division called mitosis. Sex cells – sperm and ova – are produced by a process of cell division called meiosis (see pp. 212–13).

Cytoplasm *of parent cell divides*

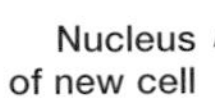

Cell Theory

In 1838, German scientists Jakob Schlieden (1804–81) and Theodor Schwann (1810–82) put forward their Cell Theory, which states that all living things are made of cells. In 1858, German physician Rudolf Virchow (1821–1902) went one stage further by including newly discovered cell division. He stated that new cells can only be made from existing ones, putting paid to the accepted idea that cells could arise spontaneously from non-living material.

Theodor Schwann

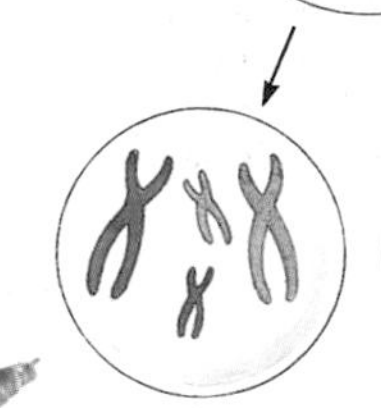

MITOSIS

This process produces two "daughter" cells that are identical to each other and to their "parent". The control centre of a body cell is called the nucleus. It contains 46 thread-like chromosomes, which hold the genetic information needed to build and run a cell. Before mitosis, each chromosome copies itself. Then, during mitosis, these copies are pulled apart, each to their own new nucleus. The result, once the cytoplasm has divided, is two genetically identical "daughter" cells.

CANCER CELLS

Normally, cell division is strictly regulated. But if its genetic material is damaged beyond repair, a cell may start to divide uncontrollably and cause a disease called cancer. Once it is out of control, division of a cancer cell (right) produces an expanding clump of abnormal cells called a tumour that can affect normal body functions. Unless the tumour is treated and destroyed, its cancer cells can spread to produce tumours in other parts of the body, eventually causing death.

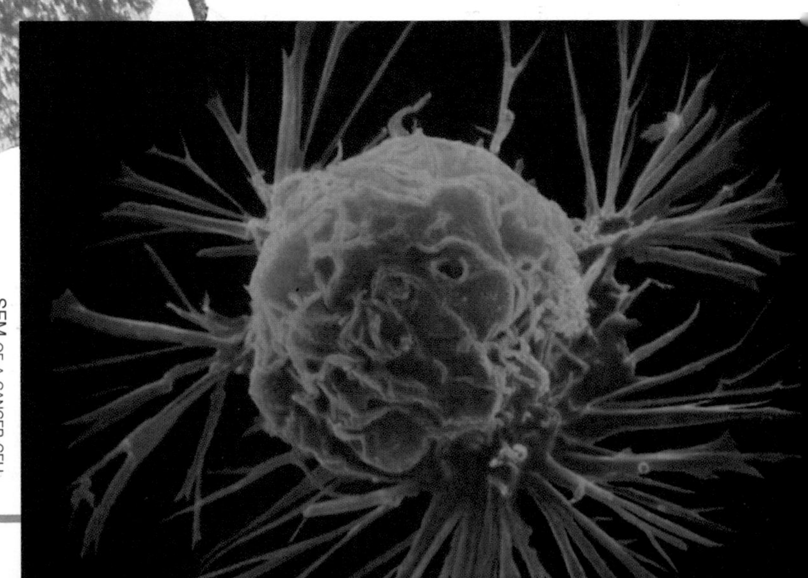

SEM OF A CANCER CELL

Cell structure

THE FACT THAT CELLS are small does not mean that they are simple. Regardless of their appearance and role, all cells share the same basic, highly organized internal structure. This structure was only revealed with the invention of the electron microscope in the 20th century. Before then, it was assumed that cells were made up of just three components: an outer cell membrane, central nucleus, and – between the two – the cytoplasm. The electron microscope revealed that, rather than being just a featureless jelly, the cytoplasm contained many different components. It is now known that these components work together rather like the different departments in a factory. Some manufacture the materials the cell needs, or recycle substances for reuse. Others generate the energy needed to power the cell's activities. All are controlled by instructions contained within the nucleus.

CYTOPLASM AND NUCLEUS
Lying between the cell membrane on the outside and the nucleus on the inside, cytoplasm consists of a clear, jelly-like fluid – mainly made up of water – called cytosol. Here, organelles, such as mitochondria, float. In the cytoplasm, microfilaments and microtubules form the cell's support system. The nucleus is the cell's control centre, containing the instructions that direct cell activities. The nucleus is surrounded by a nuclear membrane, which has holes, or pores, that allow substances to move between nucleus and cytoplasm.

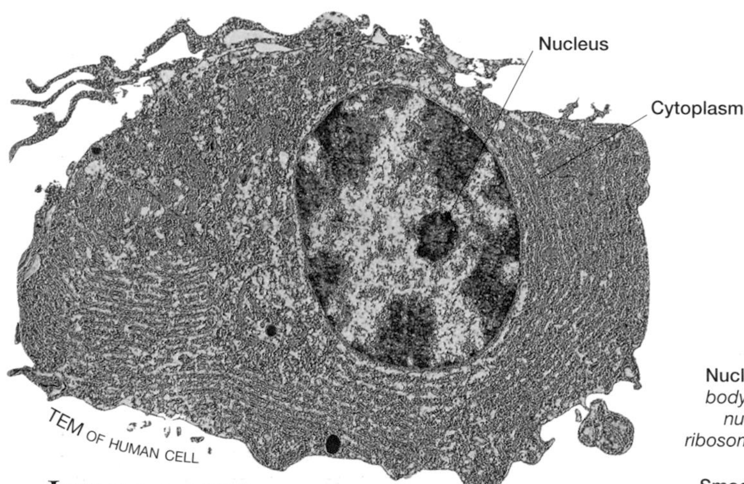

TEM OF HUMAN CELL

INSIDE A CELL

Just as each body part, or organ, has its own task, so do the tiny components, or organelles ("little organs"), inside a cell. Largest of these is the nucleus, the cell's control centre. Organelles suspended in the cytoplasm include the endoplasmic reticulum and Golgi body, which manufacture, store, and transport substances, while the mitochondria release energy from glucose and other foods. These cytoplasmic organelles serve to organize the cell's interior into compartments, preventing the cell's many chemical reactions from interfering with each other. Many organelles are surrounded by a membrane, similar in structure to the cell membrane that controls the flow of substances into and out of the cell.

Microfilaments *support and shape the cell*

Nucleolus *is a body within the nucleus where ribosomes are made*

Smooth endoplasmic reticulum *is a system of membrane-enclosed channels where lipids are made*

Secretory vesicle *is a package of substances formed by the Golgi body. It opens at the cell's surface to release its contents*

Nuclear membrane *or envelope forms the boundary of the nucleus*

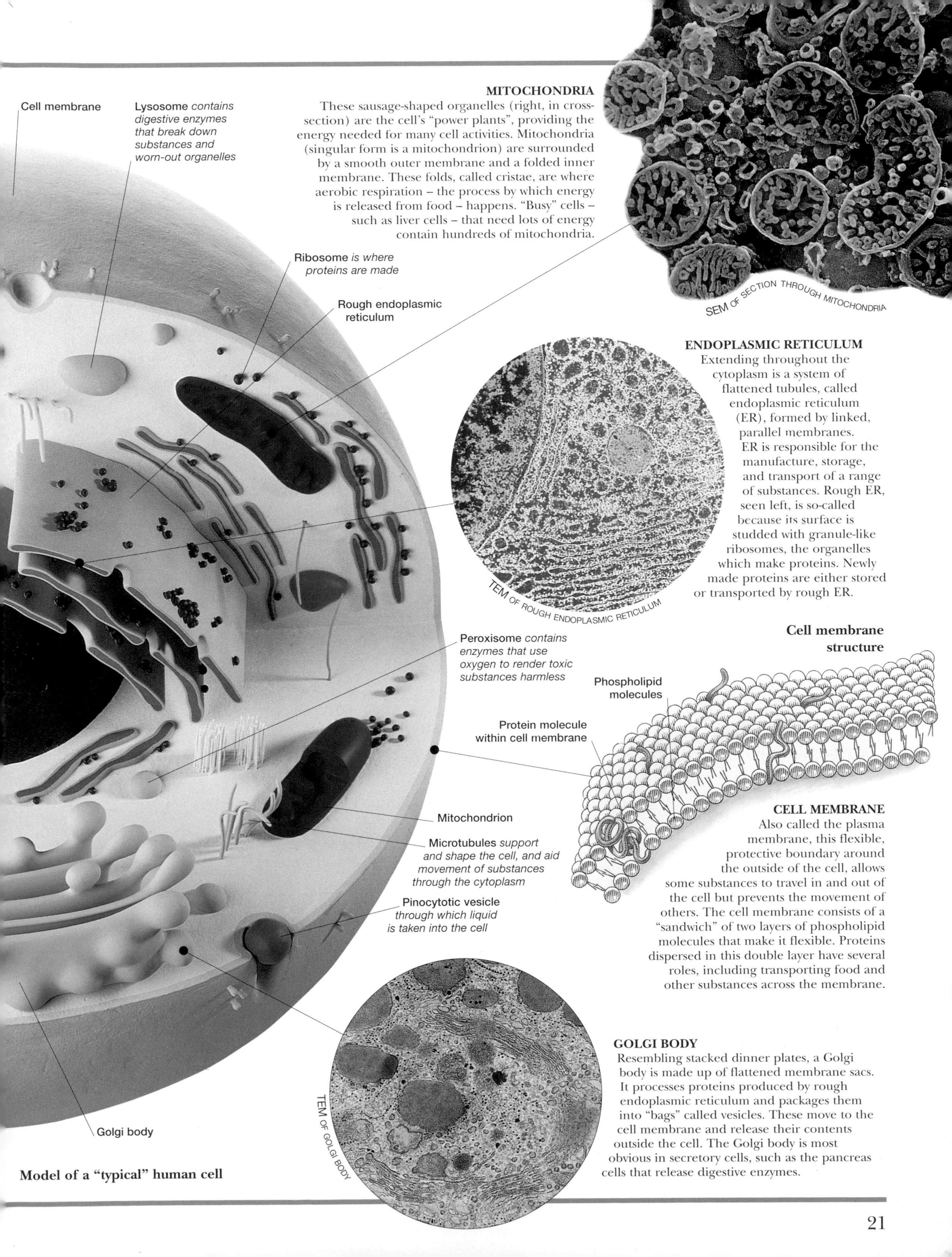

Cell membrane
Lysosome contains digestive enzymes that break down substances and worn-out organelles
Ribosome is where proteins are made
Rough endoplasmic reticulum
Peroxisome contains enzymes that use oxygen to render toxic substances harmless
Mitochondrion
Microtubules support and shape the cell, and aid movement of substances through the cytoplasm
Pinocytotic vesicle through which liquid is taken into the cell
Golgi body
SEM OF SECTION THROUGH MITOCHONDRIA
TEM OF ROUGH ENDOPLASMIC RETICULUM
Cell membrane structure
Phospholipid molecules
Protein molecule within cell membrane
TEM OF GOLGI BODY

Model of a "typical" human cell

MITOCHONDRIA

These sausage-shaped organelles (right, in cross-section) are the cell's "power plants", providing the energy needed for many cell activities. Mitochondria (singular form is a mitochondrion) are surrounded by a smooth outer membrane and a folded inner membrane. These folds, called cristae, are where aerobic respiration – the process by which energy is released from food – happens. "Busy" cells – such as liver cells – that need lots of energy contain hundreds of mitochondria.

ENDOPLASMIC RETICULUM

Extending throughout the cytoplasm is a system of flattened tubules, called endoplasmic reticulum (ER), formed by linked, parallel membranes. ER is responsible for the manufacture, storage, and transport of a range of substances. Rough ER, seen left, is so-called because its surface is studded with granule-like ribosomes, the organelles which make proteins. Newly made proteins are either stored or transported by rough ER.

CELL MEMBRANE

Also called the plasma membrane, this flexible, protective boundary around the outside of the cell, allows some substances to travel in and out of the cell but prevents the movement of others. The cell membrane consists of a "sandwich" of two layers of phospholipid molecules that make it flexible. Proteins dispersed in this double layer have several roles, including transporting food and other substances across the membrane.

GOLGI BODY

Resembling stacked dinner plates, a Golgi body is made up of flattened membrane sacs. It processes proteins produced by rough endoplasmic reticulum and packages them into "bags" called vesicles. These move to the cell membrane and release their contents outside the cell. The Golgi body is most obvious in secretory cells, such as the pancreas cells that release digestive enzymes.

Cell chemistry

INSIDE A PERSON'S CELLS a multitude of chemical reactions is taking place, regardless of whether that person is asleep or awake. During each reaction, chemical compounds are modified to meet the needs of the cell. This mass of chemical activity is far from being chaotic. The organelles of the cell – such as mitochondria – organize chemical reactions into separate compartments so that they do not interfere with each other. Furthermore, reactions are greatly accelerated by special substances called enzymes, which also control how quickly cells consume raw materials and yield new products. Key to a cell's chemistry is the release of energy from food, by the process of cell respiration. Without energy, cells would have no driving force, and life simply could not exist.

THERMOGRAM OF WOMAN'S MOUTH AND TEETH

CHEMICAL PROCESSES

Inside any cell, two basic processes – catabolism and anabolism – work side-by-side. In catabolism, energy-rich fuel molecules are broken down to release their energy to supply the cell's needs. In fact, only about 25 per cent of this energy can be used by the cell, the rest is released as heat which is used to maintain the body's temperature at 37°C (98.6°F). Anabolism takes simple building blocks to build up the more complex substances a cell needs, such as proteins and lipids. In order to work, anabolism uses the energy generated by catabolism. Together, anabolism and catabolism make up metabolism, the sum of all the chemical processes happening inside a cell.

1 molecule glucose

Glycolysis

ATP

2 molecules pyruvic acid

ENZYMES

These proteins act as catalysts, speeding up the cell's chemical reactions by thousands or millions of times without being changed or used up. Without enzymes, these reactions would take place so slowly that life could not exist. Each enzyme is specific to a particular reaction. Molecules taking part in that reaction bind to the enzyme and react to form product molecules that are then released. Some enzymes work outside cells, including the digestive enzymes that speed up the breakdown of food during digestion.

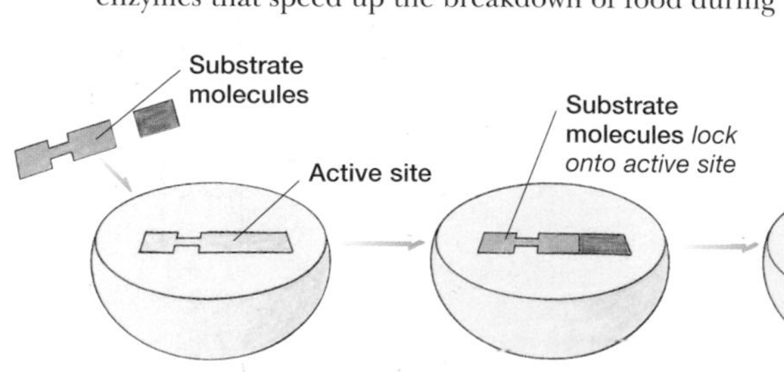
Substrate molecules
Active site
Substrate molecules lock onto active site
Product molecules move away from enzyme

1 *Molecules involved in a reaction, called substrates, fit into a part of an enzyme called the active site.*

2 *The enzyme holds the substrates together. The substrates react and form product molecules.*

3 *The products are released. The enzyme is unaffected by the reaction, and is ready to attract more molecules.*

RAW MATERIALS

Unless it has a constant stream of raw materials to supply its cells' needs, the body cannot survive. During digestion, the carbohydrates, lipids, proteins, and nucleic acids in food are broken down into simple molecules, taken into the blood, and carried to the body's cells. Once inside a cell these are assembled into new molecules by anabolism, or broken down by catabolism. Water is also vital because it provides the liquid medium inside cells in which chemical reactions take place.

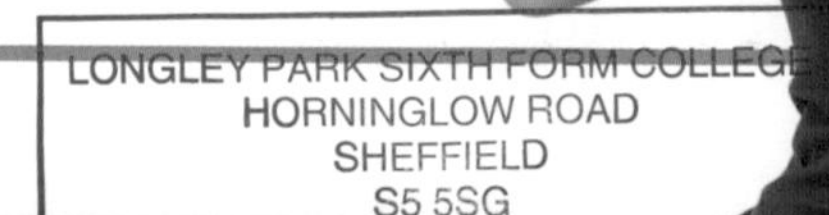

Aerobic respiration

Body cells get most of their energy from a type of cell respiration called aerobic respiration. Using oxygen, this breaks down fuel molecules such as glucose into carbon dioxide and water to release large amounts of energy. Aerobic respiration has two phases, each made up of several enzyme-catalysed reactions. Firstly, during glycolysis ("glucose splitting") in the cytoplasm, a glucose molecule is split into two molecules of pyruvic acid, with the release of a little energy. Then, inside mitochondria – the cell's "power plants" – a sequence of chemical reactions called the Krebs cycle (see p. 175) completely dismantles pyruvic acid molecules and, with the help of oxygen, releases all of their energy.

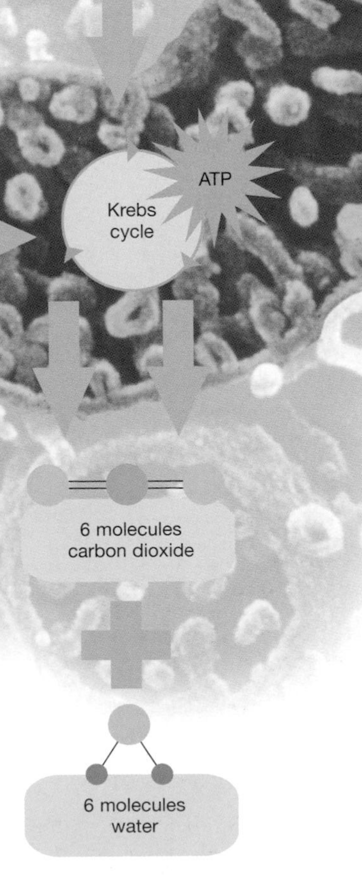

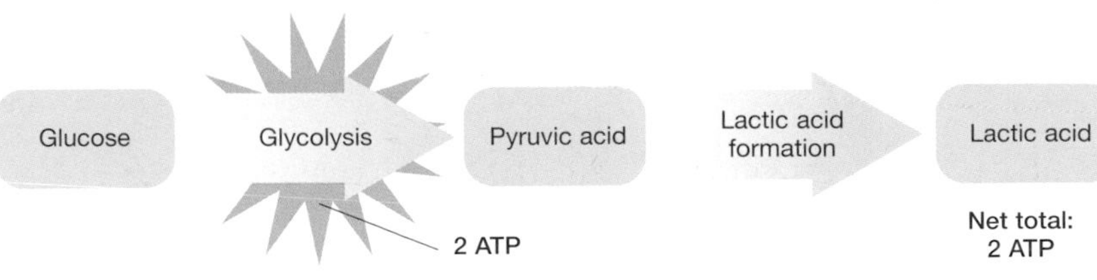

ANAEROBIC RESPIRATION
Cells can temporarily run out of oxygen, for example when muscle fibres (cells) contract rapidly during vigorous exercise. This does not stop cells releasing energy from glucose, however, even though they cannot use aerobic respiration. Instead they use anaerobic ("without air") respiration. In the cytoplasm, glucose is broken down into pyruvic acid – as during glycolysis – which is then converted into lactic acid. This yields much less energy than aerobic respiration, but happens more rapidly. Once oxygen becomes available again, lactic acid is reconverted to pyruvic acid and broken down by the Krebs cycle.

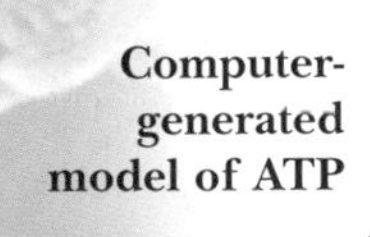

Computer-generated model of ATP

ADENOSINE TRIPHOSPHATE (ATP)
Cells cannot use the energy stored inside glucose until it has been processed by cell respiration. This releases the stored energy in small bursts and uses it to make molecules of adenosine triphosphate (ATP), the cell's main energy store and carrier. When a chemical reaction requires energy, ATP is broken down to liberate its energy store and its components are recycled to pick up more energy from cell respiration. During aerobic respiration, each molecule of glucose yields 38 molecules – "energy packets" – of ATP. The yield from anaerobic is much lower – just 2 ATP.

Chemical Components

Human cells – like those of all living things – contain chemical components that are unique to living systems. These are organic compounds, substances whose molecules are based on a "skeleton" of carbon atoms. The main types found in cells are carbohydrates, lipids, proteins, and nucleic acids.

- Carbohydrates provide cells with energy. They include simple sugars, such as glucose, that deliver a source of energy; and complex polysaccharides, such as glycogen, that form fuel stores in the liver and muscle cells.
- Lipids (fats and oils) form the membrane around the cell and its organelles, and provide a long-term energy store in adipose (fat) tissue cells.
- Proteins are complex compounds that perform many different tasks, including forming part of cell membranes, and, as enzymes, controlling cell reactions.
- Nucleic acids such as DNA store in coded form the information needed to control cells by telling them how to make proteins.

Molecular model of glycogen

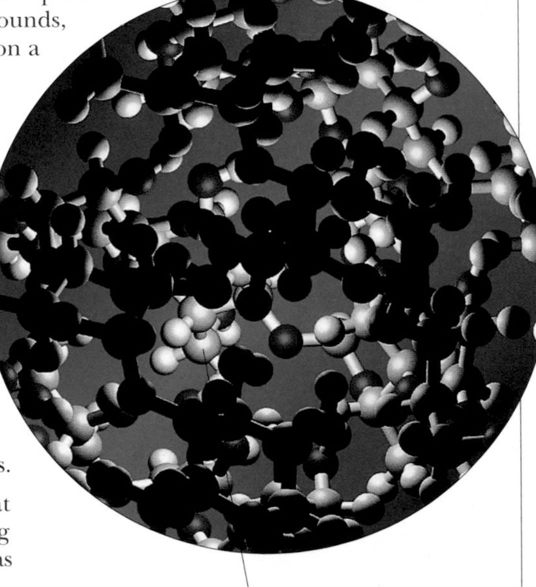

Glycogen *is a complex carbohydrate, or polysaccharide, which is made up of glucose subunits*

THE DOUBLE HELIX

ON 25TH APRIL, 1953, the world of science changed forever. Two scientists – American James Watson (b. 1928) and Briton Francis Crick (b. 1916) – announced in the journal *Nature* that they had unravelled the structure of a molecule believed to hold the key to life itself. The molecule was DNA (deoxyribose nucleic acid), which is found in the nucleus of every cell, and its structure they called the "double helix". This new understanding of DNA's structure enabled scientists to find out how it controls the activities of cells and of whole organisms, including humans.

DNA'S STRUCTURE REVEALED
This photograph shows Watson (left) and Crick next to their model of DNA. Its completion was announced in the journal *Nature* – "We wish to suggest a structure for the salt of deoxyribose nucleic acid (DNA). This structure has novel features which are of considerable biological interest."

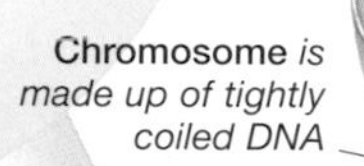
Cell
Chromosome is made up of tightly coiled DNA
"Unzipping" DNA strands separate so that the bases contained in one strand can be copied
Gene is a section of DNA that carries the instructions for making a specific protein
Guanine
Thymine
Adenine
mRNA nucleotide
Cytosine
DNA "backbone" is made of linked phosphate and deoxyribose (sugar) molecules
Transcription is where free mRNA bases pair with corresponding bases on DNA to make an mRNA strand

DNA IN ACTION
Each of the 46 chromosomes in the nucleus of a human cell is made up of a long DNA molecule, sections of which form genes. The two strands of the DNA double helix are linked by bases that always join in a specific way.

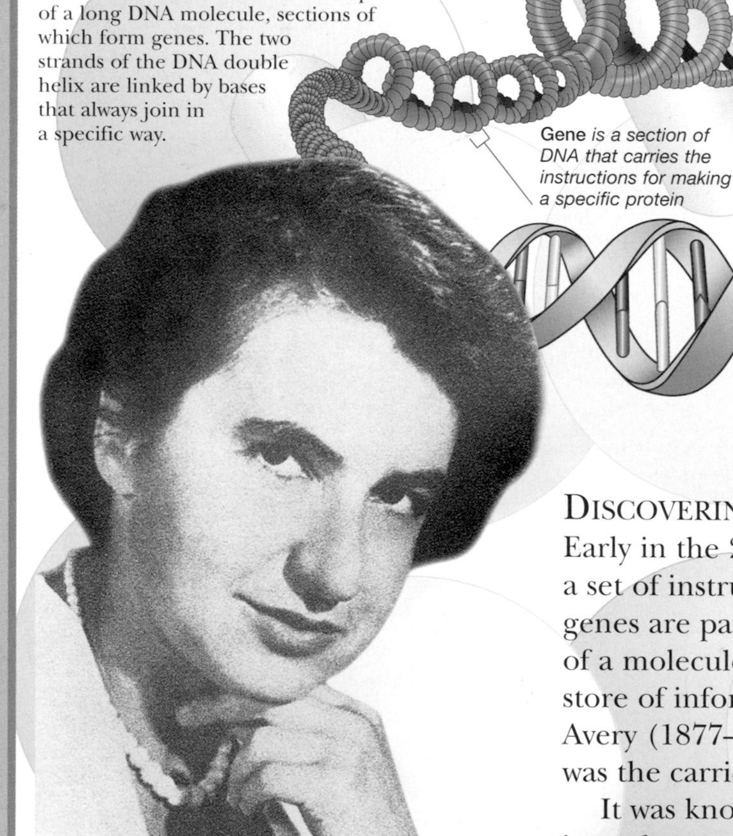

ROSALIND FRANKLIN
Born in 1920, British scientist Rosalind Franklin played a key role in the discovery of the structure of DNA. Her X-ray diffraction photographs provided Crick and Watson with vital evidence about DNA's spiral shape. Franklin died from cancer in 1958.

DISCOVERING DNA

Early in the 20th century it was realized that cells contain genes, a set of instructions for making and running an organism. Because genes are passed on from parents to offspring, they must be made of a molecule that can both replicate, or copy, itself and hold a store of information. In 1944, American bacteriologist Oswald Avery (1877–1955) showed that it was the nucleic acid DNA which was the carrier of genetic information.

It was known from work carried out during the 1930s that DNA is made up of units called nucleotides. Each nucleotide consists of a phosphate group, the sugar deoxyribose, and one of four nitrogenous bases called adenine, cytosine, guanine, and thymine. In 1953, using evidence from chemical analysis and X-ray diffraction – a method that bounces X-rays off atoms in a DNA of molecule to produce a photographic pattern that indicates its structure – Watson and Crick built their 3-D model of DNA, consisting of two parallel strands that spiral around each other.

COPYING THE MESSAGE
DNA has the unique ability to make an exact copy of itself. This TEM shows a DNA strand (shown right in yellow) that is "unzipping" to form two single "daughter" strands, each of which acts as a template to form a new double helix identical to the "parent".

DNA strand *has "unzipped" into two strands*

TEM OF REPLICATING DNA STRAND

UNDERSTANDING THE CODE

Watson and Crick's double helix resembles a twisted ladder. The uprights are made of a "backbone" of phosphate and deoxyribose, the "rungs" of paired bases, adenine with thymine, and cytosine with guanine. They soon realized that this arrangement also provided the means for replication – that the DNA double helix could "unzip", so that free nucleotides could bond to the exposed bases to produce two new DNA double helices. But how does DNA control cell activities?

Since the 1940s it had been known that genes control the production of proteins, many of which are enzymes, the biological catalysts that regulate chemical reactions inside cells. Each protein is made up of a specific sequence of amino acids,

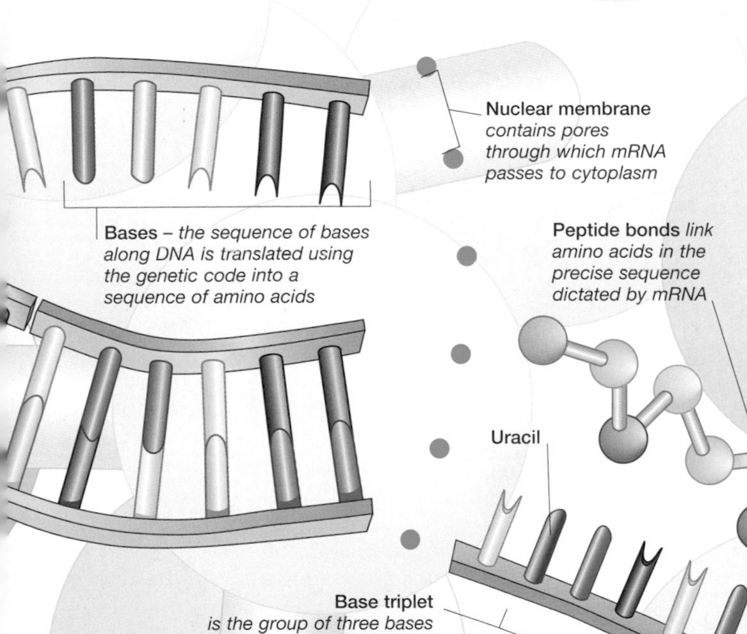
Nuclear membrane contains pores through which mRNA passes to cytoplasm
Bases – the sequence of bases along DNA is translated using the genetic code into a sequence of amino acids
Peptide bonds link amino acids in the precise sequence dictated by mRNA
Uracil
Base triplet is the group of three bases codes for a specific amino acid

Free amino acids *in the cytoplasm are "assembled" to make a particular protein*

mRNA

Ribosome *provides a site on which the mRNA message is translated into protein structure*

Newly assembled protein *detaches itself from the ribosome and folds up, its functional shape determined by the specific sequence of amino acids*

of which there are 20 different types. During the 1960s Marshall Nirenberg (b. 1927), an American biochemist, showed that within DNA an arrangement of three bases specifies one particular amino acid. This provides a genetic code in which base triplets form the "words" that instruct the cell to select the right amino acids to make a specific protein.

MAKING PROTEINS

Protein synthesis happens in the cytoplasm of the cell, but DNA molecules are too large to move out of the nucleus. So how is the message conveyed from nucleus to cytoplasm? A section of DNA (a gene) "unzips", and one strand is copied (transcription) by a smaller, single-stranded nucleic acid called messenger RNA (mRNA). This contains the same bases as DNA, except for uracil, which replaces thymine. The mRNA passes into the cytoplasm and attaches itself to a ribosome. Here, the mRNA is translated triplet by triplet so that amino acids are linked up in the correct sequence to make a specific protein.

TRANSLATION
An organelle called a ribosome passes along the strand of mRNA using the genetic code to translate it into protein structure. Amino acids are lined up in the correct sequence, and then bond together to form a protein.

Tissues

IF THE BODY'S TRILLIONS of cells all existed independently, it would be impossible to organize and operate a living human being. Instead, cells of the same or similar types are grouped together into tight-knit communities in which they work together to carry out a specific task. These groups of similar cells form tissues – a word derived from the Latin for "woven" – just as cotton threads are interwoven to make cloth. The many different types of tissue fall into four basic categories – epithelial, connective, muscular, and nervous – that interface to make up the fabric of the body. If tissues are damaged, their cells divide to repair the damage. This process, called regeneration, happens more quickly in some tissues than others. Together, different types of tissue form organs such as the heart and kidneys, as described in more detail on pp. 28–29. The roles of the four basic types of tissue can be described very simply as follows: epithelial tissues cover; connective tissues support; muscle tissues move; and nervous tissues control.

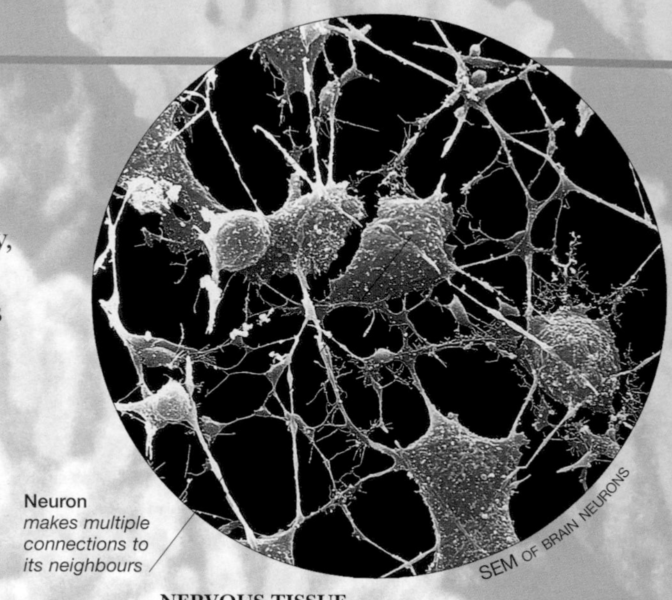

NERVOUS TISSUE
Restricted to the nervous system – the brain, spinal cord, and nerves – nervous tissue is responsible for controlling and coordinating most body processes, as well as providing humans with conscious thought and sensation. Two types of cell are found in nervous tissue. Neurons, or nerve cells, like the brain neurons above, enable communication to take place by generating and carrying electrical signals at high speed. Glial cells, or neuroglia, support and nurture the neurons.

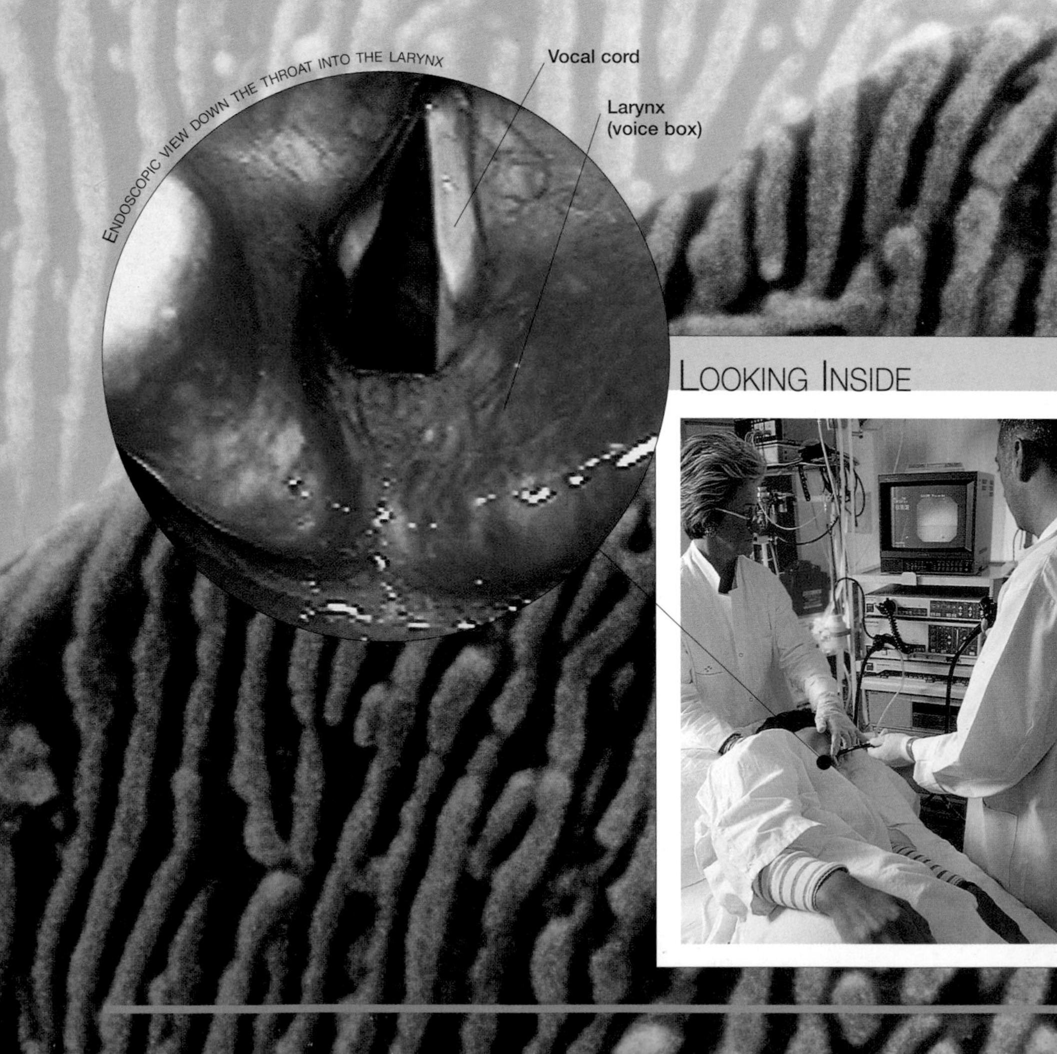

LOOKING INSIDE

Doctors can look inside the body to examine tissues and organs by using a technique called endoscopy. Endoscopes are tube-like instruments that contain long optical fibres which transmit light. The endoscope can be inserted through a natural opening, such as the mouth, as shown here, or through a small incision made in the skin. Optical fibres carry light to illuminate the tissue being observed, while a miniature camera relays images to a screen for the doctor to see. The endoscope may also include tiny scissors or forceps to take a small sample of tissue for biopsy. Tissue is then looked at under the microscope for signs of disease.

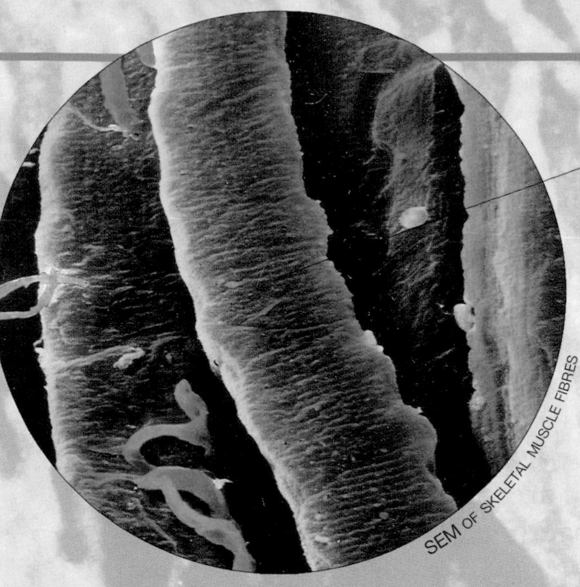

SEM OF SKELETAL MUSCLE FIBRES

MUSCLE TISSUE

The body's muscular tissues consist of cells, called fibres, that can contract, or shorten. Skeletal muscle tissue, seen in this micrograph, moves the body and maintains its posture. Smooth muscle tissue, in the walls of hollow organs, typically moves materials through the body. Cardiac muscle tissue, found in the wall of the heart, pumps blood around the body. Muscular tissues receive a rich blood supply which brings the food and oxygen needed to release energy for contraction.

CONNECTIVE TISSUES

Unlike other types of tissue, connective tissue consists of cells embedded in a matrix, or framework, which is secreted by the cells of the tissue. With major roles including binding and support, protection and insulation, connective tissues are the most abundant and diverse of all body tissues. They hold the body together by underlying epithelial tissues and packaging organs, and are also found in tendons, ligaments, and the dermis of the skin. Cartilage (left, seen inside a joint) and bone in the skeleton support and protect the body's organs. Adipose tissue, its cells packed with fat, insulates and cushions organs. Blood, a connective tissue with a liquid matrix, transports materials around the body.

Chondrocyte *is a mature cartilage cell*

Cartilage matrix *is rich in collagen fibres*

LM OF SECTION THROUGH HYALINE CARTILAGE

EPITHELIAL TISSUE

Also called epithelium, epithelial tissue consists of a continuous sheet of cells, which may be one cell or many layers thick. It forms the outer layer of skin, and the inner linings of the digestive system – including the folded lining of the oesophagus, shown here – and of the respiratory, urinary, and reproductive systems, blood vessels and the heart. By covering and lining, epithelial tissues protect the body's surfaces, form a barrier to micro-organisms, and provide an interface through which all substances entering or leaving the body must pass.

LM OF SECTION THROUGH NASAL MUCOUS MEMBRANE

SEM OF LINING OF OESOPHAGUS

MEMBRANES

Together, epithelial and connective tissues combine to form the membranes that line hollow organs and body cavities. The micrograph (above) shows a section through the membrane that lines the nasal cavity. Epithelial tissue (blue-pink) secretes sticky mucus that traps dust particles in the air. The connective tissue (green-gold) is reinforced by tough collagen fibres and stretchy elastin fibres, which underpin and stabilize the epithelial tissue above it.

Imaging techniques

TODAY, DOCTORS AND researchers have access to many different methods of looking inside a living body. This allows them to explore tissues and organs to search for disease, or to find out how the body works. This has not always been the case. Until 40 years ago, X-rays – which do not clearly reveal softer body tissues – were about the only means of seeing inside a living person without having to perform surgery. Since the 1970s, however, modern technology, especially advances in computers, has produced a variety of powerful imaging techniques, including CT and MRI scanning. Like X-rays, these techniques are non-invasive (do not require surgery), but they produce images – many examples of which feature throughout this encyclopedia – that are considerably more detailed than anything obtained before.

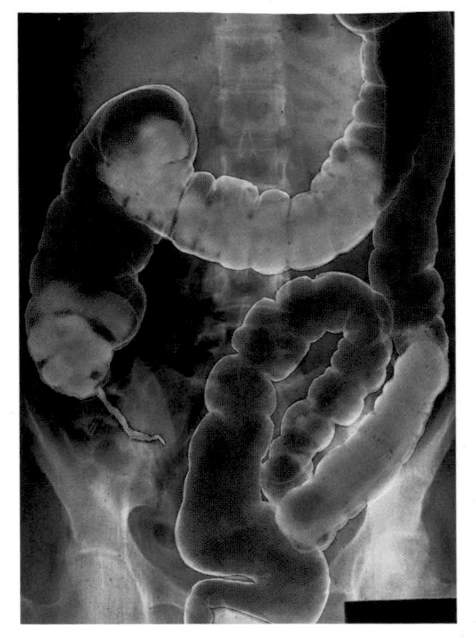

X-RAYS (RADIOGRAPHY)
With this technique, a high-energy form of radiation is passed through the body and projected onto a photographic film to produce an X-ray photograph, or radiograph. The film remains white where hard tissues, such as bone, have absorbed X-rays, but turns grey or black where X-rays have passed through soft tissues, such as muscle. In this contrast radiograph, barium sulphate – a substance that absorbs X-rays – has been introduced into the large intestine (made of soft tissues) so its outline (orange) can be clearly seen. The colours are false and were added afterwards.

COMPUTED TOMOGRAPHY (CT)
Combining X-rays with a computer, CT scanning produces much more detailed images than ordinary X-ray photographs. As a person lies inside a scanner (below), it rotates around him sending narrow beams of X-rays through his body and into a detector. A computer analyzes information from the detector to produce a "slice" through the organs in that part of the body. These slices can be built up to produce a 3-D image, like this one (right) of the skull and brain.

DIFFERENT METHODS

Imaging techniques vary in the way they work and in their applications. X-rays and CT scanning use high-energy radiation. Ordinary X-rays are typically used for looking at bones. Contrast X-rays use special substances to reveal hollow structures such as the intestines or blood vessels. CT scanning uses computers to produce detailed images, usually of the head or abdomen. PET and radionuclide scanning both use radioactive substances to reveal chemical activity in tissues rather than detailed structure. MRI scanning and ultrasound do not use radiation. MRI scans produce detailed images of any body tissue. Ultrasound can show both structure and movement, and is commonly used to observe the development of the fetus.

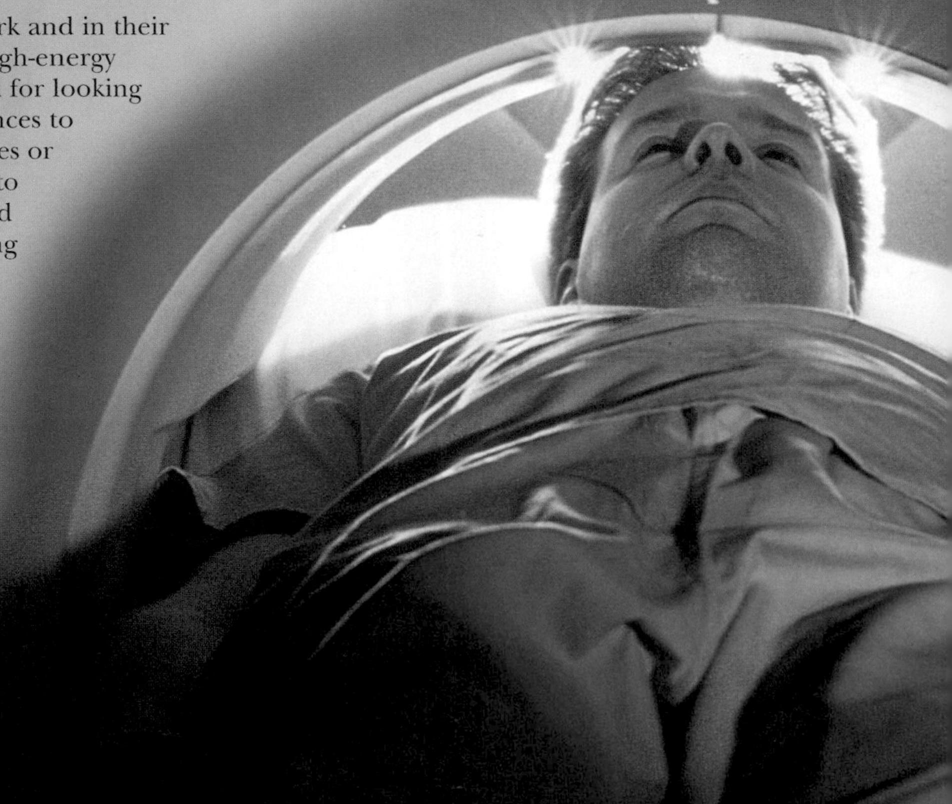

Patient undergoing CT scan

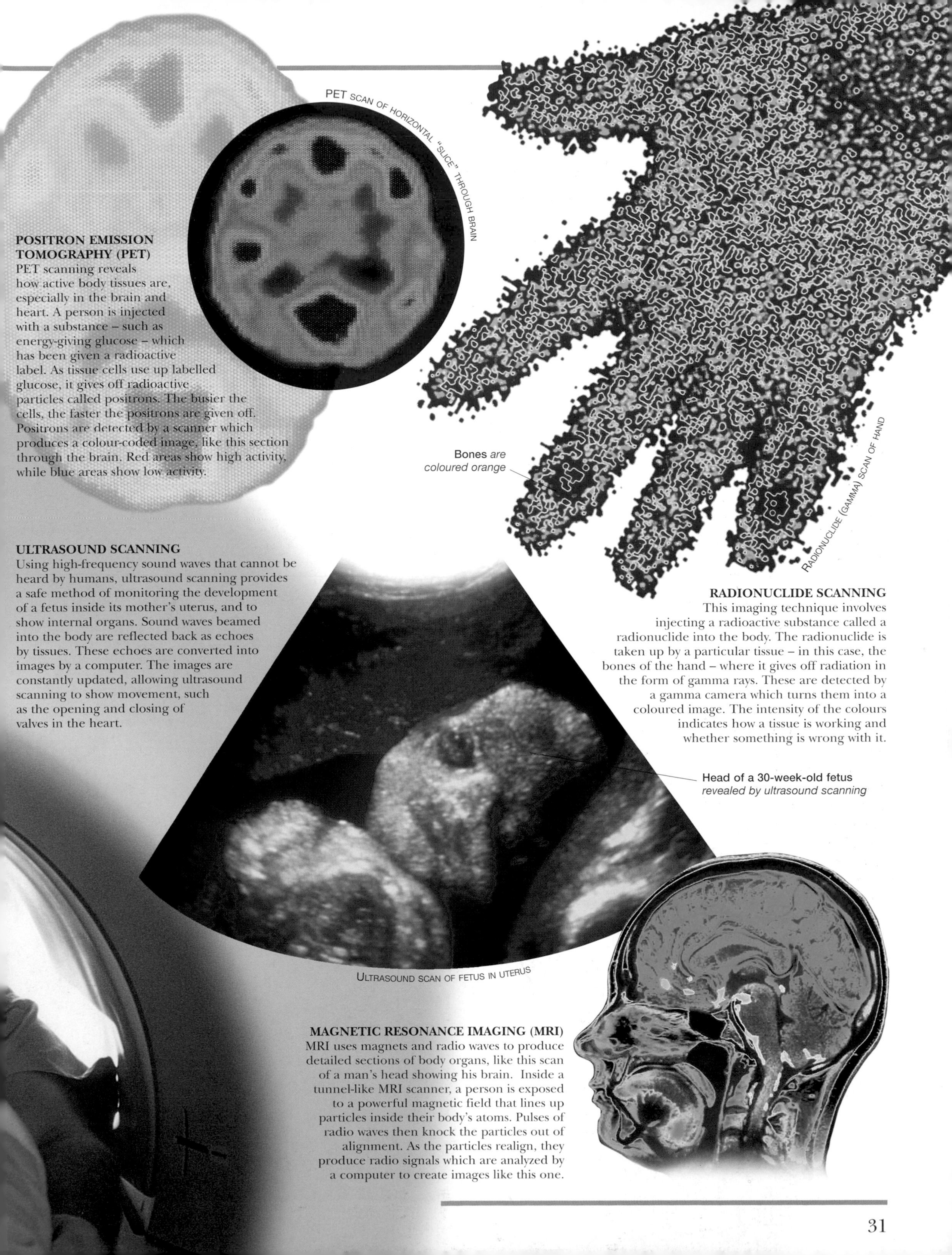

POSITRON EMISSION TOMOGRAPHY (PET)
PET scanning reveals how active body tissues are, especially in the brain and heart. A person is injected with a substance – such as energy-giving glucose – which has been given a radioactive label. As tissue cells use up labelled glucose, it gives off radioactive particles called positrons. The busier the cells, the faster the positrons are given off. Positrons are detected by a scanner which produces a colour-coded image, like this section through the brain. Red areas show high activity, while blue areas show low activity.

ULTRASOUND SCANNING
Using high-frequency sound waves that cannot be heard by humans, ultrasound scanning provides a safe method of monitoring the development of a fetus inside its mother's uterus, and to show internal organs. Sound waves beamed into the body are reflected back as echoes by tissues. These echoes are converted into images by a computer. The images are constantly updated, allowing ultrasound scanning to show movement, such as the opening and closing of valves in the heart.

RADIONUCLIDE SCANNING
This imaging technique involves injecting a radioactive substance called a radionuclide into the body. The radionuclide is taken up by a particular tissue – in this case, the bones of the hand – where it gives off radiation in the form of gamma rays. These are detected by a gamma camera which turns them into a coloured image. The intensity of the colours indicates how a tissue is working and whether something is wrong with it.

MAGNETIC RESONANCE IMAGING (MRI)
MRI uses magnets and radio waves to produce detailed sections of body organs, like this scan of a man's head showing his brain. Inside a tunnel-like MRI scanner, a person is exposed to a powerful magnetic field that lines up particles inside their body's atoms. Pulses of radio waves then knock the particles out of alignment. As the particles realign, they produce radio signals which are analyzed by a computer to create images like this one.

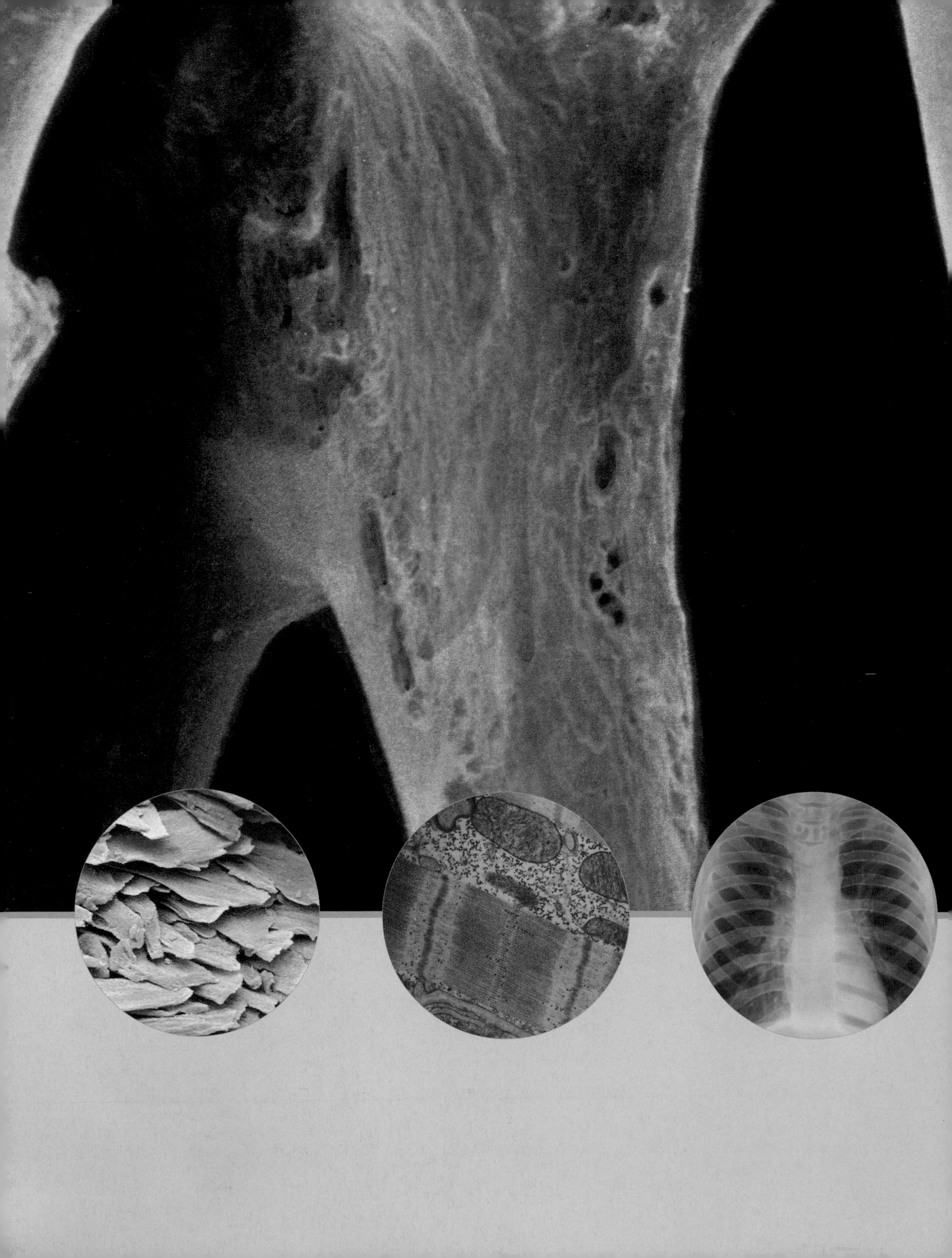

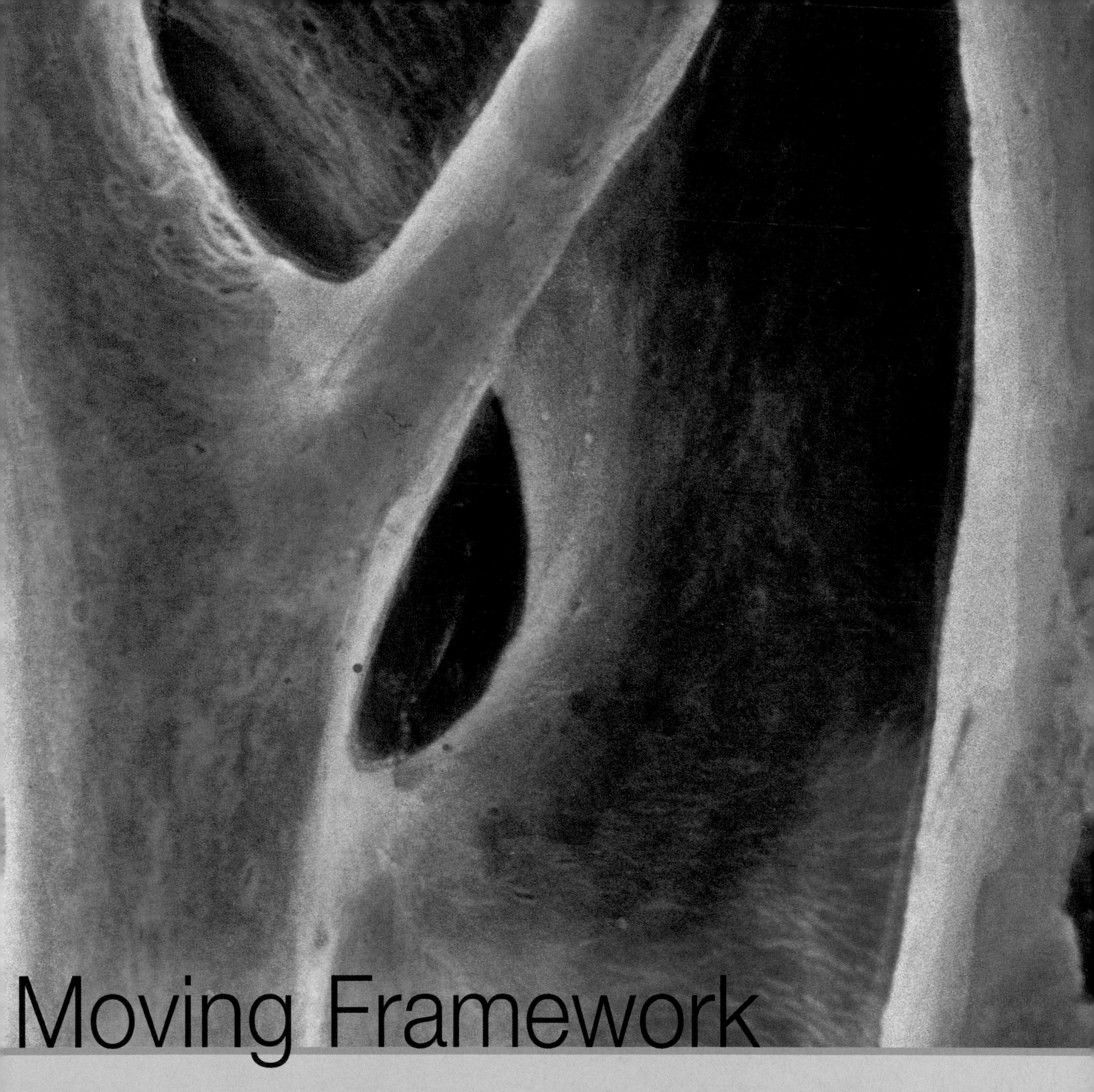

Moving Framework

THREE SYSTEMS COVER, move, shape, and support the body. Skin is a living, stretchy overcoat that shields the delicate tissues inside the body from the harsh conditions outside. Bones form the skeleton, a structure strong enough to support the body's weight and prevent its collapse, yet light and flexible enough to let the body move. Muscles shape the body, and, by pulling bones, produce a multitude of movements from raising an eyebrow to running a marathon.

INTEGUMENTARY system

SKIN, HAIR, AND NAILS make up the integumentary system. The skin, which covers the entire surface of the body and forms a vital barrier between the body and its surroundings, is the body's largest organ. In an average adult, it weighs about 5 kg (11 lb). It provides protection against injury, infection by micro-organisms, and damage by harmful rays in sunlight. The skin is also a sense organ that can detect touch, warmth, cold, and pain. In addition, it helps to control body temperature and produces vitamin D, which is necessary for healthy bones. Hair and nails grow directly from the skin to provide additional covering and protection.

LIVING LAYERS

The skin has two layers. The upper layer, the epidermis, is made up of cell layers (left, shown separated) that become flatter and tougher towards the surface. The lower layer, the dermis, contains strong, flexible fibres, blood vessels, nerves, and sensory receptors. Lodged in the dermis are sebaceous glands and coiled sweat glands. Hairs grow from follicles that extend upwards from the dermis through the epidermis. Beneath the dermis lies subcutaneous tissue, a fatty layer that insulates the body and stores energy.

Cornified layer *consists of flat, dead cells that are constantly worn away*
Clear layer *is most apparent in the thick skin that covers the soles and palms*
Granular layer *is where cells flatten and fill with tough keratin as they move towards the surface cells*
Basal layer *produces new cells to replace those lost from the surface*
Spiny layer *cells are linked by spine-like connections*
Dermal papillae *attach dermis to the epidermis*
Nerve *relays messages between skin receptors and the brain*
Arrector pili muscle *pulls hair upright to produce goose bumps*
Sebaceous gland *produces oily sebum which keeps skin and hair soft and flexible*
Subcutaneous fat *helps to insulate the body*
Touch receptor *detects light touch*
Sweat gland *releases sweat onto the skin's surface to cool the body*
Pressure receptor *detects pressure and vibrations*
Blood vessel *helps to regulate body temperature*
Hair follicle *is a cavity in the skin from which hair grows*

Section through the skin

PROTECTIVE OVERCOAT
The epidermis provides a waterproof coating that protects the body from drying out or becoming waterlogged. Water is repelled from the skin's surface by the keratin that fills epidermal cells and by oily sebum produced by the sebaceous glands. As a person showers, skin keeps the water out. Meanwhile, receptors that are sensitive to heat or cold detect the temperature of the water, and receptors that are sensitive to touch detect the flow of water against the skin.

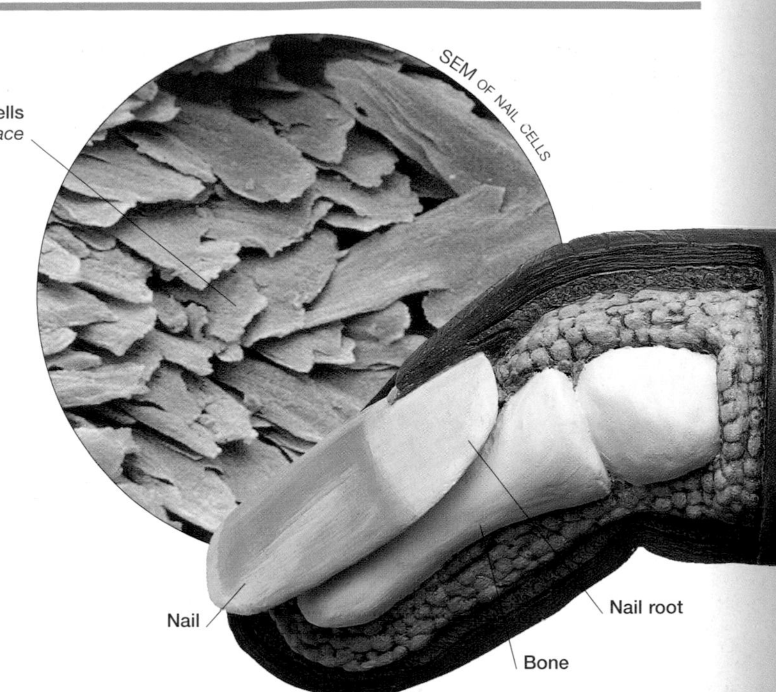

Section through a finger showing position of nail

NAILS

The sensitive tips of the fingers and toes have hard plates called nails that provide protection and support. Each nail grows from a living root at its base, which is overlapped by a fold of skin, and consists mainly of flattened dead cells filled with the tough protein keratin. Nails grow faster in warm weather than cold, and fingernails grow three or four times faster than toenails.

INTEGUMENTARY SYSTEM FUNCTIONS

Protection	*Against the harmful effects of physical injury, chemicals, heat, sunlight, infection, and excessive water; also against water loss.*
Temperature regulation	*Keeps body temperature stable by sweating and varying size of blood vessels in the skin; vessels narrow to conserve heat and widen to lose heat; hair limits heat loss from the head.*
Sensation	*Detects touch/pressure, pain, warmth, and cold.*
Vitamin D synthesis	*Produces vitamin D in the presence of ultraviolet rays from the sun.*
Excretion	*Eliminates small amounts of waste substances from the body in sweat.*
Gripping	*Provides a surface with good grip for handling objects and to prevent slipping; nails make handling easier.*
Absorption	*Can take small amounts of certain substances into the body from the surface.*

"NAKED" APES

Humans appear naked compared to great apes, our closest animal relatives. However, apes and humans have similar numbers of body hairs. Whereas the ape's body is coated with long, coarse hair (terminal hair), the human body is mostly covered with short, fine hair (vellus). Humans have terminal hair on the scalp and in a few other areas only. A hairy coat helps keep the ape warm; humans rely mainly on clothing instead.

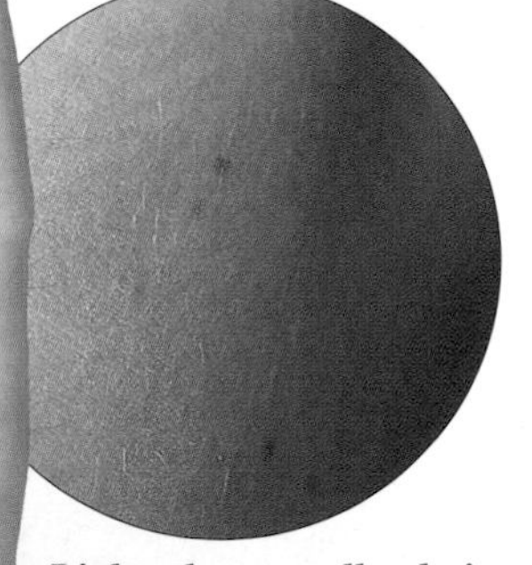

Light, downy vellus hair

Coarse terminal hair

Skin surface

COVERING AN AREA of up to 2 sq m (21.5 sq ft), the skin's surface is far from smooth. In fact, it is marked by numerous criss-crossing lines and by creases, grooves, ridges, and bumps. It is kept soft and supple by a thin coating of oil, called sebum. Scattered over the surface are the openings of millions of hair follicles and sweat ducts. Sweating is one of the ways in which the skin helps to control body temperature. The surface of the skin is also home to a variety of bacteria, visible only under a microscope. Sometimes large numbers flourish, causing spots or rashes.

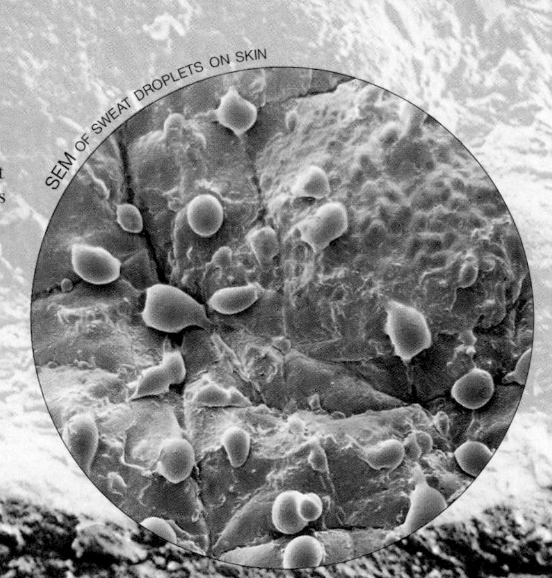
SEM OF SWEAT DROPLETS ON SKIN

COOLING DOWN
If the body overheats, droplets of sweat ooze from sweat glands on to the skin's surface. Water in the sweat evaporates and draws heat away from the body, cooling it down. The sweat glands in the armpits produce a large amount of sweat, which many adults use antiperspirants to control. Cooling is also achieved by blood vessels in the skin, which widen to speed up heat loss from the body.

Goose bumps

GOOSE BUMPS
When the body is cold, blood vessels in the skin narrow to conserve heat and small bumps, called goose bumps, appear. Goose bumps are formed when a tiny muscle attached to the base of each hair shortens, pulling the hairs upright and lifting the surrounding skin. In animals with thick coats, air becomes trapped between the upright hairs, creating a blanket that helps keep the body warm.

EPIDERMAL RIDGES
This close-up view shows the hundreds of tiny ridges on the surface of the skin on the palm of a hand. Sweat ducts open in rows along the crests of each ridge. The undersides of hands and feet are the only areas covered with these ridges, which are separated by fine parallel grooves and form curved patterns on the skin. They are also the only areas that have no hair or oil glands. A ridged, hairless surface provides good grip for handling objects and prevents slipping on surfaces, such as when walking or climbing.

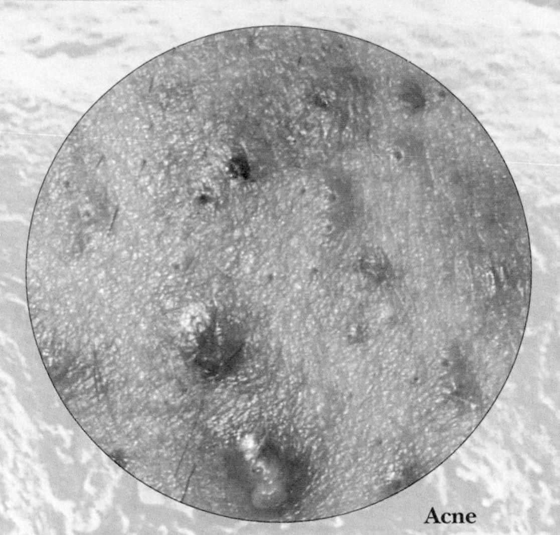

Acne

SPOTS AND PIMPLES
Bacteria or other micro-organisms can infect the skin, resulting in spots and pimples. Skin problems are common during adolescence, when changes in hormone levels cause sebaceous glands to produce excess sebum which becomes trapped in hair follicles. Skin bacteria thrive in the sebum, causing inflammation of surrounding tissues, and often giving rise to acne, as shown above on the face of a teenage boy.

SEM OF BACTERIA ON SKIN

SKIN BACTERIA
This magnified view shows some of the millions of harmless bacteria that normally live on the skin. These bacteria help to keep the skin healthy by preventing harmful varieties from growing. The oil on the surface of the skin also helps eliminate harmful bacteria. Fungi are often present in deep folds in the skin, such as between the toes or in the groin.

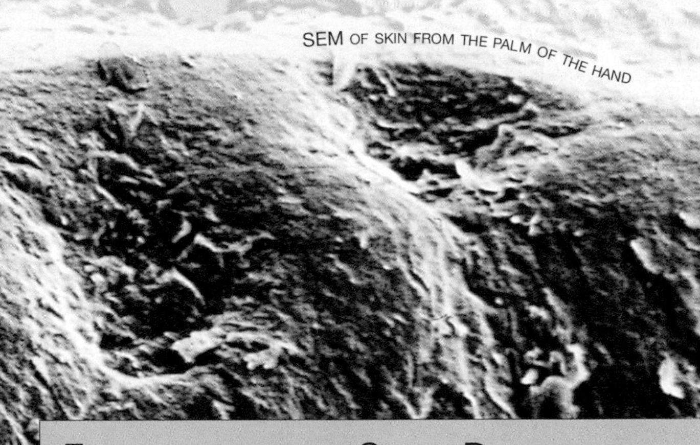

SEM OF SKIN FROM THE PALM OF THE HAND

FINGERPRINTS AND CRIME DETECTION

The swirling epidermal ridges at the ends of the fingers and thumbs produce patterns that are unique to each individual. Well supplied with sweat glands, these ridges leave behind sweat patterns, better known as fingerprints, when they touch smooth surfaces. The unique nature of fingerprints makes them useful in crime detection. Investigators look for similarities between fingerprints found at a crime scene and those of suspects. During the 1880s, interest in the use of fingerprints grew, and London's Metropolitan Police set up a fingerprint department in 1901 to help identify criminals. In 1902, a burglar named Jackson became the first criminal to be convicted on the basis of fingerprint evidence. In the United States, the first fingerprint file was established by J. Edgar Hoover, who became director of the FBI in 1924.

A fingerprint, with its unique pattern of whorls and arches

Skin features

TOGETHER, THE TWO layers of the skin – the epidermis and dermis – combine to give it depth. This can vary considerably – the skin is very thin and delicate in some areas, and much thicker and tougher in others. Regardless of its thickness, skin everywhere undergoes a continuous process of renewal. Dead, flattened cells are constantly worn away to be replaced by new cells generated by the division of living cells in the lowest layer of the epidermis. Scattered within this multiplying layer are cells that produce melanin, the brown pigment that helps to give skin its colour.

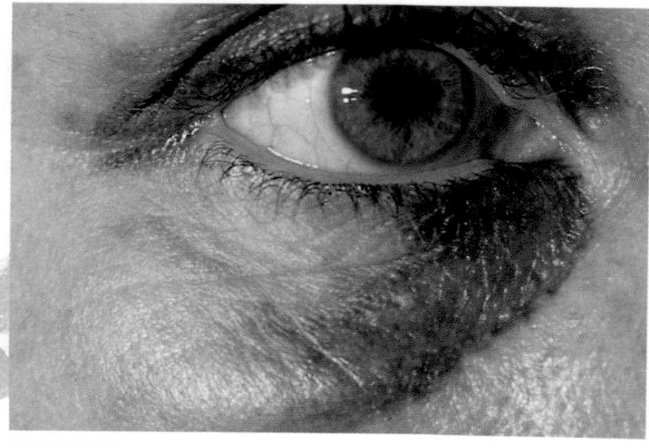

BRUISES
Bruising – which is usually caused by a knock or fall – occurs when blood leaks from damaged blood vessels. It can be very noticeable around the eye because here the skin is not only very thin but also loosely anchored, so blood can easily collect – producing a black eye. Bruises usually look dark purple or blue at first, then turn brown, green, or yellow as they fade.

TWO LAYERS

The SEM below shows the two skin layers: the epidermis above, and the thicker dermis beneath. The epidermis contains flat, overlapping cells that are filled with a tough protein called keratin and which form a protective, waterproof covering. Below, the dermis contains collagen and elastin fibres which give the skin strength and flexibility, allowing it to stretch and return to its normal shape. The dermis is also supplied with nerves and blood vessels.

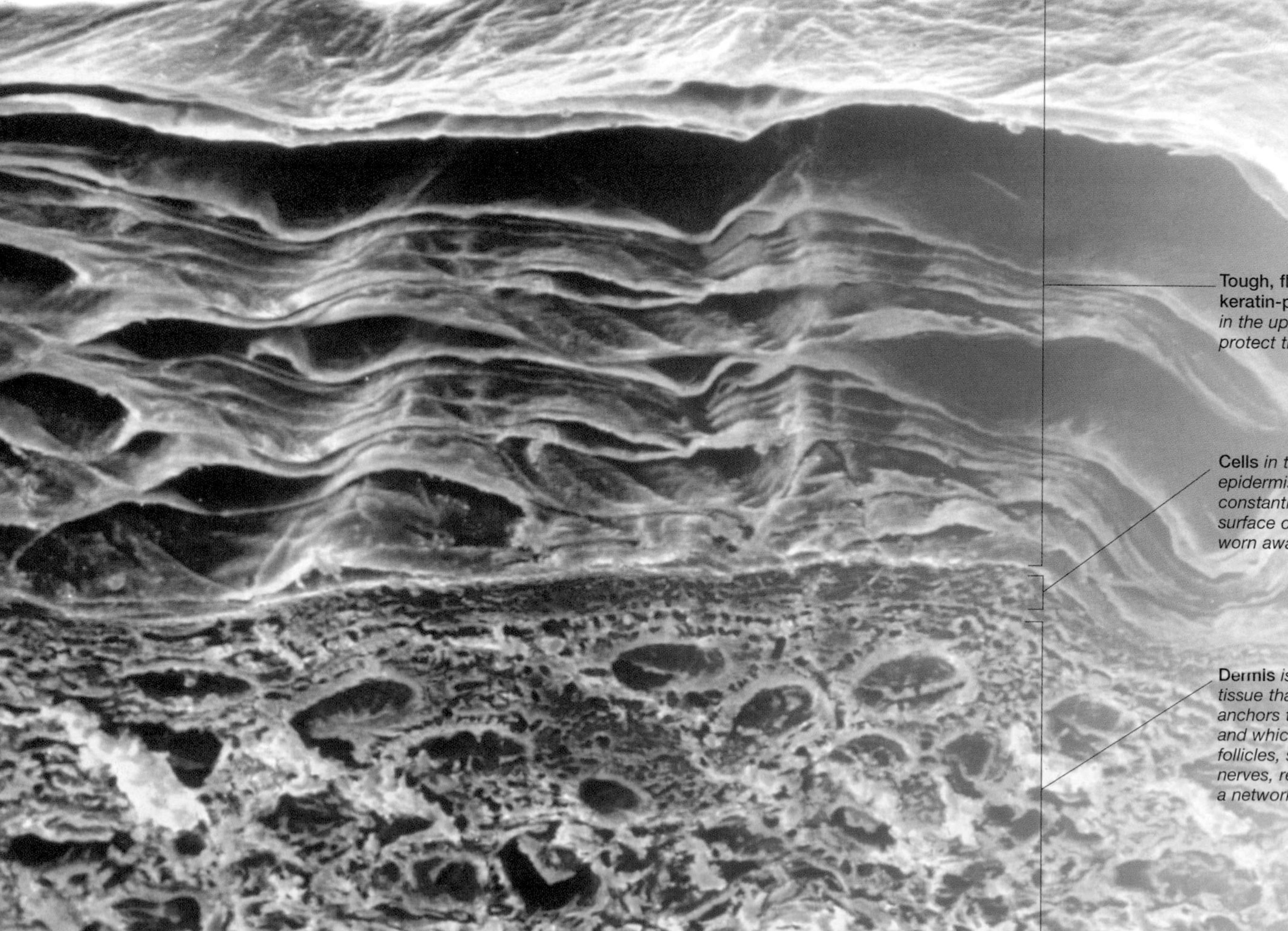

Tough, flattened keratin-packed cells *in the upper epidermis protect the layers below*

Cells *in the lower epidermis divide constantly and replace surface cells that are worn away*

Dermis *is a connective tissue that supports and anchors the epidermis and which contains hair follicles, sweat glands, nerves, receptors, and a network of blood vessels*

Sole of the foot: 4 mm (0.16 in)

Eyelid: 0.5 mm (0.02 in)

HOW THICK?

Skin varies in thickness over the surface of the body. It is very thin on the eyelids and lips, and very thick on the soles of the feet and palms of the hands. The epidermis of the skin tends to become thicker and harder if it experiences a lot of wear and tear. For example, people who often walk barefoot have tougher soles of the feet.

SUNTAN

Sunlight stimulates the skin to make more melanin – which is why skin darkens, or tans, in the sun. The extra melanin provides increased protection against harmful ultraviolet rays from the sun. Too much ultraviolet radiation can cause sunburn. It can also make the skin appear dry and wrinkled, and increases the risk of developing skin cancer in later life. Wearing a hat, and applying sunscreens regularly, helps to reduce the sun's effects on the skin.

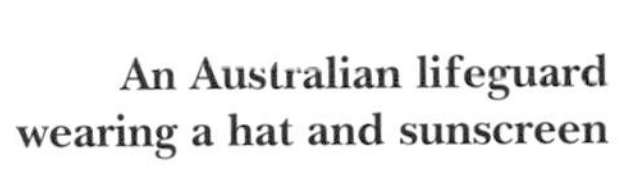

An Australian lifeguard wearing a hat and sunscreen

SKIN FLAKES
Tens of thousands of dead, flattened cells are shed from the skin every minute. They constantly rub off or peel away from the skin surface, like flakes of old paint, to be replaced by new cells that push up from the lower epidermis towards the surface. These fallen flakes of dead skin, together with other particles and fibres, form household dust.

SEM OF SKIN FLAKES

MANY SHADES

Skin colour ranges from nearly black through varying shades of brown to pale pink, and depends on how much melanin the skin contains. Melanin is a dark pigment made by cells called melanocytes in the lower epidermis, and serves to protect the skin against damage by ultraviolet rays. The amount each person has is determined by their ancestry. People with dark skin have a lot of melanin, people with fair skin have less.

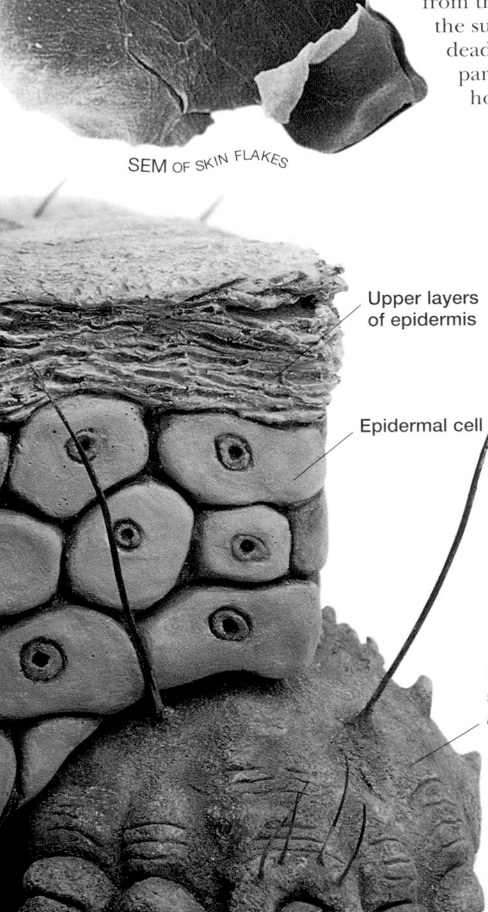

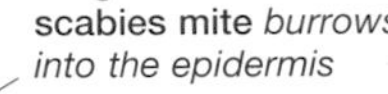

Pregnant female scabies mite *burrows into the epidermis*

ITCH MITES
The itch mite has been a human parasite for thousands of years. The tiny female mite burrows deep into the epidermis, where she lays her eggs, and a few weeks later an intensely itchy rash – called scabies – develops on the trunk and limbs. Passed on via close contact, such as holding hands with an infected person, the scabies mite can only be killed by special medicated lotions and creams.

Hair

ALMOST EVERY PART OF THE body – apart from the palms of the hands, soles of the feet, lips, and nipples – is covered by hair. Hairs are long filament-like structures made up of dead cells that grow out of the skin. Short, fine vellus hair covers much of the body, while longer, thicker terminal hair is found in the eyebrows, eyelashes, nose hairs, and on the scalp. After puberty, terminal hair appears in the armpits and pubic regions of both sexes, as well as on the faces and chests of males. There are about 100,000 hairs on the scalp, of which about 100 are lost daily to be replaced by new growth. Scalp hair serves to insulate the head and protects it from harmful sunlight radiation.

HEAD OF HAIR

Whether hair is straight, curly, or wavy depends on the shape of its shaft. In cross- section, shafts of straight hair are round, those of wavy hair are oval, and those of curly hair flat. Hair colour depends on how much of each of the three melanin variants – yellow, red, and brown-black – hair contains. The relative amounts of each determines whether hair is blond, red, brown, or black. In older people, a slow-down in melanin production produces grey hair.

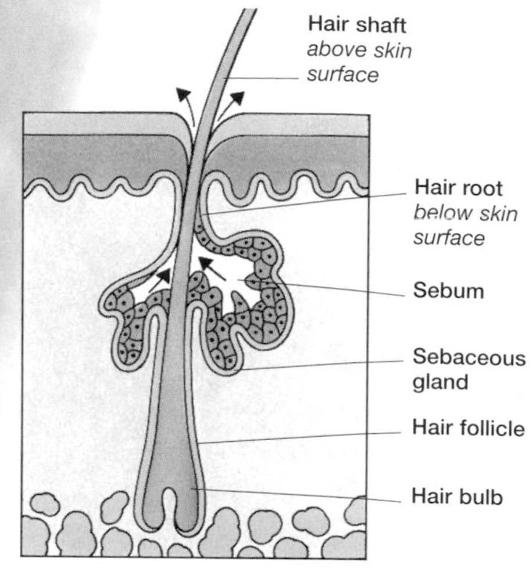

Hair follicle cross-section

HAIR FOLLICLES

Hairs grow from tiny pits in the skin called follicles. Cells in the hair bulb at the bottom of the follicle divide to form the hair and push it upwards. As cells move upwards, they fill with keratin and die, which is why a haircut is painless. Oily sebum, released by sebaceous glands into hair follicles, keeps the hair shaft moist and flexible.

Hair shaft

HAIR GROWTH

The terminal hairs on the scalp and, in the case of adult males, on the face, grow at a rate of about 10 mm (0.4 in) a month. Each hair has a growth phase lasting several years, then a resting phase, before it is pushed out by a new hair. People control the growth by cutting their hair, and many men also shave their beards. Here, beard hairs are regrowing after a shave.

SEM OF SHAVED BEARD HAIR

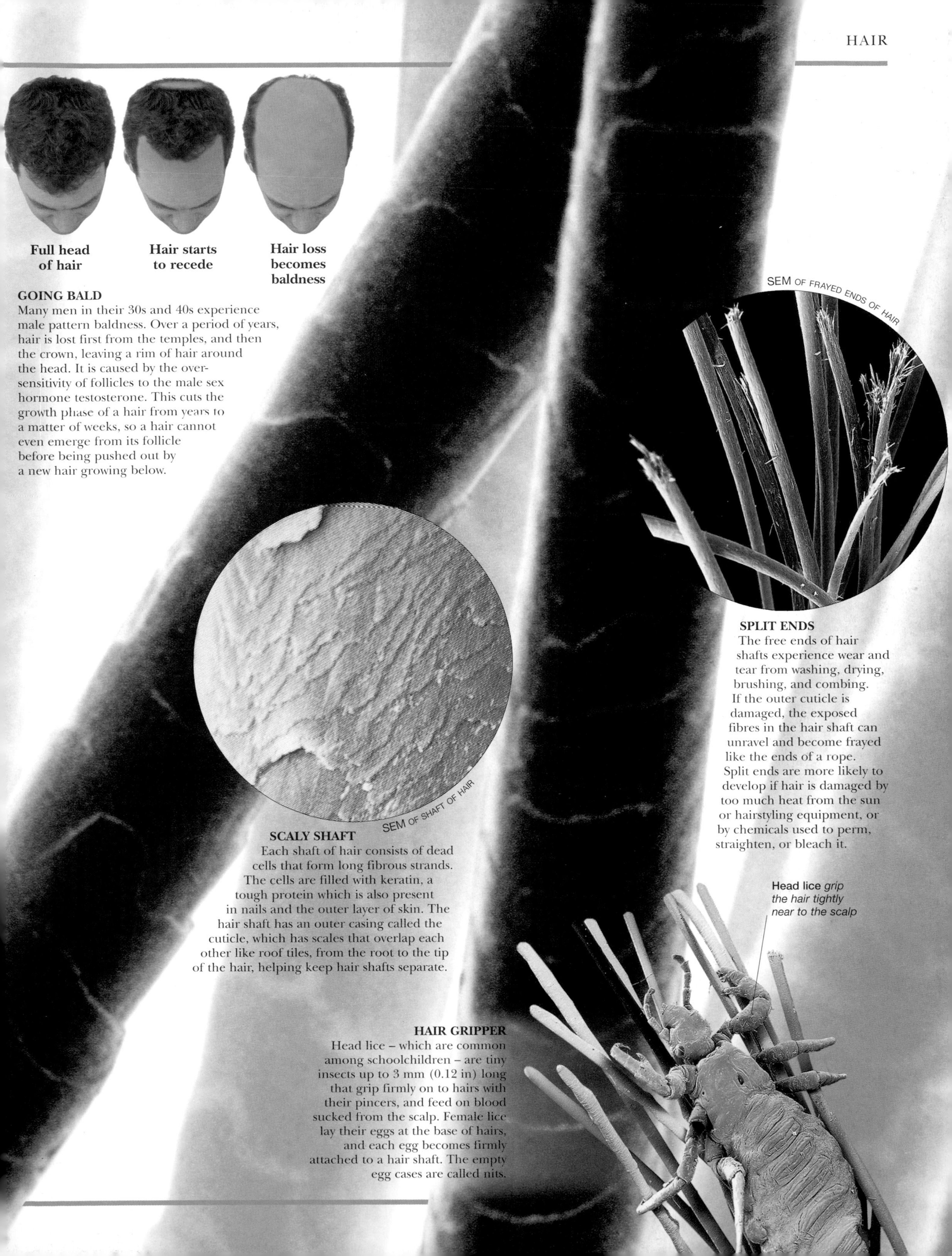

GOING BALD

Many men in their 30s and 40s experience male pattern baldness. Over a period of years, hair is lost first from the temples, and then the crown, leaving a rim of hair around the head. It is caused by the over-sensitivity of follicles to the male sex hormone testosterone. This cuts the growth phase of a hair from years to a matter of weeks, so a hair cannot even emerge from its follicle before being pushed out by a new hair growing below.

SPLIT ENDS

The free ends of hair shafts experience wear and tear from washing, drying, brushing, and combing. If the outer cuticle is damaged, the exposed fibres in the hair shaft can unravel and become frayed like the ends of a rope. Split ends are more likely to develop if hair is damaged by too much heat from the sun or hairstyling equipment, or by chemicals used to perm, straighten, or bleach it.

SCALY SHAFT

Each shaft of hair consists of dead cells that form long fibrous strands. The cells are filled with keratin, a tough protein which is also present in nails and the outer layer of skin. The hair shaft has an outer casing called the cuticle, which has scales that overlap each other like roof tiles, from the root to the tip of the hair, helping keep hair shafts separate.

HAIR GRIPPER

Head lice – which are common among schoolchildren – are tiny insects up to 3 mm (0.12 in) long that grip firmly on to hairs with their pincers, and feed on blood sucked from the scalp. Female lice lay their eggs at the base of hairs, and each egg becomes firmly attached to a hair shaft. The empty egg cases are called nits.

SKELETAL system

THE STRONG INNER framework of the body is formed by the skeletal system. Made up of separate bones that are linked together at joints, the skeletal system not only gives the body its shape, but also provides anchorage for the muscles that move it. It supports and protects vital organs, such as the brain, heart, and lungs. Although bones themselves are rigid, they are linked by joints that give the skeleton a great deal of flexibility. Before birth, the skeleton is made mostly of cartilage, and as a result is less rigid. As the body grows, cartilage is gradually replaced by bone, though some remains in the joints and also in the nose and ears.

Human skeleton seen from the front

Clavicle *(collar bone) extends from sternum to scapula, with which it forms the pectoral girdle*

Scapula *(shoulder blade) has a hollow into which the rounded head of the humerus fits*

Sternum *(breast bone) is connected to the ribs by strips of cartilage*

Ribs *surround and protect the heart and lungs*

Humerus *(upper arm bone) is the longest bone in the arm and extends from shoulder to elbow*

Backbone *(spine) is a strong, flexible chain of bones called vertebrae*

Pelvic (hip) girdle *supports abdominal organs and anchors leg bones*

Radius and ulna *(forearm bones) run from the elbow to the wrist*

LIVING FRAMEWORK

The skeleton is made up of 206 separate bones, which differ in shape, size, and name. The skull, backbone, ribs, and sternum form the central part of the skeleton. The bones of the arms and legs hang symmetrically on either side, attached respectively by the pectoral girdle (clavicle and scapula) and the pelvic girdle. Bones are not dry and lifeless, as may be thought, but are active organs containing living cells surrounded by protein fibres and mineral crystals. They act as mineral stores and are constantly exchanging calcium with blood. Inside some bones, marrow produces red and white blood cells.

Compact bone

Osteocyte

TEM OF AN OSTEOCYTE

BONE CELL

This highly magnified image shows a mature bone cell called an osteocyte. It lies within a fluid-filled space called a lacuna, which is surrounded by compact bone. The cell has a large oval nucleus, which is visible in the lower part of the cell. Osteocytes send thin branches out into the surrounding bone to link with the branches of other osteocytes. Their role is to maintain bone, exchanging nutrients and waste with the blood. Other types of bone cell, called osteoblasts and osteoclasts, make bone and break it down respectively

Metacarpal is one of 27 hand bones that form the most flexible part of the skeleton

Femur (*thigh bone*) *is the largest bone in the body. It has a rounded end that fits into the pelvic girdle; the other end has a wide, grooved surface that forms part of the knee*

RADIONUCLIDE (GAMMA) SCAN OF THE KNEE

18th-century engraving by Deuchar of Holbein's *Dance of Death*

RADIONUCLIDE (GAMMA) SCAN OF A HEALTHY SKELETON

SYMBOL OF DEATH

After death, once the flesh has rotted away, the bones are all that remain. Consequently, the skeleton has, for centuries, been used around the world as a symbol of death and disease. One example of this is the skull and crossbones which was used on the flags of pirates' ships to signal danger to others.

FLEXIBLE SUPPORT

The image on the right shows a scan of the knees – the joints between each femur (thigh bone) and tibia (shin bone). Joints are the parts of the skeleton where two or more bones meet. Held in place by strong bands of tissue called ligaments, they allow the bones to move and give the skeleton its flexibility. Each joint has its own range of movements, though most, like the knees, can move freely. In order to reduce friction, the ends of the bones at joints are covered with smooth cartilage.

Tibia (*shin bone*) *bears most of the weight in the lower leg; its sharp front edge forms the shin*

BONE SCAN

The image on the right is of a radionuclide (gamma) scan of the whole skeleton in a living person. Images such as these, which are produced in hospitals or clinics with special scanning equipment, can be very useful in medicine because they help doctors discover whether the bones are diseased. Doctors also use other tests, such as X-rays, magnetic resonance imaging (MRI), and ultrasound, to examine the skeletal system.

SKELETAL SYSTEM FUNCTIONS

Function	Description
Support	*Provides supportive framework for body tissues and organs; gives the body its shape.*
Protection	*Provides protection for internal organs: ribs protect heart and lungs; skull protects the brain; spine protects the spinal cord; pelvis protects the uterus and bladder.*
Movement	*Provides a strong yet light framework and anchorage for muscles; joints allow flexibility.*
Blood cell production	*Produces different types of blood cells in the red marrow of certain bones.*
Mineral storage	*Acts as a reservoir for minerals, particularly calcium and phosphorus.*

Tarsal *is one of 26 foot bones that supports the body during walking and standing*

BACKGROUND: SEM OF BONE MARROW

Skull

THE MOST COMPLEX PART of the skeleton, the skull shapes the head and face, protects the brain, and houses the special sense organs. It is made of 22 separate bones, 21 of which are locked together by immovable joints to form a structure of extraordinary strength. The only moveable skull bone is the mandible, or lower jaw. Skull bones are divided into two sets. The cranial bones form the domed upper part, or cranium, which surrounds, supports, and protects the brain and the organs of hearing. The facial bones form the framework of the face and jaw, and provide attachment sites for the muscles that produce facial expressions. Together, both cranial and facial bones form the orbits, or eye sockets, and the nasal cavity.

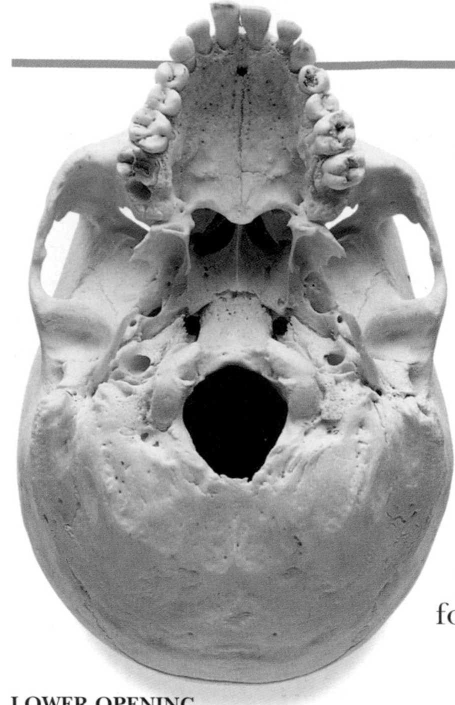

LOWER OPENING
At the base of the skull is a large circular hole – the foramen magnum (above, centre). The lowest part of the brain passes through this opening and continues downwards as the spinal cord. Other smaller holes are for the passage of nerves and blood vessels.

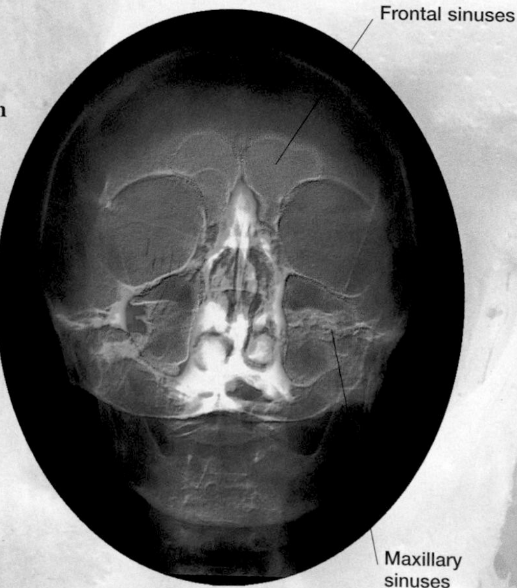

X-ray of the skull, seen from the front, showing two pairs of sinuses

AIR SPACES
Some of the bones surrounding the nasal cavity contain hollow, air-filled spaces called sinuses. The sinuses lighten the skull's weight, and act as an echo chamber, giving a slight "ring" to the voice. They are lined with a moist membrane and connect through small openings with the inside of the nasal cavity.

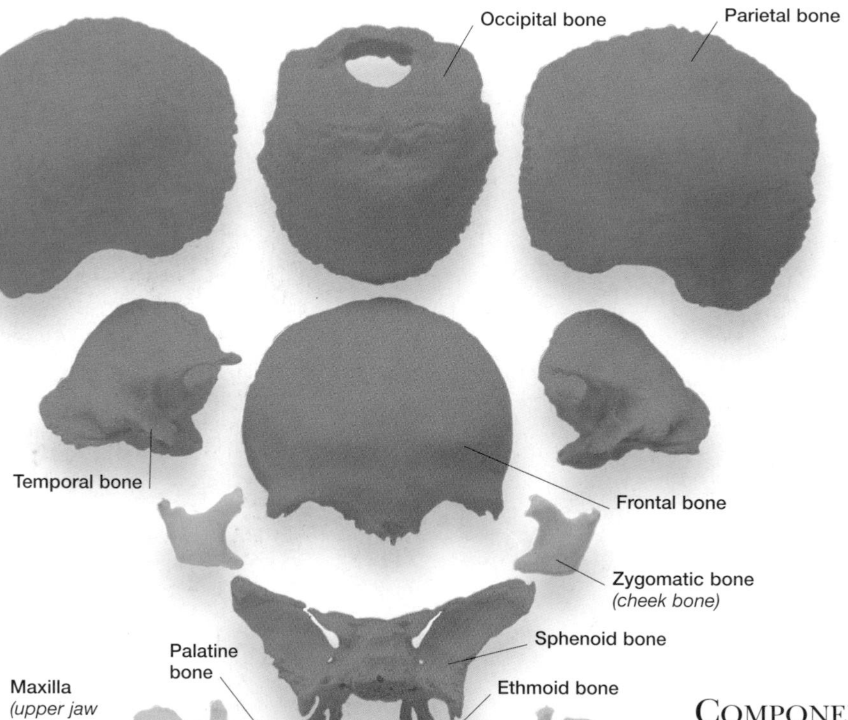

Bones of the skull

Cranial bones (8)

Facial bones (14)

COMPONENT PARTS

Eight bones form the cranium around the brain. The frontal bone is at the front, the two parietal bones form the sides and top, the occipital bone the back and – with the sphenoid – the base, the two temporal bones the side, and the ethmoid part of the nasal cavity. Each temporal bone has an opening to the inner parts of the ear which are encased within. The remaining 14 facial bones form the skeleton of the face. The zygomatic bones are the cheek bones. The palatine bones, nasal bones, inferior conchae, vomer, and lacrimal bones (not shown here) surround the nasal cavity. The maxillae (upper jaw bones) and the mandible (lower jaw bone) contain sockets for the teeth.

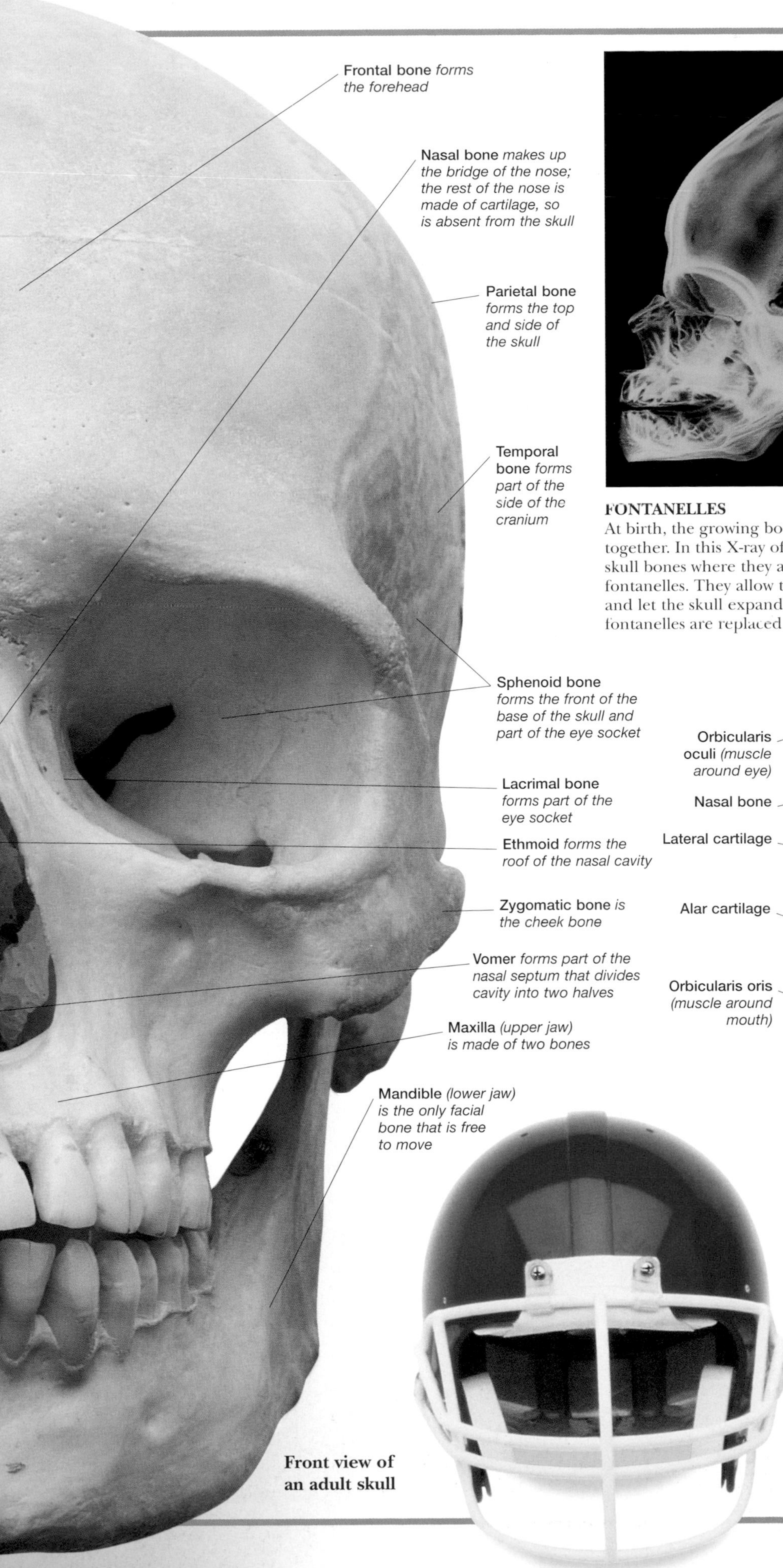

Frontal bone *forms the forehead*
Nasal bone *makes up the bridge of the nose; the rest of the nose is made of cartilage, so is absent from the skull*
Parietal bone *forms the top and side of the skull*
Temporal bone *forms part of the side of the cranium*
Sphenoid bone *forms the front of the base of the skull and part of the eye socket*
Lacrimal bone *forms part of the eye socket*
Ethmoid *forms the roof of the nasal cavity*
Zygomatic bone *is the cheek bone*
Vomer *forms part of the nasal septum that divides cavity into two halves*
Maxilla *(upper jaw) is made of two bones*
Mandible *(lower jaw) is the only facial bone that is free to move*

Front view of an adult skull

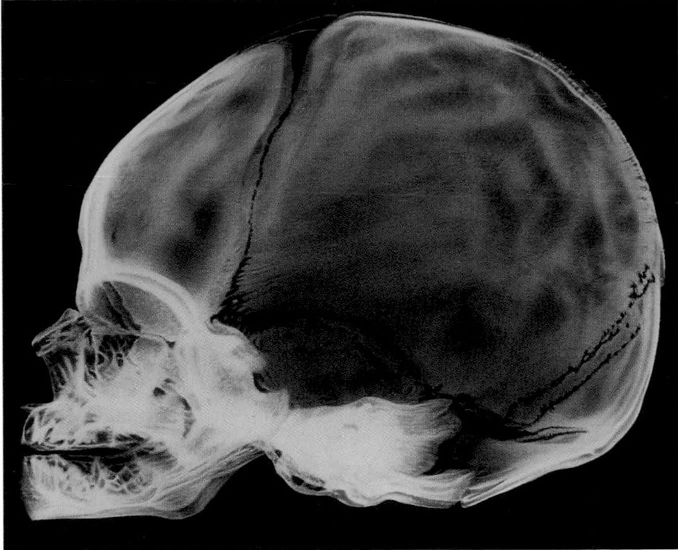

FONTANELLES
At birth, the growing bones of a baby's skull are not yet fixed firmly together. In this X-ray of a baby's skull, "gaps" are visible between skull bones where they are linked by areas of membranes called fontanelles. They allow the skull to be "squeezed" during birth and let the skull expand as the brain grows. By 12 to 18 months, fontanelles are replaced by bone.

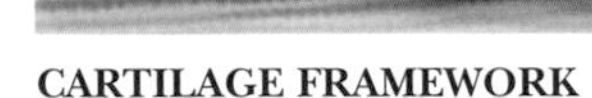

Orbicularis oculi *(muscle around eye)*
Nasal bone
Lateral cartilage
Alar cartilage
Orbicularis oris *(muscle around mouth)*

CARTILAGE FRAMEWORK
Cartilage is a tough, flexible tissue that is softer than bone. The nose (above) is shaped by several pieces. Lateral and alar cartilages form the sides of the nose, while septal cartilages form a central partition between the two nostrils. Cartilage is also present at the bone ends in mobile joints, at the front ends of the ribs, and in the ears.

EXTRA PROTECTION
Heavy blows to the head can be dangerous because they may damage the brain. Helmets provide extra protection against the risk of head injuries during hazardous activities, and have been worn for centuries by soldiers as part of their armour. They are also worn by cyclists and motor cyclists, by people participating in sports such as cricket and American football, and by those working on building sites or in unsafe environments.

Backbone and ribs

THE BACKBONE (also known as the vertebral column or spine) and ribs, together with the skull, form the axial skeleton. These 80 bones run down the mid-line of the body and form its core. The strong yet flexible backbone holds the head and trunk upright, and allows them to bend and twist. It consists of 24 vertebrae, with a further nine that are fused into the sacrum and coccyx. Vertebrae are irregular bones, each with a centrum that bears the body's weight, and extensions or processes that form joints with other vertebrae or provide attachment points for ligaments and muscles. Twelve pairs of thin ribs curve from the backbone round to the sternum (breastbone), protecting the organs of the thorax and aiding breathing.

SHAPED LIKE AN "S"

Four curves – cervical, thoracic, lumbar, and sacral – give the backbone its characteristic "S" shape when viewed from the side. This shape – aided by the discs between vertebrae – gives the backbone the springiness to absorb shocks during movement, strengthens it, and positions the body directly over the legs and feet. Seven cervical vertebrae at the neck support the head. The top two, the atlas and axis, allow the head to nod and shake. Twelve thoracic vertebrae form joints with the ribs. Five large lumbar vertebrae bear most of the body's weight. The sacrum, made of five fused vertebrae, anchors the pelvic girdle before tapering into the four fused vertebrae of the coccyx.

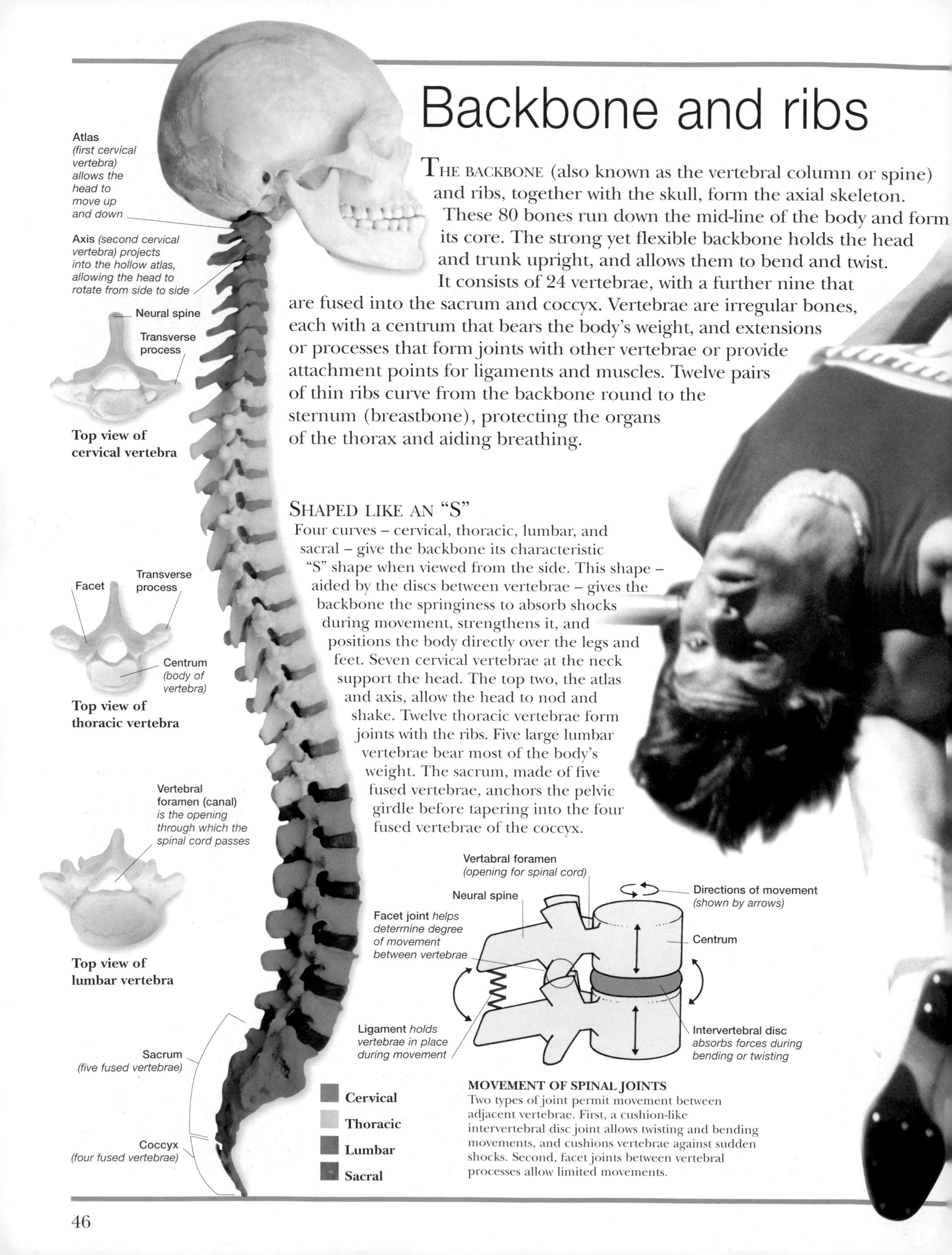

MOVEMENT OF SPINAL JOINTS
Two types of joint permit movement between adjacent vertebrae. First, a cushion-like intervertebral disc joint allows twisting and bending movements, and cushions vertebrae against sudden shocks. Second, facet joints between vertebral processes allow limited movements.

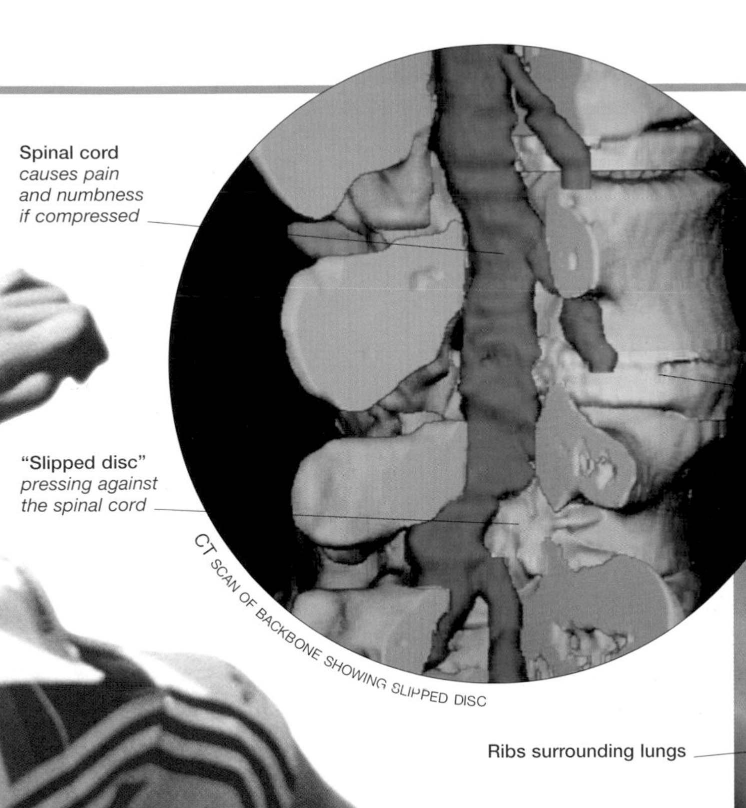

CT SCAN OF BACKBONE SHOWING SLIPPED DISC

SLIPPED DISC

Each intervertebral disc between the vertebrae consists of a pad of fibrous cartilage with a jelly-like centre. Sometimes the fibrous coat breaks open, and part of the disc's core protrudes. If this happens, the core may put pressure on a spinal nerve or, as seen in this CT scan, the core (yellow) actually presses on the spinal cord (blue). This disc prolapse – known more commonly as a slipped disc – causes back pain, and may also cause weakness and pain in the arms and legs.

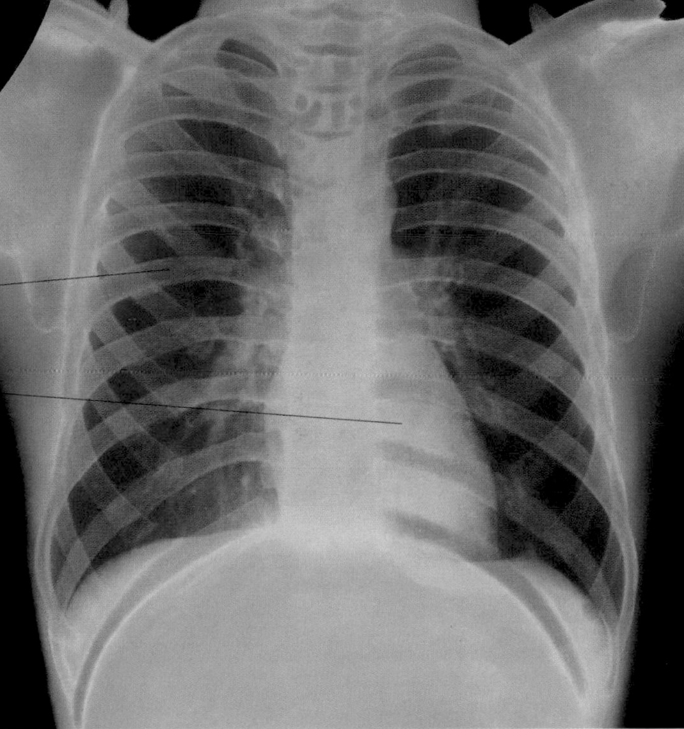

CHEST X-RAY OF AN 11-YEAR-OLD BOY

ORGAN PROTECTOR

This chest X-ray shows the cage-like structure formed by the ribs that protects the heart, lungs, and the organs of the upper abdomen. The 12 pairs of flat, curved bones extend from the backbone, where the rear end of each rib forms a joint with one of the thoracic vertebrae, around the wall of the thorax to meet the sternum at the front. Flexible costal cartilage connects the upper ten ribs to the sternum. The up and down movements of the ribcage during breathing move air in and out of the lungs.

SPINE FLEXIBILITY

The joints between neighbouring vertebrae that make up the bony chain running down the back allow only limited movement. But, added together, these small movements make the backbone as a whole very flexible. The backbone can bend forwards and backwards, and from side to side, and permits rotation with the body twisting on its axis. The body can bend further forwards (flexion) than it can backwards (extension) because the shape of vertebrae limits backward movement. Of all the vertebrae, the cervical vertebrae allow the greatest flexibility, a feature that becomes obvious in human neck movements.

This high-jumper demonstrates the flexibility of the spine

Limbs and girdles

WHENEVER SOMEONE jumps in the air, kicks a ball, swings a tennis racket, or writes a letter, he or she is making full use of their limbs and girdles. The limbs are the arms and legs, while the girdles are rings of bones that attach each limb pair to the axial, or central, part of the skeleton. Arm and leg bones share the same fundamental structure. Each has the same number of bones organized into three major segments – a single upper bone, a pair of lower bones, and a collection of small hand or foot bones – linked by freely movable joints. The pectoral (shoulder) girdle, consisting on each side of a scapula (shoulder blade) and a clavicle (collar bone), forms the link between the axial skeleton and the upper arm bones, while the pelvic (hip) girdle does the same job for the upper leg bones. The pectoral girdle is the weaker but more mobile of the two, and the pelvic is stronger but more rigid. Together, limbs and girdles make up that section of the bony framework called the appendicular skeleton.

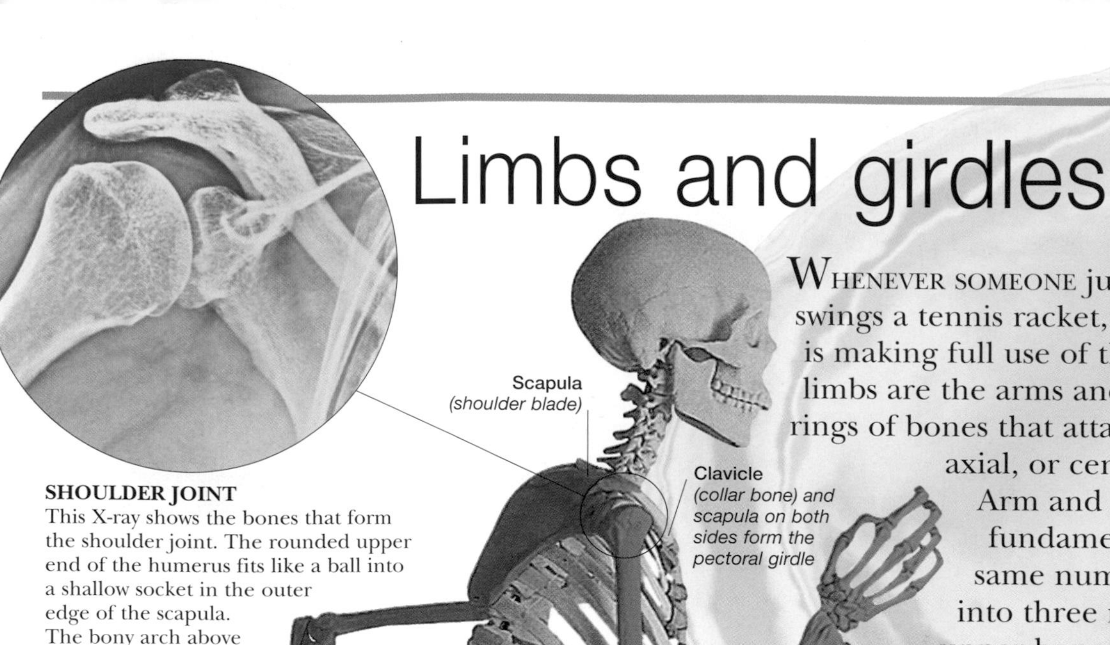

Scapula *(shoulder blade)*
Clavicle *(collar bone) and scapula on both sides form the pectoral girdle*
Elbow *is formed by two joints between the radius and humerus and ulna and humerus*
Hand *is made up of 27 small bones*
Ulna *hooks into the humerus and runs down to the wrist on the little finger side*
Humerus *(upper arm) is the longest bone in the arm*
Radius *(lower arm) crosses over the ulna when the palm faces down*
Pelvic girdle, *formed from two hip bones, anchors the femurs to the rest of the skeleton*
Appendicular skeleton
Axial skeleton
Femur *(thigh bone) has a rounded top that fits into a hip bone, and a wide, lower end that forms part of the knee*
Tibia *(shin bone) bears most of the weight in the lower leg, and has a sharp front, the shin*
Fibula *is the smaller of the lower leg bones; helps to swivel the foot*
Knee joint *is a hinge joint between the femur and tibia*
Foot *is made up of 26 bones, and forms a flexible joint with the tibia and fibula at the ankle*

SHOULDER JOINT
This X-ray shows the bones that form the shoulder joint. The rounded upper end of the humerus fits like a ball into a shallow socket in the outer edge of the scapula. The bony arch above the humerus is also part of the scapula, which joins with the clavicle (collar bone).

APPENDICULAR SKELETON

The 126 bones of the appendicular skeleton (blue) "append", or hang on to, the axial skeleton. Thanks to the looseness of the pectoral girdle, arm bones have an incredible range of movement which, coupled with the flexibility of the hand bones, makes them ideal for manipulating objects, a key skill for humans. Leg bones are thicker and firmly attached to the immobile pelvic girdle, reflecting their roles of moving and supporting the body's weight, but reducing their overall mobility compared with arm bones.

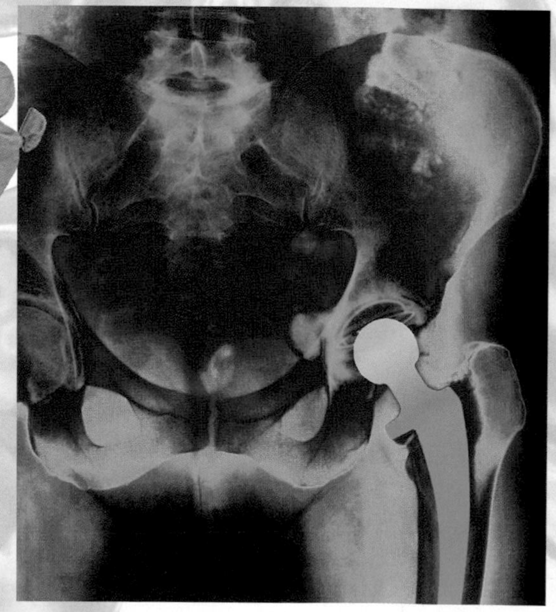

ARTIFICIAL JOINTS
This false colour X-ray shows an artificial left hip joint. Hips, knees, and sometimes other joints that have been damaged by arthritis or injury, often need to be replaced in older people. During joint replacement surgery, the ends of the damaged bones are removed and replacement parts (yellow) made of metal or other synthetic materials are inserted instead.

Looking at Bones

Anatomy, the study of the structure of the human body, has been a subject of great interest since ancient times. In the past, it was made difficult because many cultures forbade the dissection, or cutting up, of dead bodies to look inside. For that reason, the easiest way to study anatomy was to look at the bones, the only part of the body to remain long after death. This illustration from 1363 shows a lesson in skeletal anatomy given by the French anatomist Guy de Chauliac.

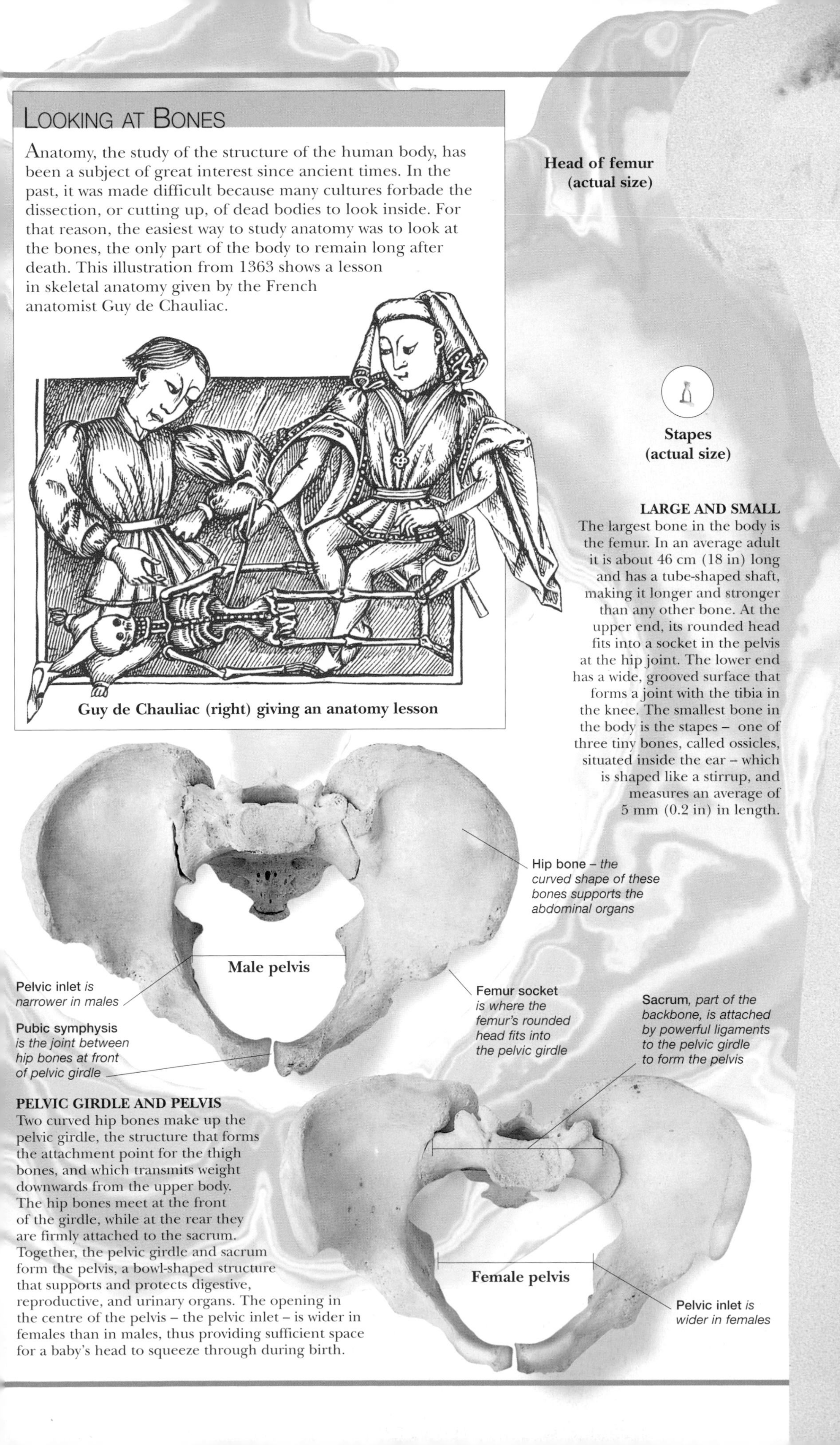

Head of femur
(actual size)
Stapes
(actual size)
Hip bone – the curved shape of these bones supports the abdominal organs
Male pelvis
Pelvic inlet is narrower in males
Pubic symphysis is the joint between hip bones at front of pelvic girdle
Femur socket is where the femur's rounded head fits into the pelvic girdle
Sacrum, part of the backbone, is attached by powerful ligaments to the pelvic girdle to form the pelvis
Female pelvis
Pelvic inlet is wider in females

Guy de Chauliac (right) giving an anatomy lesson

LARGE AND SMALL

The largest bone in the body is the femur. In an average adult it is about 46 cm (18 in) long and has a tube-shaped shaft, making it longer and stronger than any other bone. At the upper end, its rounded head fits into a socket in the pelvis at the hip joint. The lower end has a wide, grooved surface that forms a joint with the tibia in the knee. The smallest bone in the body is the stapes – one of three tiny bones, called ossicles, situated inside the ear – which is shaped like a stirrup, and measures an average of 5 mm (0.2 in) in length.

PELVIC GIRDLE AND PELVIS

Two curved hip bones make up the pelvic girdle, the structure that forms the attachment point for the thigh bones, and which transmits weight downwards from the upper body. The hip bones meet at the front of the girdle, while at the rear they are firmly attached to the sacrum. Together, the pelvic girdle and sacrum form the pelvis, a bowl-shaped structure that supports and protects digestive, reproductive, and urinary organs. The opening in the centre of the pelvis – the pelvic inlet – is wider in females than in males, thus providing sufficient space for a baby's head to squeeze through during birth.

Hands and feet

THE HINGED ATTACHMENTS at the lower ends of the limbs are called hands and feet. The hands are attached at the wrist joints, and the feet at the ankle joints. Although both hands and feet have five flexible digits, the hands are more versatile. They can perform an enormous range of movements. In particular, they can form a grip that allows them to hold and handle objects. The feet are mobile platforms for the body. They support its weight and also act as springboards to propel the body forward during walking, running, or jumping.

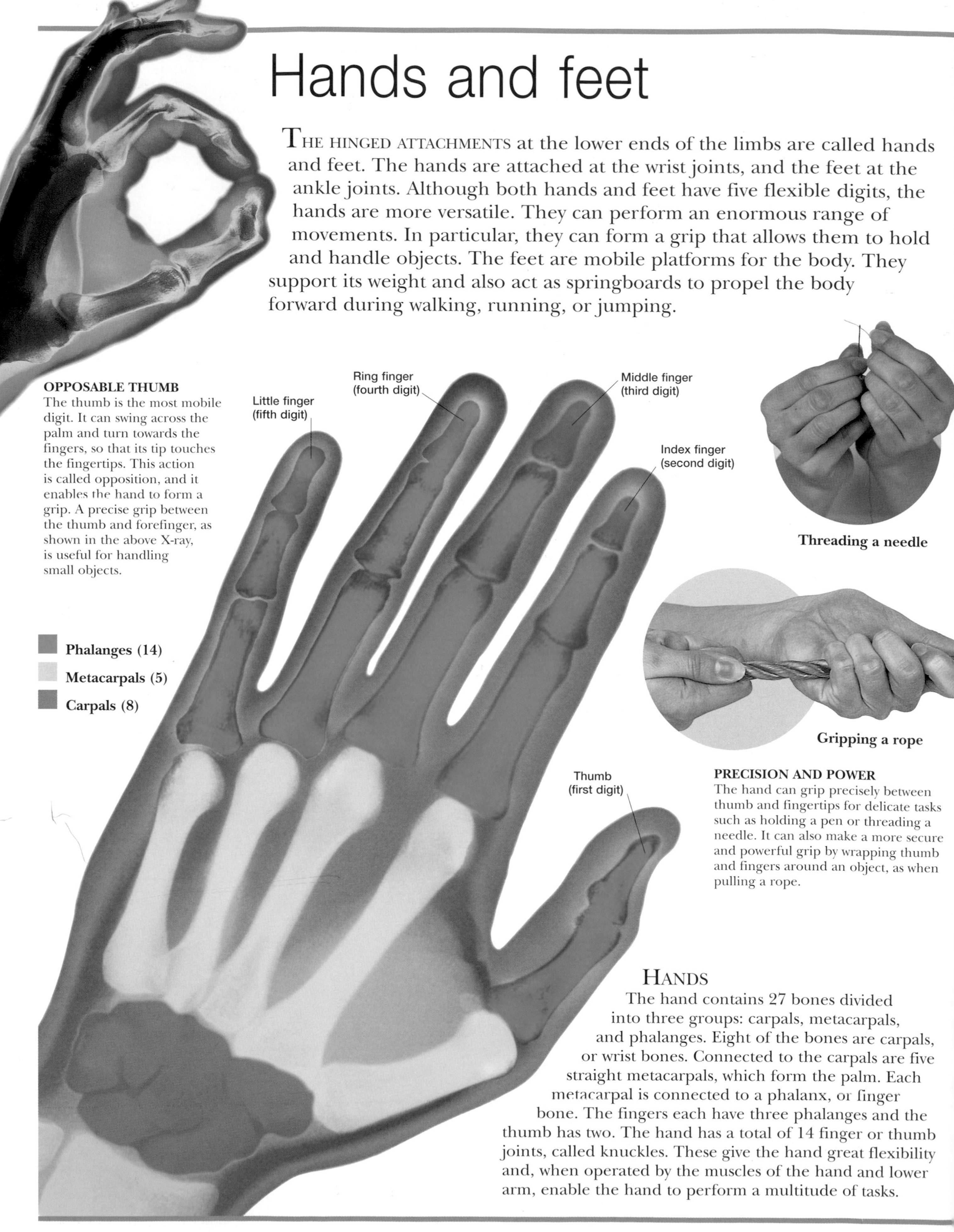

OPPOSABLE THUMB
The thumb is the most mobile digit. It can swing across the palm and turn towards the fingers, so that its tip touches the fingertips. This action is called opposition, and it enables the hand to form a grip. A precise grip between the thumb and forefinger, as shown in the above X-ray, is useful for handling small objects.

Threading a needle

Gripping a rope

PRECISION AND POWER
The hand can grip precisely between thumb and fingertips for delicate tasks such as holding a pen or threading a needle. It can also make a more secure and powerful grip by wrapping thumb and fingers around an object, as when pulling a rope.

HANDS

The hand contains 27 bones divided into three groups: carpals, metacarpals, and phalanges. Eight of the bones are carpals, or wrist bones. Connected to the carpals are five straight metacarpals, which form the palm. Each metacarpal is connected to a phalanx, or finger bone. The fingers each have three phalanges and the thumb has two. The hand has a total of 14 finger or thumb joints, called knuckles. These give the hand great flexibility and, when operated by the muscles of the hand and lower arm, enable the hand to perform a multitude of tasks.

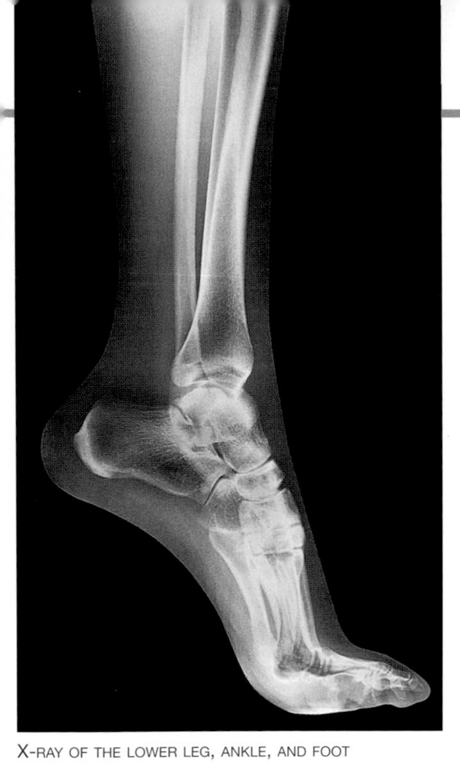

X-RAY OF THE LOWER LEG, ANKLE, AND FOOT

JUMP OFF
The feet provide support and stability to prevent the body from falling over while standing or moving on flat or uneven surfaces. The feet are also strong flexible levers that push the body off the ground during walking, climbing, running, and jumping. This X-ray shows the bones of the foot, ankle joint, and lower leg as the foot pushes the body up. The toes are the only part that remain in contact with the ground.

FEET

Each foot contains 26 bones arranged in three groups: tarsals, metatarsals, and phalanges. There are seven tarsals, or ankle bones. The largest of these is the calcaneus, or heel bone, which projects backwards behind the ankle joint. The calcaneus is connected to the talus, which forms a joint with the bones of the lower leg. In front of the tarsals are five metatarsal bones, or sole bones, linked to the 14 phalanges, two of which are in the big toe, and three in each other toe. The feet are less flexible than the hands because they are bound together by strong ligaments.

Great (big) toe (first digit)

Second toe (second digit)

Third toe (third digit)

Fourth toe (fourth digit)

Little toe (fifth digit)

Talus *forms a joint with the tibia and fibula at the ankle*

Calcaneus *(heel bone)*

Phalanges (14)

Metatarsals (5)

Tarsals (7)

FLEXIBLE ARCH
Footprints made by bare feet reveal that the soles are not flat and that only part of the foot touches the ground. Flexible curves, called arches, raise part of the foot off the ground. The arches are held up by ligaments and by tendons pulled by their muscles. Arches help spread out body weight and provide a springiness that absorbs shocks during running and walking. Feet with reduced arches are called flat feet.

BONES IN EVOLUTION

MUCH OF OUR KNOWLEDGE of our ancestry comes from the study of the human skeleton. Bones and teeth are the only parts of the body that form fossils. This usually occurs when human remains become buried in mud, and bone tissue is replaced by minerals. After death, bones become separated, so it is extremely rare to find more than scattered fossilized bones and skull fragments. But painstaking searches in east Africa, followed by careful measurements and comparisons of bones, have allowed palaeontologists to piece together the story of our evolution from our ape-like ancestors.

PROCONSUL AFRICANUS
Skeletons of ape-like *Proconsul africanus* (above), which lived about 20 million years ago, show long pelvic bones and arms typical of primates that walk using all four limbs. By 5 million years ago, hominids have strong leg bones and a broad pelvis, suggesting they walked on two legs.

DOWN FROM THE TREES

Humans belong to a group of mammals called primates, and share a common ancestor with lemurs, monkeys, and humans. The hominids, a family of upright-walking humans to which we belong, arose some 6 million years ago, at a time when the climate in Africa became drier and grasslands began to replace forests. As our ape-like ancestors spent more time foraging and hunting on the ground, a species that walked upright evolved. It seems likely that the ability to see approaching danger over tall grasses, the advantage of having hands free to manipulate objects, and the necessity to cool the body in a hot climate were all important factors in the transition from four legs to two.

Large, projecting jaw
Clavicle
Humerus
Vertebra
Carpal
Femur (thigh bone)
Knee joint
Tibia (upper piece)

AUSTRALOPITHECUS AFARENSIS
About 40 per cent of the skeleton of "Lucy" has been found. She belonged to the species *Australopithecus afarensis*, the first known primate to have walked upright. Lucy was about 143 cm (4 ft 9 in) tall and is named after the Beatles' song "Lucy in the Sky with Diamonds", which was popular among the palaeontologists who found her.

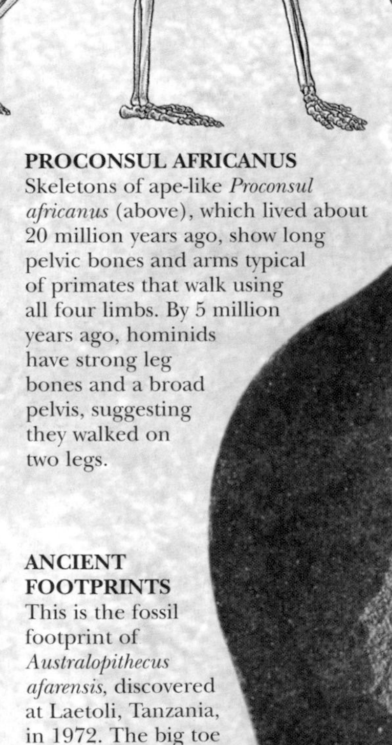

ANCIENT FOOTPRINTS
This is the fossil footprint of *Australopithecus afarensis*, discovered at Laetoli, Tanzania, in 1972. The big toe (top right) hardly diverges from the rest of the foot, unlike that of primates such as chimpanzees that walk on all fours and use this toe like a thumb. This fossil demonstrates conclusively that these early humans walked upright on two legs.

EARLY FAMILY HISTORY

Fossil bones of ape-like *Proconsul africanus*, which lived 20 million years ago, provide palaeontologists with a point of reference for comparing more recent hominid fossils. *Proconsul* had the arched backbone, long pelvic bones, and long fore limbs that are typical of a primate that walks on all four limbs, with gripping hands and feet for climbing. The earliest hominids that walked upright were the Australopithecines (meaning "southern ape"). In 1974, American Donald Johanson (b. 1943) found the fossilized partial skeleton of *Australopithecus afarensis* in the Afar region of Ethiopia.

HOMO HABILIS
Fossil bones of *Homo habilis* (meaning "handy man") have been found among the stone tools that he used. This species was only about 135 cm (4 ft 6 in) tall and had a more rounded head, narrower jaw, and longer face than *Australopithecus*.

***Homo habilis* used simple tools**

"Lucy", as the specimen became known, was 3 million years old. She had the curved, S-shaped spine and short pelvis that is typical of a primate that walks upright. Other key features of upright-walking primates include big toes aligned with all of the others, and feet with pronounced arches. In 1972, Mary Leakey (1913–96) found 3.5-million-year-old footprints at Laetoli in Tanzania that had been left by feet just like this, preserved in hardened volcanic ash. These provide the earliest direct evidence of primates walking upright.

HUMANS APPEAR

Fossil bones of *Homo habilis*, the earliest member of our own genus, were found by Mary and Louis Leakey (1903–72) in 1960, in the Olduvai Gorge of Tanzania. This species lived about 2 million years ago, and it had a small cranium, slightly protruding face, and long, thick leg bones. Fossil bones of *Homo erectus*, which evolved about 1.8 million years ago, have been found throughout Europe and Asia, suggesting that this was the first species to have travelled out of Africa and colonized other continents. Fossil bones show that Neanderthals then evolved as a distinct species about 250,000 years ago, and modern humans evolved alongside them until Neanderthals became extinct, about 35,000 years ago.

HOMO ERECTUS
Homo erectus was taller than *Homo habilis* and colonized Europe and Asia. Chinese fossils dating from 360,000 years ago have been found in caves amongst ash, charcoal, and charred bones – showing that this species had learned to use fire.

Eyebrow ridges *are very distinctive in Neanderthals*

Skull of *Homo neanderthalensis*

HOMO NEANDERTHALENSIS
Studies of fossil hominid skulls show that skull size and brain volume have increased, teeth have become smaller, jaws less jutting, eyebrow ridges less prominent, and the chin more pronounced. Neanderthals had broad skulls and pronounced eyebrow ridges, but could easily pass unnoticed in the street today.

PALAEONTOLOGIST AT WORK
Palaeontologists rarely find more than fragments of skeletons. These are collected, photographed, carefully measured, and used to reconstruct the appearance of our earliest ancestors. Computer technology is often used to analyse and reconstruct skeletons.

Bone structure

BONES ARE REMARKABLE pieces of engineering, combining lightness with a strength that rivals steel. They are also slightly springy, which helps them to withstand sudden jolts and knocks. This unusual mix of features comes from their underlying structure, which consists of living cells and protein fibres, wrapped around layers of hard mineral salts. Bones are at their densest and toughest near the outside, where they bear the most stress. On the inside they have a sponge-like design that helps to reduce weight. Long bones – like the ones in the arms and the legs – also have a large central cavity. This is packed with marrow, a substance that stores fat and makes blood cells.

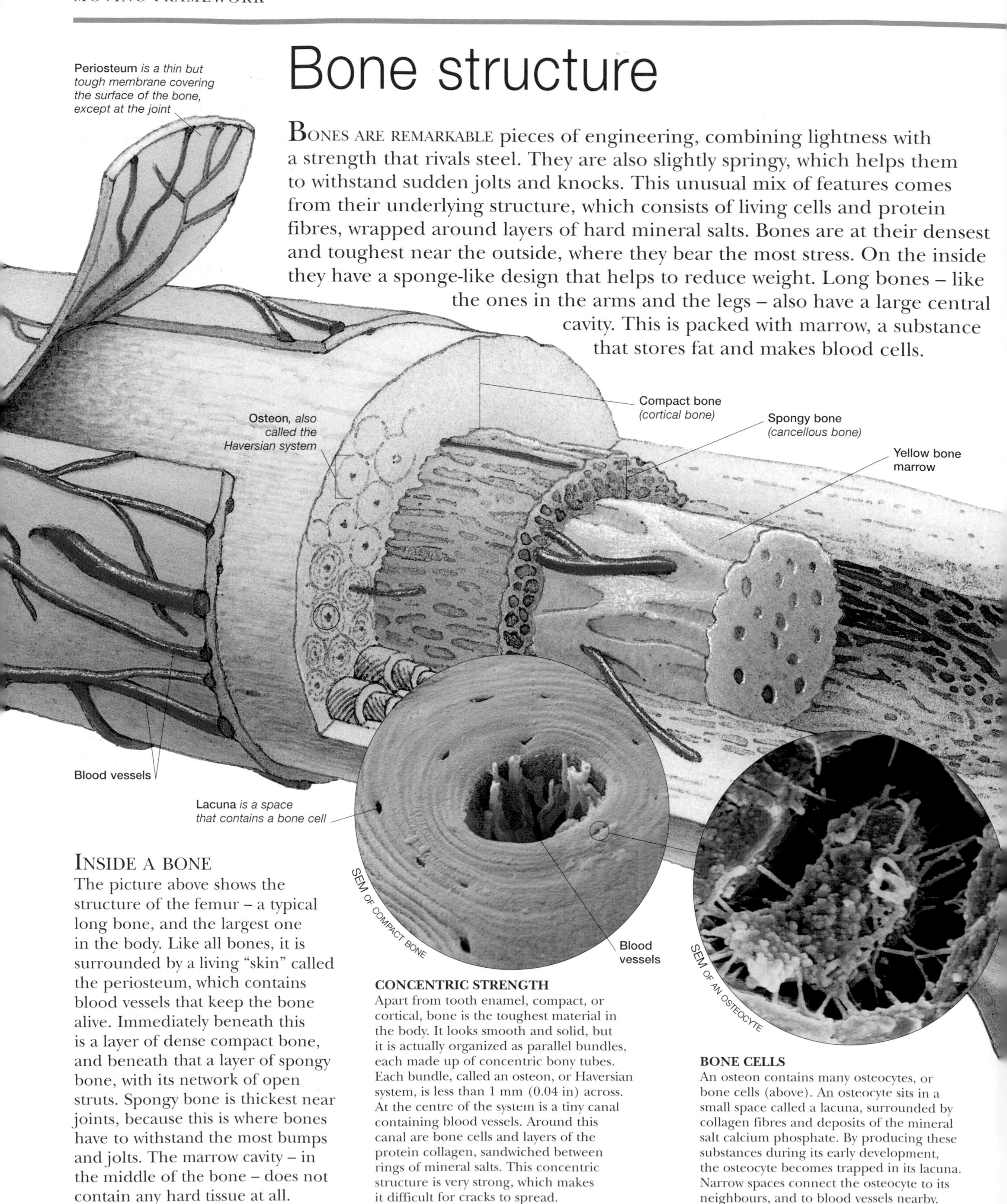

INSIDE A BONE

The picture above shows the structure of the femur – a typical long bone, and the largest one in the body. Like all bones, it is surrounded by a living "skin" called the periosteum, which contains blood vessels that keep the bone alive. Immediately beneath this is a layer of dense compact bone, and beneath that a layer of spongy bone, with its network of open struts. Spongy bone is thickest near joints, because this is where bones have to withstand the most bumps and jolts. The marrow cavity – in the middle of the bone – does not contain any hard tissue at all.

CONCENTRIC STRENGTH
Apart from tooth enamel, compact, or cortical, bone is the toughest material in the body. It looks smooth and solid, but it is actually organized as parallel bundles, each made up of concentric bony tubes. Each bundle, called an osteon, or Haversian system, is less than 1 mm (0.04 in) across. At the centre of the system is a tiny canal containing blood vessels. Around this canal are bone cells and layers of the protein collagen, sandwiched between rings of mineral salts. This concentric structure is very strong, which makes it difficult for cracks to spread.

BONE CELLS
An osteon contains many osteocytes, or bone cells (above). An osteocyte sits in a small space called a lacuna, surrounded by collagen fibres and deposits of the mineral salt calcium phosphate. By producing these substances during its early development, the osteocyte becomes trapped in its lacuna. Narrow spaces connect the osteocyte to its neighbours, and to blood vessels nearby.

SEM OF SPONGY BONE AFFECTED BY OSTEOPOROSIS

OSTEOPOROSIS

As people grow older, they can lose calcium and protein in their bones, causing the bones to become lighter and more likely to break. This condition, known as osteoporosis, affects both women and men, though women are particularly at risk because of changes in their hormone levels following the menopause (see p. 242). The photograph on the left shows spongy bone that has been weakened in this way. Compared with healthy spongy bone, shown below, its struts are full of spaces, or pores.

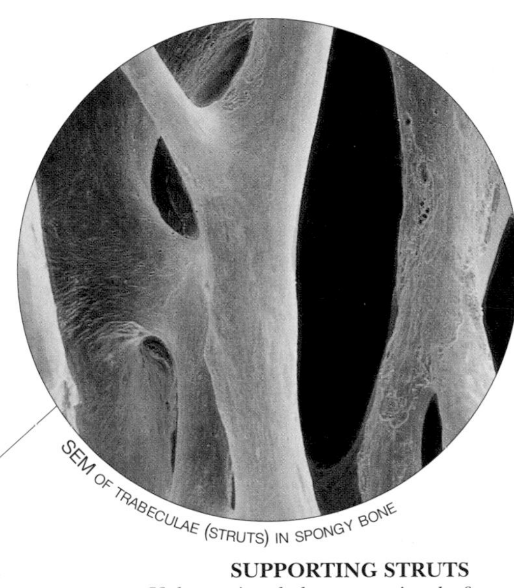
SEM OF TRABECULAE (STRUTS) IN SPONGY BONE

SUPPORTING STRUTS

If the entire skeleton consisted of compact bone, it would be much too heavy to move. Fortunately, spongy, or cancellous, bone helps to reduce its weight. Despite its name, it is not really spongy. Instead, it is made up of a network of rigid struts, which takes some of the load that the bone has to bear. These struts are separated by a maze of tiny spaces, which are filled with bone marrow.

SEM OF RED BONE MARROW

BONE MARROW

Marrow is an important tissue because it is responsible for making blood cells. At birth, all bones contain blood-cell producing red marrow, but, during the teenage years, some of it is replaced by yellow marrow, which consists mainly of fat. In adults, red marrow is found mainly in spongy bone. Yellow marrow is concentrated in the central cavity of long bones, where it acts mainly as a storage tissue.

Structure of a femur (thigh bone)

Growth and repair

UNLIKE SCAFFOLDING or steel girders, bones change all the time. They grow in step with the rest of the body, can repair minor fractures, and even some major breaks. Even when they have reached adult size, they are constantly but invisibly renewed. How do they do it? The answer lies in the cells that all bones contain. Some of these cells, called osteoblasts, deposit protein and mineral salts, forming new bone. At the same time, other cells, called osteoclasts, do the reverse. By adjusting the balance between these two processes, the body makes bones grow. It also ensures that bones keep strong, by renewing or "remodelling" parts that get the most wear and tear.

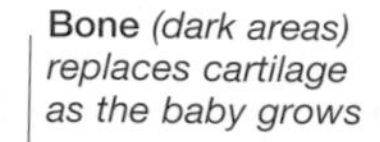
Bone *(dark areas) replaces cartilage as the baby grows*
Joints *are one of the last areas to be ossified*

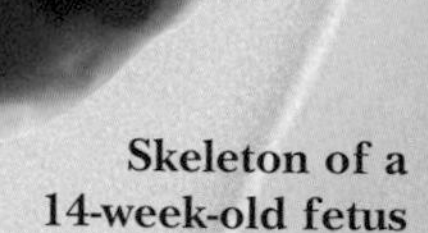

Skeleton of a 14-week-old fetus

EARLY DEVELOPMENT

When the skeleton first forms, it consists solely of cartilage. During a process called ossification, osteoblast cells lay down mineral salts in this framework, gradually turning most of it into bone. This image shows a fetus that is 14 weeks old. The dark areas show that its bones are partially ossified, but its joints still consist of cartilage. Ossification continues after birth, and it lasts throughout adult life. For example, the flap of cartilage below the breastbone, called the xiphoid process, often ossifies at the age of 40 or older.

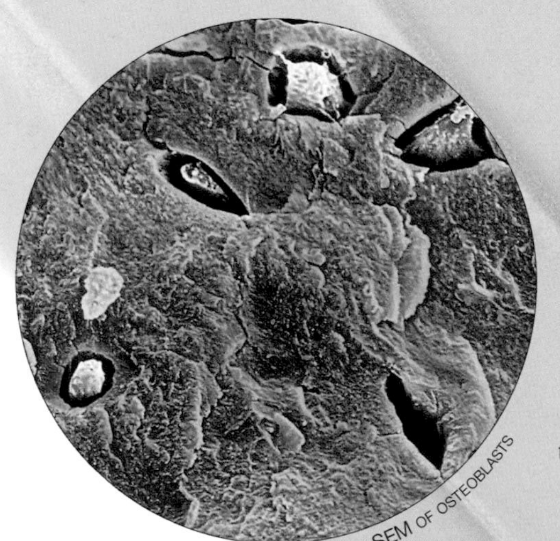

SEM OF OSTEOBLASTS

BUILDING BONES
This close-up view shows active osteoblasts (bone-making cells) on the surface of a bone. They lie just underneath the bone's outer "skin" or periosteum, and build up the bone from the outside. As a bone grows, they gradually become embedded in the bone itself. Once this has happened, they retire from bone-making, and change into mature bone cells or osteocytes.

GROWING BONES
The X-rays below show how hand bones develop between infancy and early adulthood. At first, the bones are only partially ossified, and their ends consist of cartilage. But as the cartilage grows, the ossified zone also expands, until finally the adult bone is complete. When bones grow, most of the growth occurs at their ends. Existing bone tissue often has to be broken down so that the bone can reach its adult shape.

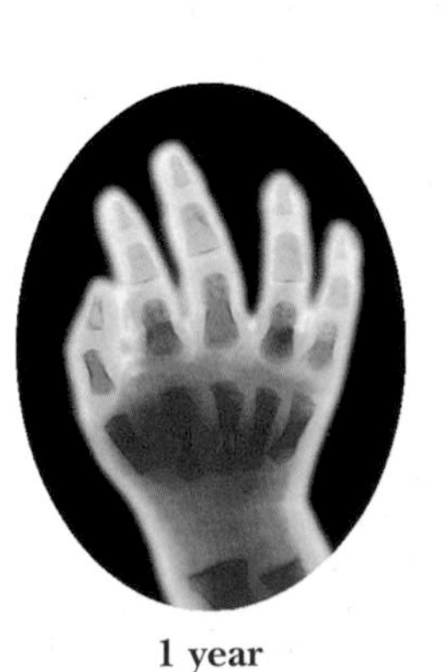

1 year

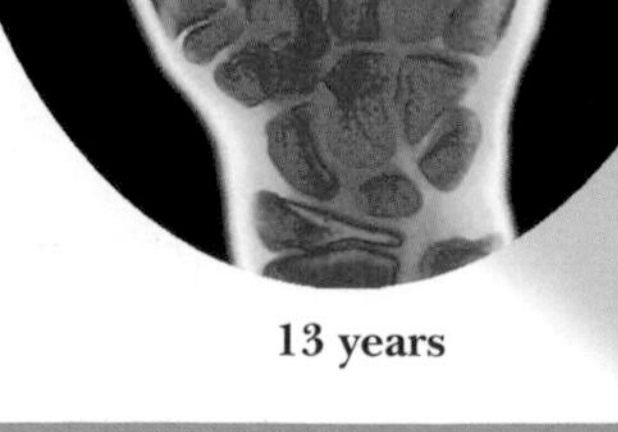

3 years

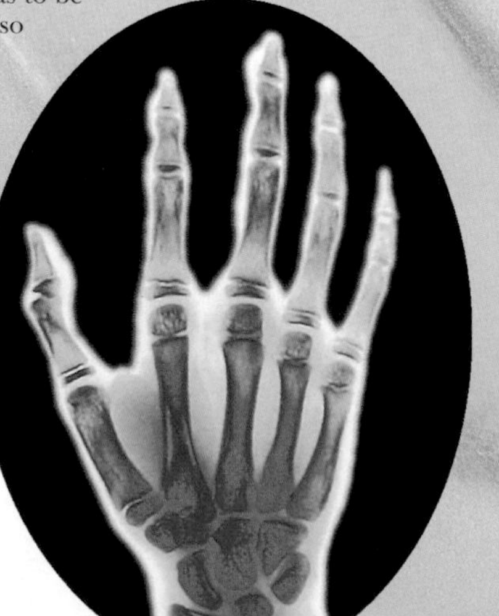

13 years

20 years

Compound fracture *– the broken end of the radius is sticking through the skin and may need surgery*

COLOURED X-RAY OF A PINNED FRACTURE OF THE TIBIA AND FIBULA BONES

BONE FRACTURES

Although bones are tough, they sometimes break – particularly if they have been weakened by disease. The fractured ends may stay in place, but in a bad break they sometimes separate, as in the broken forearm shown here. If this happens, the fracture has to be realigned or "reduced" to produce a proper repair. In simple fractures, the broken ends remain beneath the skin, but in compound fractures they stick out and surgery may be required.

X-RAY OF FRACTURE OF RADIUS AND ULNA

Ulna *is one of the forearm bones (the other is the radius)*

AIDING REPAIR

When a bone has been broken, the fractured ends need to stay together until the repair is complete. To ensure this happens, the fracture is often immobilized in a plaster or plastic cast. With severe fractures, the bones are fixed with metal pins or screws (as above with this broken tibia), which keep them correctly aligned.

SETTING BONES

Until the modern health care era, serious fractures could lead to permanent disabilities. One way to reduce this risk was to call in a bone-setter, who realigned broken bones so that they healed in the correct position. The art of bone-setting is thousands of years old, and arose in different cultures all over the world. Medieval bone-setters often doubled up as blacksmiths, and, apart from a strong pair of hands, used little equipment. Later, more sophisticated techniques developed. The scene below, from the 1600s, shows a European bone-setter at work. Helped by an assistant tugging on a set of pulleys, the setter is about to reunite the ends of the patient's broken arm.

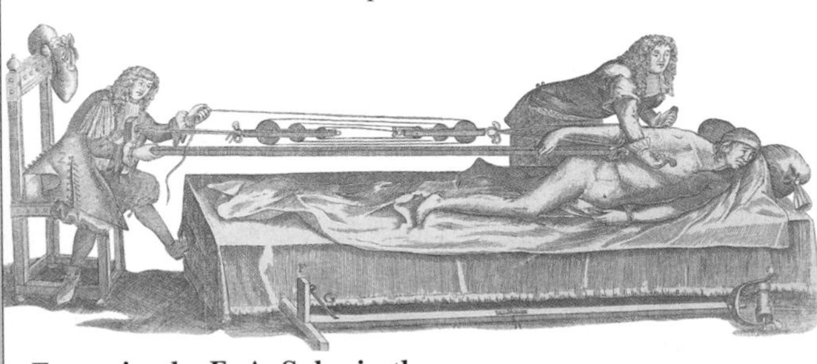

Engraving by E. A. Sohn in the scientific journal *Acta Eruditorum*

HOW BONES HEAL

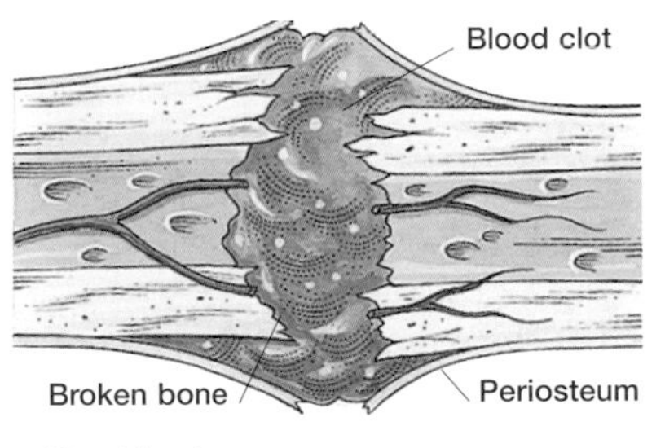

Blood clot formation
Within 6 hours of being fractured, a bone has already started the task of self-repair. A blood clot forms between the ends, sealing off any broken blood vessels within the bone. Cells in the periosteum start to divide, so that it can grow back around the site of the break.

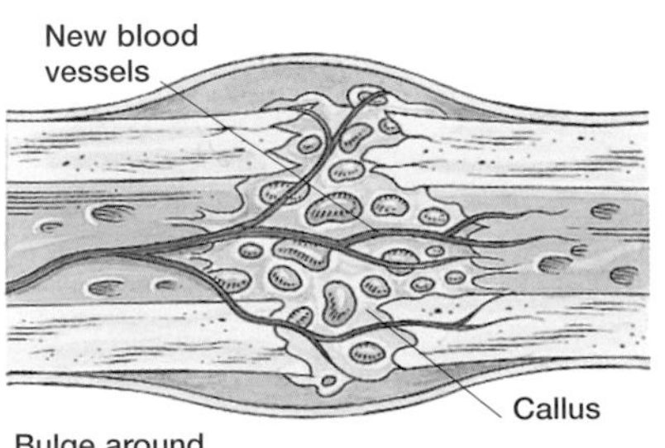

Callus formation
After 3 weeks, the periosteum is complete, and the blood clot has been replaced by areas of fibrous tissue, forming a mass called a callus. Blood vessels grow through the callus, and osteoblasts start to lay down new bone tissue inside it, connecting the bone's broken ends.

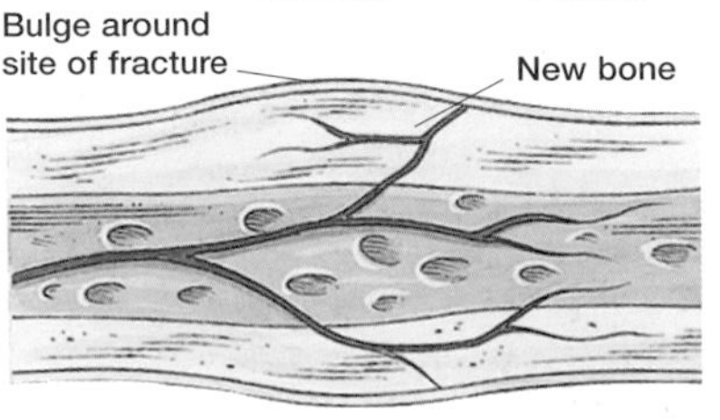

New bone formation
By 3–4 months, the fracture is almost fully repaired. The callus has been replaced by new bone, with spongy bone on the interior, and compact bone around the outside. Osteoclasts complete the operation by absorbing any remaining fragments of dead bone.

Joints

IF THE BONES IN THE SKELETON were solidly fixed together, nobody would be able to move. However, thanks to more than 400 separate joints, our skeletons are rigid but flexible at the same time. A joint is any part of the skeleton where two or more bones fit together. Some are big enough and strong enough to take the whole of the body's weight, while the smallest of them – between the tiny bones in the ear – are almost too small to see. As well as varying in size, joints also vary in the way they are built. Some hold bones together very firmly, so that they can hardly move at all. But in synovial joints, the most common type of joint, the bones slide over each other as smoothly as parts in a well-oiled machine.

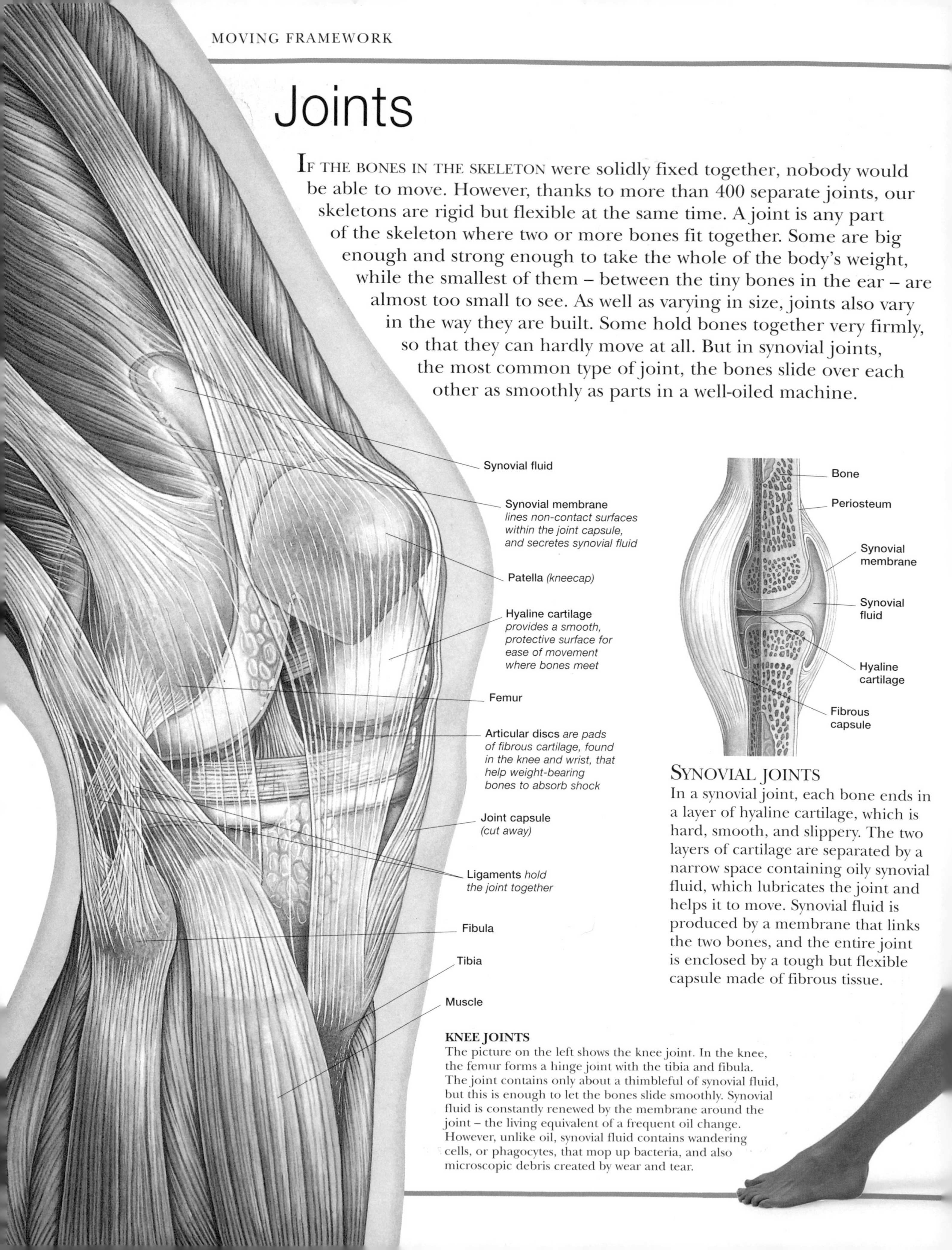
Synovial fluid
Synovial membrane *lines non-contact surfaces within the joint capsule, and secretes synovial fluid*
Patella *(kneecap)*
Hyaline cartilage *provides a smooth, protective surface for ease of movement where bones meet*
Femur
Articular discs *are pads of fibrous cartilage, found in the knee and wrist, that help weight-bearing bones to absorb shock*
Joint capsule *(cut away)*
Ligaments *hold the joint together*
Fibula
Tibia
Muscle
Bone
Periosteum
Synovial membrane
Synovial fluid
Hyaline cartilage
Fibrous capsule

SYNOVIAL JOINTS

In a synovial joint, each bone ends in a layer of hyaline cartilage, which is hard, smooth, and slippery. The two layers of cartilage are separated by a narrow space containing oily synovial fluid, which lubricates the joint and helps it to move. Synovial fluid is produced by a membrane that links the two bones, and the entire joint is enclosed by a tough but flexible capsule made of fibrous tissue.

KNEE JOINTS

The picture on the left shows the knee joint. In the knee, the femur forms a hinge joint with the tibia and fibula. The joint contains only about a thimbleful of synovial fluid, but this is enough to let the bones slide smoothly. Synovial fluid is constantly renewed by the membrane around the joint – the living equivalent of a frequent oil change. However, unlike oil, synovial fluid contains wandering cells, or phagocytes, that mop up bacteria, and also microscopic debris created by wear and tear.

STRAPPED TOGETHER
Mobile joints are held together by straps of extra-tough tissue called ligaments. In some synovial joints, such as the hip and knee, there are ligaments inside the joint capsule, but the largest and strongest ones are on the outside. Ligaments connect the adjoining bones, and are made of densely packed fibres of collagen, arranged in parallel rows. These fibres allow a joint to move, but because they are difficult to stretch, they stop the bones being pulled apart. The hip joint has very strong ligaments, which is one reason why it rarely dislocates.

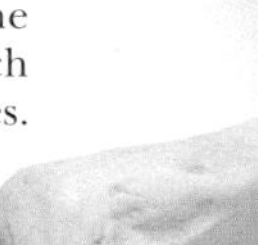

Pelvic (hip) bone

Ligament

Femur *(thigh bone)*

Hip joint showing ligaments

X-RAY OF FINGER

SPRAINS AND DISLOCATION
Synovial joints are tough, but compared to other parts of the skeleton they are quite easily damaged. Sudden wrenches sometimes result in a sprain – a painful injury involving damage to muscles and ligaments. But if the blow is really hard, it may force the two bones apart. This is known as a dislocation, and it is treated by carefully moving the bones back into position. The X-ray above shows a dislocated knuckle – one of the most common kinds of dislocation.

Fused joint

Sutures between skull bones

LOCKED TOGETHER
In the skull, most bones are locked together by joints called sutures. The wavy edges of skull bones fit tightly together like the pieces of a jigsaw. By adulthood, sutures are fully formed and cannot move, giving the skull its great strength.

SEMI-MOVEABLE JOINTS
In about a third of the body's joints, there is no synovial space, which means that the adjoining bones have much less freedom to move. In the backbone, for example, the joints between neighbouring vertebrae are only slightly flexible, but because there are many of them, the whole backbone can bend.

LOOKING INSIDE JOINTS
Many things can go wrong with joints, particularly complicated ones such as the knee. During strenuous exercise, the knee's internal ligaments can become torn, and pieces of cartilage may break away, stopping it moving altogether. To examine this kind of injury, doctors often use a special kind of endoscope (see p. 27) called an arthroscope. This slides into the joint capsule through a small incision, allowing the doctor to see the inside of the joint and to carry out repairs.

ARTHROSCOPIC IMAGE OF KNEE CARTILAGE

A surgeon using an arthroscope to see inside a knee joint

Joints and movement

THE HUMAN BODY CAN CARRY OUT an incredible variety of movements, from threading a needle or typing on a keyboard, to turning somersaults or leaping through the air. Each of these movements involves many different joints, acting together to produce a single action. The way individual joints move depends on their shape. Some joints, such as the knee, are shaped so that they move in just one direction, or plane. Others, such as the shoulder, are much more flexible, and allow movement in many directions. Joints can also vary slightly from one person to another. People who are "double-jointed" – for example in their fingers – have unusually flexible joints, although they have the same total number as everyone else.

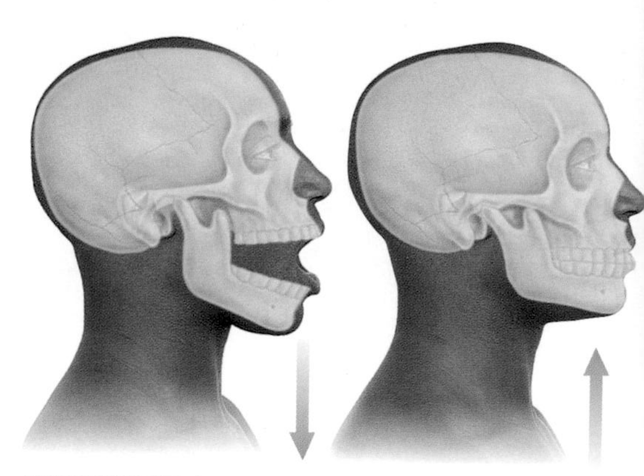

DEPRESSION AND ELEVATION
Used anatomically, depression means a downward movement, while elevation is an upward one. The lower jaw, or mandible, is depressed when the mouth is opened and elevated when it is closed.

TYPES OF JOINT

This skeleton shows six main types of synovial joints (see p. 58), with simplified diagrams showing how they work. Each joint allows a different set of movements between two neighbouring bones. In most of these joints, the bones either hinge against each other, or they rotate. The exception is the plane joint, shown on the far right. Here, the bones glide past each other in the same plane.

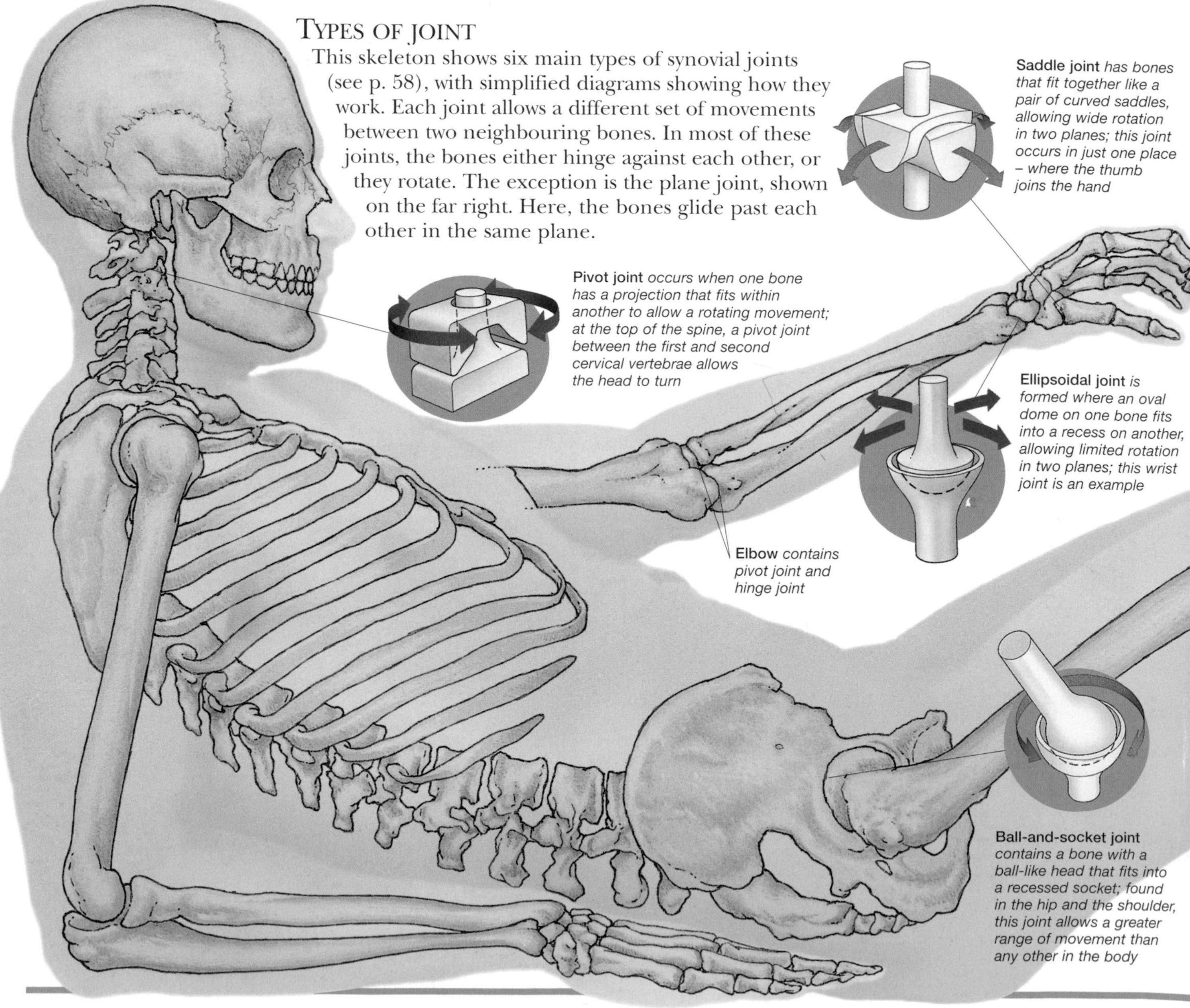

Saddle joint *has bones that fit together like a pair of curved saddles, allowing wide rotation in two planes; this joint occurs in just one place – where the thumb joins the hand*

Pivot joint *occurs when one bone has a projection that fits within another to allow a rotating movement; at the top of the spine, a pivot joint between the first and second cervical vertebrae allows the head to turn*

Ellipsoidal joint *is formed where an oval dome on one bone fits into a recess on another, allowing limited rotation in two planes; this wrist joint is an example*

Elbow *contains pivot joint and hinge joint*

Ball-and-socket joint *contains a bone with a ball-like head that fits into a recessed socket; found in the hip and the shoulder, this joint allows a greater range of movement than any other in the body*

ROTATION AND CIRCUMDUCTION
During rotation, a bone swivels around on its own axis. This happens in the neck when the head is turned, and also in the elbow when the hand is turned palm-upwards. Circumduction is the kind of movement that takes place when an entire arm or leg is moved in a circle.

FLEXION AND EXTENSION
When a joint is flexed, the angle between its two bones decreases. This is what happens, for example, when the elbow is bent. When the joint is extended, the angle is increased, so that the bones are more stretched out. Flexion and extension apply to any joints that hinge – either in one plane or two.

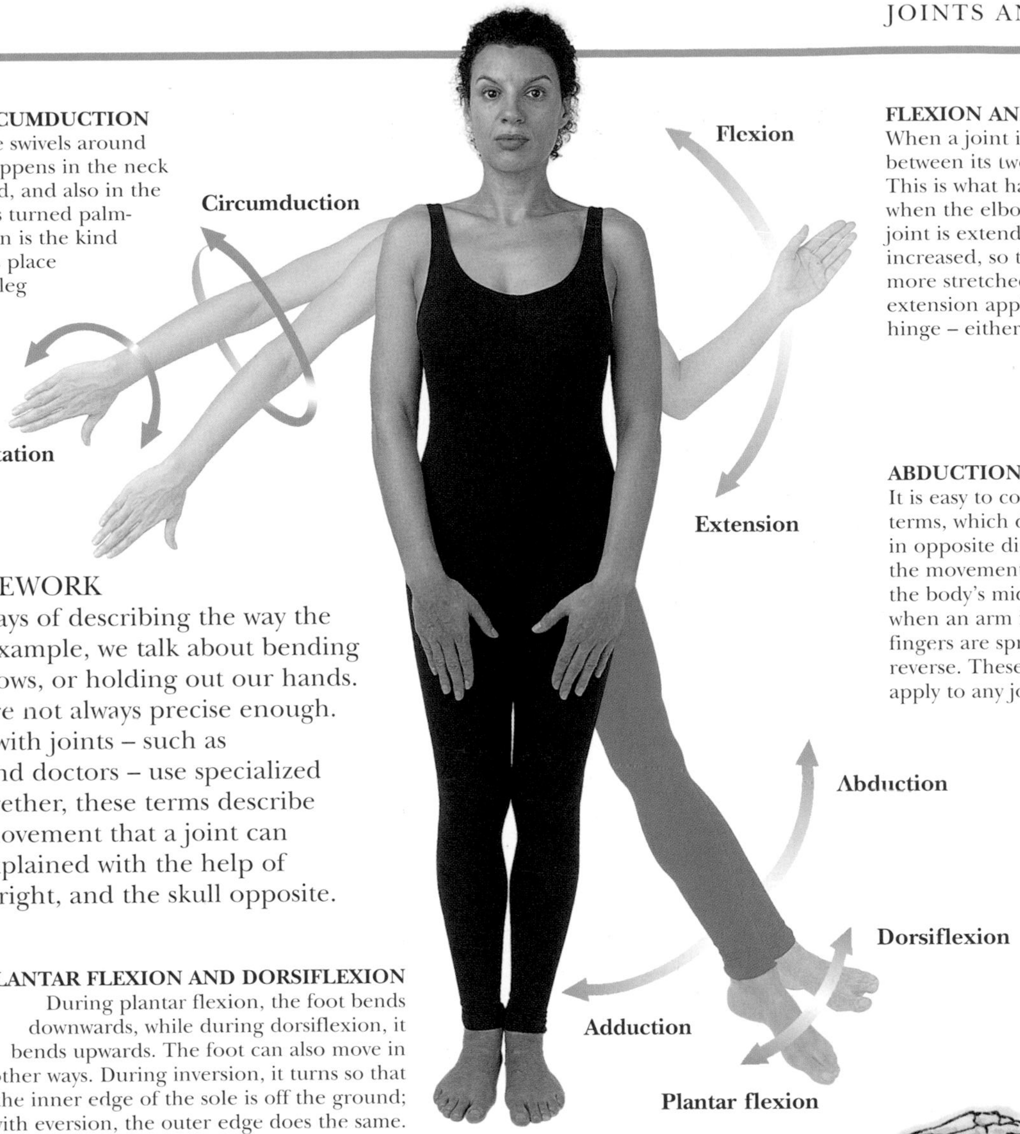

Circumduction

Flexion

Rotation

Extension

ABDUCTION AND ADDUCTION
It is easy to confuse these two terms, which describe movement in opposite directions. Abduction is the movement of a bone away from the body's midline – for example, when an arm is raised, or when the fingers are spread. Adduction is the reverse. These two movements apply to any joints that hinge.

MOVING FRAMEWORK

There are many ways of describing the way the body moves. For example, we talk about bending our knees and elbows, or holding out our hands. But these words are not always precise enough. People who work with joints – such as physiotherapists and doctors – use specialized terms instead. Together, these terms describe every individual movement that a joint can make. They are explained with the help of the model on the right, and the skull opposite.

Abduction

Dorsiflexion

PLANTAR FLEXION AND DORSIFLEXION
During plantar flexion, the foot bends downwards, while during dorsiflexion, it bends upwards. The foot can also move in other ways. During inversion, it turns so that the inner edge of the sole is off the ground; with eversion, the outer edge does the same.

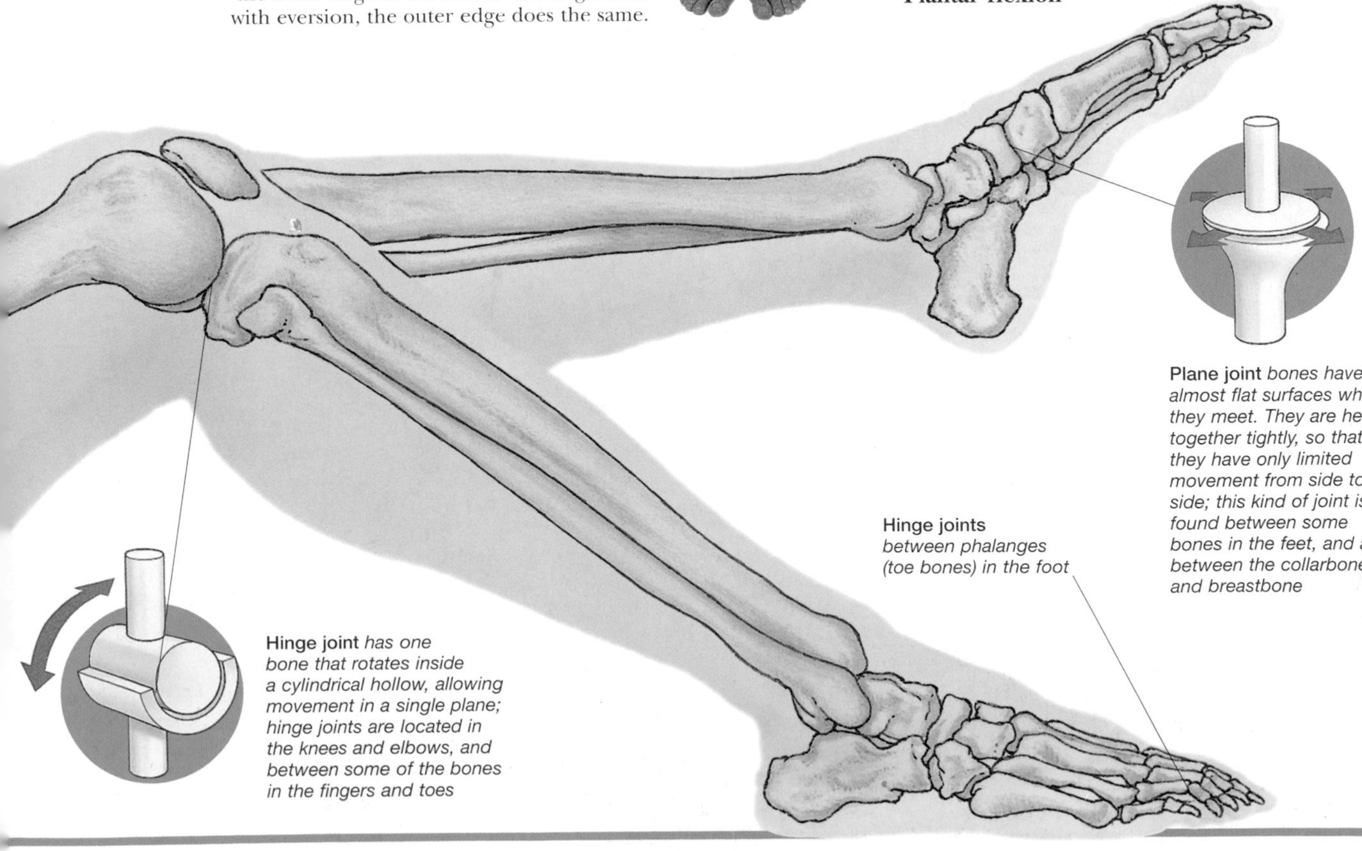

Adduction

Plantar flexion

Plane joint *bones have almost flat surfaces where they meet. They are held together tightly, so that they have only limited movement from side to side; this kind of joint is found between some bones in the feet, and also between the collarbone and breastbone*

Hinge joints *between phalanges (toe bones) in the foot*

Hinge joint *has one bone that rotates inside a cylindrical hollow, allowing movement in a single plane; hinge joints are located in the knees and elbows, and between some of the bones in the fingers and toes*

MUSCULAR system

ALL BODY MOVEMENTS – whether lifting a finger, running for a bus, or eating lunch – are produced by muscles. Many muscles are attached to bones, which they pull to move and support the skeleton. Others work automatically and unseen within internal organs, often pushing fluids or food through the body. Whatever their role, all muscle tissues share common features. Their cells, called fibres, possess the unique ability of using energy to contract, or get shorter, thereby producing a pulling force. Contraction of fibres is triggered by electrical nerve signals from the brain. Fibres are elastic enough to stretch yet still return to their original length.

Skeletal muscles are used to produce these high kicks

SMOOTH MUSCLE
The fibres that make up smooth muscle are short, unstriped, and have tapering ends. They are packed closely together in muscle sheets in the walls of internal organs, such as the small intestine, along which they push food. These sheets are often arranged in two layers – one circular and one longitudinal (lengthways) – with opposite effects. Smooth muscle is also called involuntary muscle because it cannot be controlled consciously – it is triggered by the autonomic nervous system (the part of the nervous system that automatically controls internal functions) and hormones (see p. 118).

LM OF SMOOTH MUSCLE FIBRES

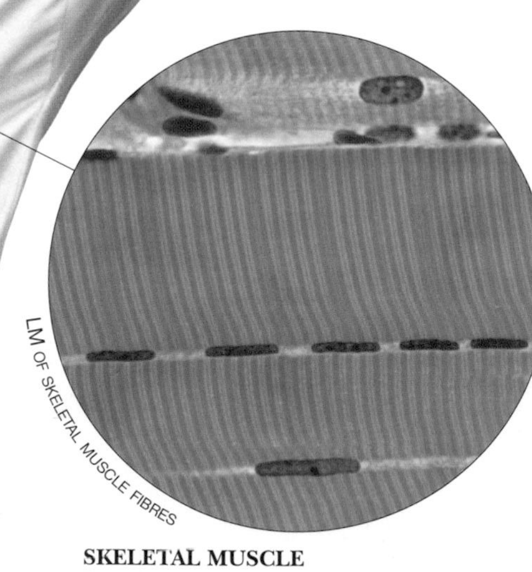

SKELETAL MUSCLE
Running in parallel, the fibres of skeletal muscle are long and cylindrical, sometimes reaching up to 30 cm (12 in) in length. Their striped, or striated, appearance gives skeletal muscle its alternative name of striate muscle. It is also known as voluntary muscle because a conscious, or voluntary, decision is made by the nervous system to trigger contraction and move the body.

MUSCLE TYPES

There are three distinct types of muscle tissue in the body – skeletal, cardiac, and smooth. Skeletal muscle tissue forms the muscles that move and support the skeleton. It contracts rapidly and powerfully but tires easily. Cardiac muscle tissue is found only in the wall of the heart. Here, for a lifetime, it contracts regularly, automatically, and without tiring, to pump blood around the body. Smooth muscle tissue is found mainly in the walls of hollow organs, such as the bladder. It contracts slowly and rhythmically, performing actions such as pushing urine out of the bladder. The three types can be easily distinguished by the appearance of their fibres (cells).

LM OF CARDIAC MUSCLE FIBRES

CARDIAC MUSCLE

The branched, striated fibres of cardiac muscle form an interconnected network in the heart wall. Cardiac muscle contracts automatically, with its own built-in rhythm, some 100,000 times each day. It cannot be controlled consciously. However, its rate of contraction is regulated by the autonomic nervous system, according to whether the body is active or at rest.

Skeletal muscle *is covered by a connective tissue sheath called the epimysium*

Tendon *makes a secure attachment between the muscle and a bone*

Bone *is pulled by the tendon when a muscle contracts*

Periosteum *is the membrane that covers the bone's surface*

BONE CONNECTORS

Skeletal muscles are anchored to bones – and sometimes to each other – by strong cords or sheets called tendons that enable muscles to pull bones without tearing. Tendons have enormous tensile strength (resistance to breaking). They are packed with fibres of the tough protein collagen, which are arranged in parallel bundles. The connective tissue sheath that encloses the muscle extends to the tendon. The tendon then extends to the bone, where its fibres penetrate the periosteum (surface membrane) and embed themselves firmly in the bone's outer layer, completing the connection between muscle and bone.

HEAT GENERATORS

By making use of heat generated inside it, the body maintains a constant internal temperature of 37°C (98.6°F). About 85 per cent of this heat is produced by muscle contraction. Of the energy converted by muscles, only 25 per cent causes movement. The remaining 75 per cent is "waste" energy, released as heat. This thermogram reveals how body heat is lost through the skin – lighter colours mark areas of greatest heat radiation.

MUSCULAR SYSTEM FUNCTIONS

Movement	*Skeletal muscles produce a wide range of movements including running, picking up objects, and changing facial expressions. Cardiac muscle in the heart pumps blood to all body tissues. Smooth muscle produces movement in internal organs, such as pushing food along the intestines.*
Posture maintenance	*Certain skeletal muscles are kept in a partially contracted state to hold the body upright.*
Joint stability	*When they contract, some skeletal muscles help to stabilize highly mobile joints such as shoulders.*
Heat generation	*Because they are not totally efficient, muscles generate heat as a by-product when they contract. Heat generation, or thermogenesis, is vitally important in maintaining normal body temperature. In cold conditions, the body uses involuntary contractions (shivering) to generate additional heat.*

Skeletal muscles

In any description of the muscular system, the skeletal muscles take centre stage. Making up nearly half of the body's mass, skeletal muscles can perform a wide range of movements, from blinking an eyelid to wielding a sledge-hammer. They are primarily attached to bones by tough, fibrous tendons. Typically, each muscle links two bones across a flexible joint so that muscle contraction, or shortening, either results in movement or assists in holding the body upright. A few muscles – such as those that produce facial expressions – work by tugging on the skin.

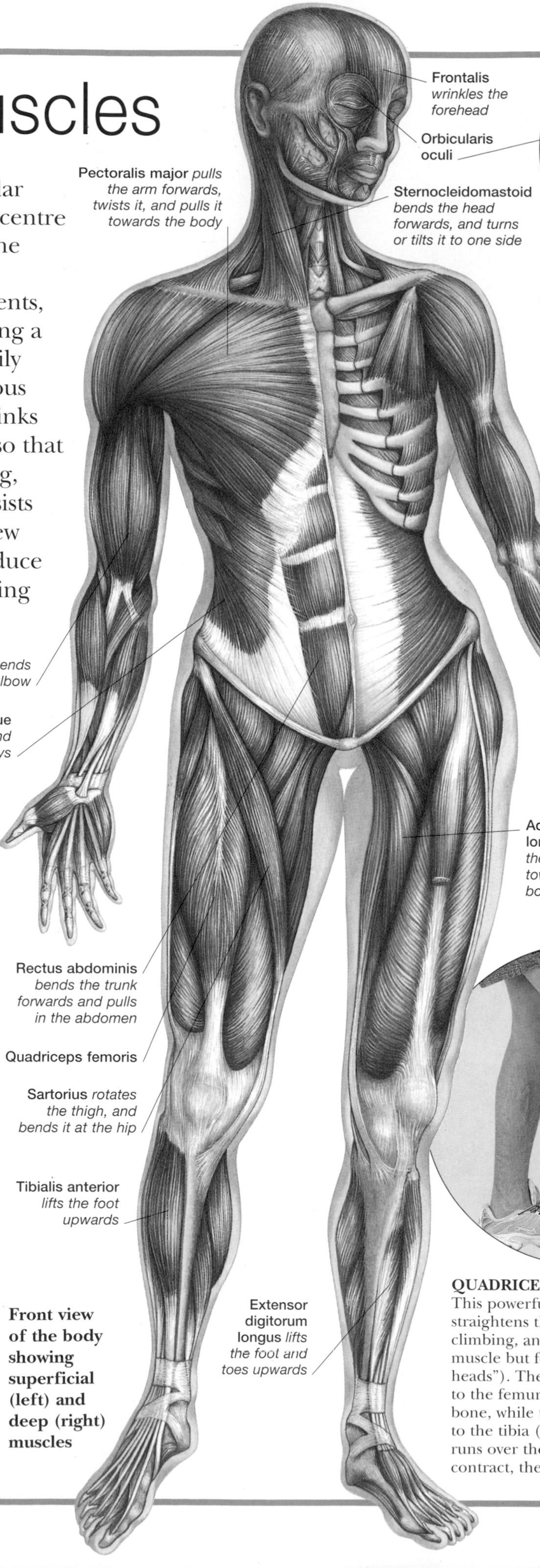

Frontalis *wrinkles the forehead*

Orbicularis oculi

Pectoralis major *pulls the arm forwards, twists it, and pulls it towards the body*

Sternocleidomastoid *bends the head forwards, and turns or tilts it to one side*

Biceps brachii *bends the arm at the elbow*

External oblique *twists the trunk and bends it sideways*

Adductor longus *pulls the leg inwards towards the body's midline*

Quadriceps femoris

Rectus abdominis *bends the trunk forwards and pulls in the abdomen*

Quadriceps femoris

Sartorius *rotates the thigh, and bends it at the hip*

Tibialis anterior *lifts the foot upwards*

Extensor digitorum longus *lifts the foot and toes upwards*

Front view of the body showing superficial (left) and deep (right) muscles

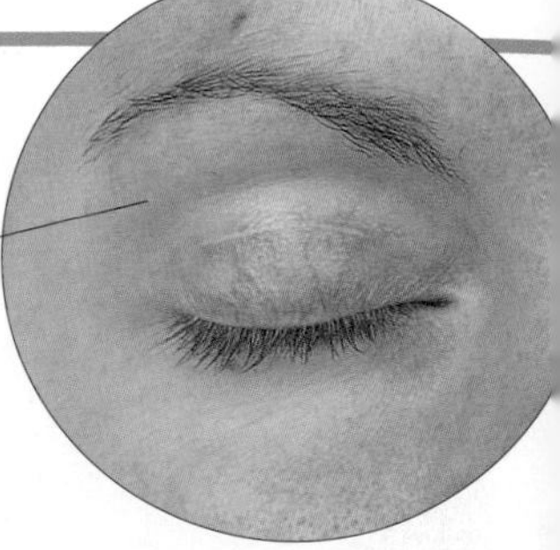

ORBICULARIS OCULI

Forming a ring around the eye, the orbicularis oculi (meaning "circular" and "of the eye") protects it from injury and intense light by causing blinking and squinting. The muscle is attached to the bony eye socket and to the eyelids. When the orbicularis oculi contracts, the ring gets smaller, and the eyelids move to close the eye. A similar type of muscle, the orbicularis oris, surrounds the mouth and closes the lips.

NAMING MUSCLES

Every skeletal muscle is given a Latin name according to one or more of its features, as described below. Some muscle names cover several features. The extensor carpi radialis longus, for example, extends (straightens) the wrist ("carpi"), lies close to the radius (lower arm) bone, and is longer than other wrist extensors.

Muscle features and descriptions

Location Example: the frontalis runs over the frontal bone of the skull.

Relative size using terms such as maximus (largest), minimus (smallest), longus (long), and brevis (short). Example: the gluteus maximus is the biggest gluteal (buttock) muscle.

Shape Example: the two trapezius muscles form a trapezoid (four-sided) shape.

Action using terms such as flexor (bends a joint) and extensor (straightens a joint). Example: the flexor carpi ulnaris bends the hand at the wrist.

Origin and insertion Example: the sternocleidomastoid has origins (where bones do not move) on the breastbone – sternum – and collar bone – clavicle ("cleido") – and insertions (where bones do move) on the mastoid process of the skull's temporal bone.

Number of origins using terms such as biceps ("two heads"). Example: the biceps brachii (arm) has two origins on the scapula, or shoulder blade.

QUADRICEPS FEMORIS

This powerful thigh muscle, which straightens the knee when running, climbing, and kicking, is actually not one muscle but four (quadriceps means "four heads"). Their upper ends are attached to the femur (thigh bone) or pelvic (hip) bone, while their lower ends are anchored to the tibia (shin bone) by a tendon that runs over the knee. When the muscles contract, the lower leg is pulled forwards.

BODY MOVERS

Peel away the body's skin and fat, and the glistening redness of its skeletal muscles is revealed. There are more than 650 named skeletal muscles, ranging enormously in shape and size from the powerful gluteus maximus in the buttocks to the tiny stapedius, just 1 mm (0.04 in) long, tucked inside the ear. Skeletal muscles are arranged in layers that overlap each other. Muscles lying close to the skin's surface are referred to as superficial, while those lying in layers beneath them are called deep.

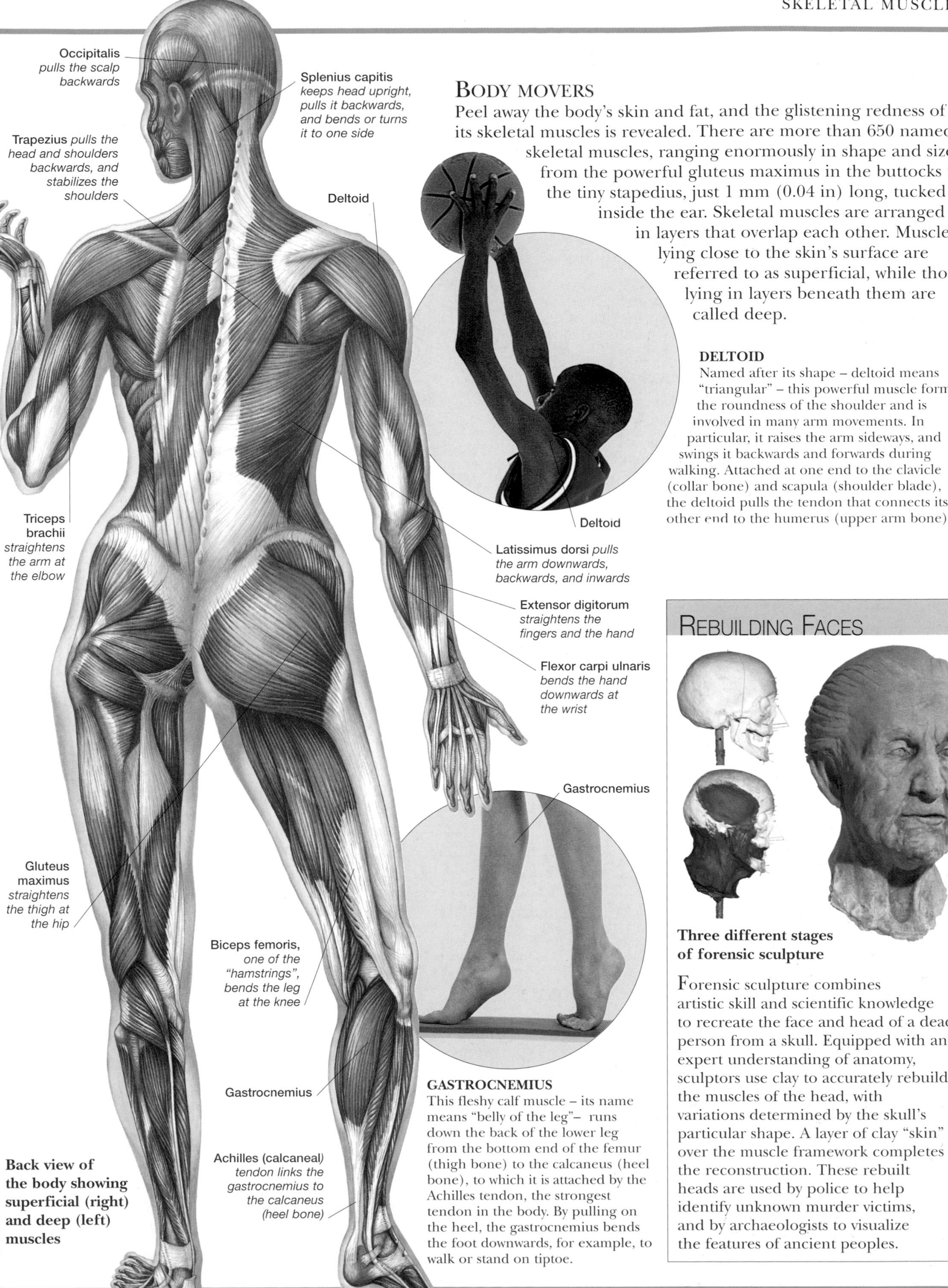
Occipitalis pulls the scalp backwards
Splenius capitis keeps head upright, pulls it backwards, and bends or turns it to one side
Trapezius pulls the head and shoulders backwards, and stabilizes the shoulders
Deltoid
Deltoid
Triceps brachii straightens the arm at the elbow
Latissimus dorsi pulls the arm downwards, backwards, and inwards
Extensor digitorum straightens the fingers and the hand
Flexor carpi ulnaris bends the hand downwards at the wrist
Gastrocnemius
Gluteus maximus straightens the thigh at the hip
Biceps femoris, one of the "hamstrings", bends the leg at the knee
Gastrocnemius
Achilles (calcaneal) tendon links the gastrocnemius to the calcaneus (heel bone)

Back view of the body showing superficial (right) and deep (left) muscles

DELTOID

Named after its shape – deltoid means "triangular" – this powerful muscle forms the roundness of the shoulder and is involved in many arm movements. In particular, it raises the arm sideways, and swings it backwards and forwards during walking. Attached at one end to the clavicle (collar bone) and scapula (shoulder blade), the deltoid pulls the tendon that connects its other end to the humerus (upper arm bone).

GASTROCNEMIUS

This fleshy calf muscle – its name means "belly of the leg"– runs down the back of the lower leg from the bottom end of the femur (thigh bone) to the calcaneus (heel bone), to which it is attached by the Achilles tendon, the strongest tendon in the body. By pulling on the heel, the gastrocnemius bends the foot downwards, for example, to walk or stand on tiptoe.

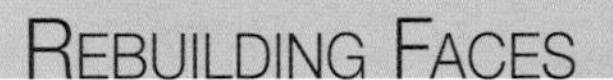

REBUILDING FACES

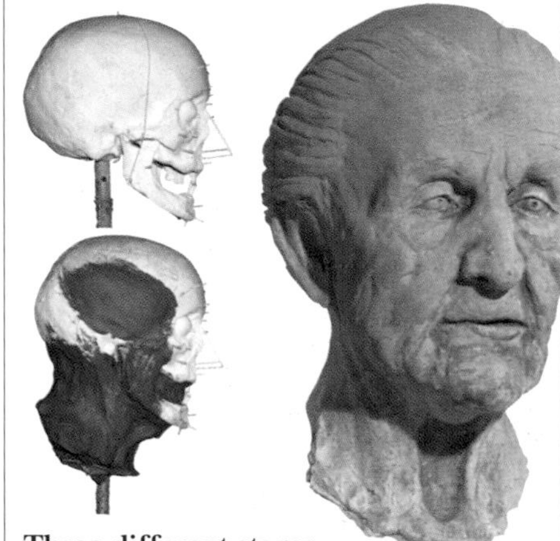

Three different stages of forensic sculpture

Forensic sculpture combines artistic skill and scientific knowledge to recreate the face and head of a dead person from a skull. Equipped with an expert understanding of anatomy, sculptors use clay to accurately rebuild the muscles of the head, with variations determined by the skull's particular shape. A layer of clay "skin" over the muscle framework completes the reconstruction. These rebuilt heads are used by police to help identify unknown murder victims, and by archaeologists to visualize the features of ancient peoples.

How muscles contract

UNDERSTANDING HOW muscle fibres are constructed is key to working out how muscles contract. Each long, cylindrical muscle fibre is filled with smaller fibres, called myofibrils, which are packed with a highly ordered array of protein filaments. The arrival of a nerve message from the brain causes these filaments to interact, making the muscle fibre – and muscle – shorten. The more signals that arrive, the more a muscle contracts, until it reaches about 70 per cent of its resting length. The whole process involves the transformation of chemical energy stored in nutrients, such as glucose, into kinetic (movement) energy. In the absence of nervous stimulus, the muscle fibre relaxes.

Muscle *is an organ that contracts to move, support, or stabilize part of the body*

Fascicle *is one of the bundles of fibres that make up a muscle. It is surrounded by a connective tissue sheath, the perimysium*

Muscle fibre *is one of the long, thin cells that makes up a muscle*

Capillary *supplies muscle fibres with blood*

Sarcomere *is a section of myofibril between Z lines*

Myofibril *is one of the parallel, rod-like strands that pack the inside of a muscle fibre*

Z line

FROM FIBRE TO FILAMENT

Hundreds of bundled muscle fibres run in parallel along the length of a muscle. Each fibre is packed with rod-like myofibrils which contain two types of protein filament – thick (myosin) and thin (actin). These filaments are arranged in repeated patterns called sarcomeres, that give muscle fibres their striped appearance. Extending from myosin filaments are small "heads" that, in resting muscle, extend towards acting filaments.

Mitochondrion

TEM OF SECTION THROUGH SKELETAL MUSCLE FIBRE

Myofibril

Sarcomere

Myosin filament

Relaxed muscle fibre

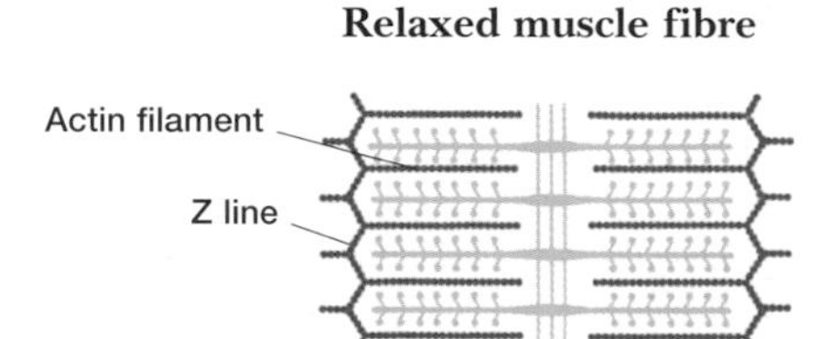

Contracted muscle fibre

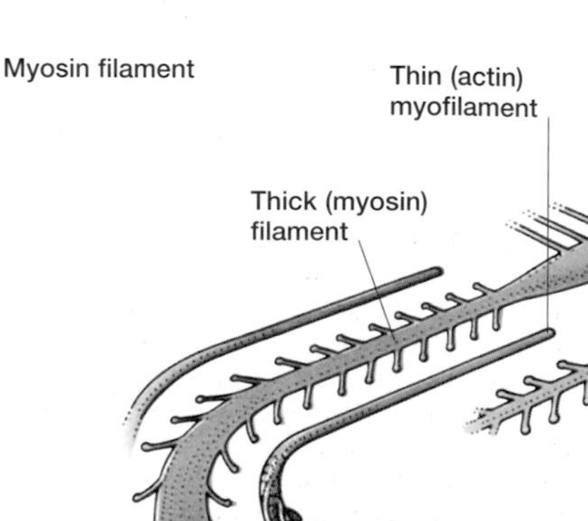

Head of myosin molecule *is "charged" with ATP and interacts with actin during contraction*

ENERGY PROVIDERS

Energy-rich nutrients cannot be used directly for muscle contraction – they must first be converted into ATP (adenosine triphosphate). This substance stores energy, carries it to where it is needed, and releases it on demand. ATP is produced by aerobic respiration inside the mitochondria squeezed in between myofibrils. During respiration, glucose – delivered to muscle fibres by the blood or extracted from their glycogen store – is broken down using oxygen, releasing its energy to make ATP.

SLIDING FILAMENTS

To make contractions happen, filaments in myofibrils slide over each other. In each sarcomere, myosin filaments are located centrally while the actin filaments that surround and overlap them are attached to the Z line. When a muscle fibre is stimulated to contract, myosin heads bind to actin and, using energy supplied by ATP, repeatedly swivel towards the centre of the sarcomere, pulling actin filaments inwards and making the sarcomere shorter until the stimulus stops.

STIMULUS TO CONTRACT

Skeletal muscle is also called voluntary muscle because its contraction is caused by a person's conscious decision to do something, for example, bending an arm. This decision is communicated by electrical signals called nerve impulses that flash along nerves from the central nervous system (brain and spinal cord) to muscles. In reality, everyday movements, such as walking, generally happen without a person having to consciously think about their minute-by-minute operation.

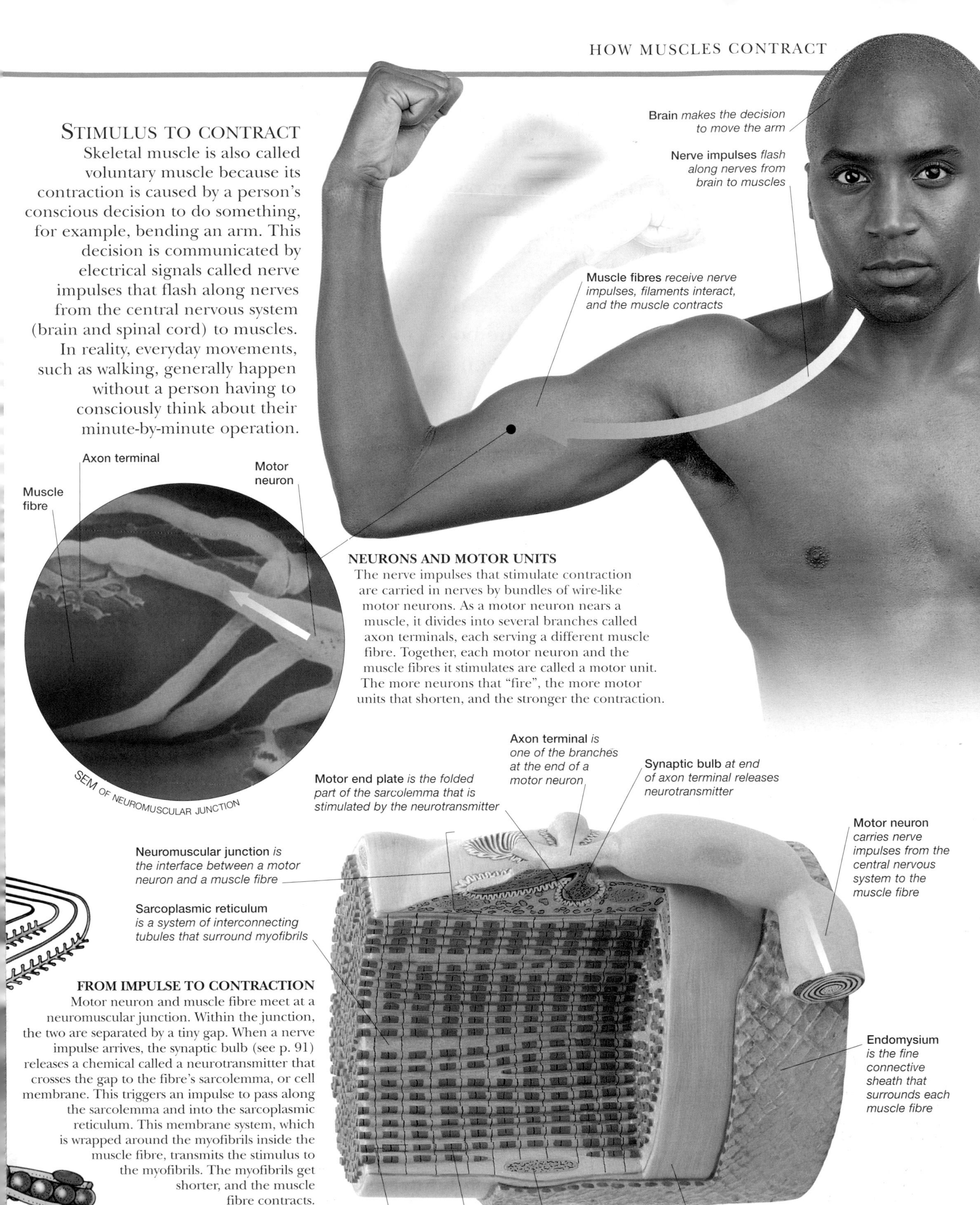

Brain makes the decision to move the arm
Nerve impulses flash along nerves from brain to muscles
Muscle fibres receive nerve impulses, filaments interact, and the muscle contracts
Axon terminal
Motor neuron
Muscle fibre
SEM OF NEUROMUSCULAR JUNCTION
Axon terminal is one of the branches at the end of a motor neuron
Synaptic bulb at end of axon terminal releases neurotransmitter
Motor end plate is the folded part of the sarcolemma that is stimulated by the neurotransmitter
Motor neuron carries nerve impulses from the central nervous system to the muscle fibre
Neuromuscular junction is the interface between a motor neuron and a muscle fibre
Sarcoplasmic reticulum is a system of interconnecting tubules that surround myofibrils
Endomysium is the fine connective sheath that surrounds each muscle fibre
Sarcolemma is the cell membrane of a muscle fibre
Myofibril
Sarcomere
Nucleus

NEURONS AND MOTOR UNITS

The nerve impulses that stimulate contraction are carried in nerves by bundles of wire-like motor neurons. As a motor neuron nears a muscle, it divides into several branches called axon terminals, each serving a different muscle fibre. Together, each motor neuron and the muscle fibres it stimulates are called a motor unit. The more neurons that "fire", the more motor units that shorten, and the stronger the contraction.

FROM IMPULSE TO CONTRACTION

Motor neuron and muscle fibre meet at a neuromuscular junction. Within the junction, the two are separated by a tiny gap. When a nerve impulse arrives, the synaptic bulb (see p. 91) releases a chemical called a neurotransmitter that crosses the gap to the fibre's sarcolemma, or cell membrane. This triggers an impulse to pass along the sarcolemma and into the sarcoplasmic reticulum. This membrane system, which is wrapped around the myofibrils inside the muscle fibre, transmits the stimulus to the myofibrils. The myofibrils get shorter, and the muscle fibre contracts.

Cutaway view of a muscle fibre and neuromuscular junction

Movement and posture

IF ITS MUSCLES WERE suddenly inactivated, the body would not only be immobilized but would also collapse. As well as moving the body, skeletal muscles hold it upright and maintain posture. To perform both roles, muscles pull the bones, to which they are attached by tendons, across a joint. When a muscle contracts, one of the bones to which it is attached – the insertion – moves, while the other attachment point – the origin – remains fixed. Since they can only pull and not push, muscles work in antagonistic pairs to produce opposing movements. Flexor muscles in the forearm, for example, bend fingers, while their antagonists, the extensor muscles, straighten them. Muscles that work together to produce the same movement are called synergists.

Triceps *contracts to straighten the arm*

Biceps *is fully contracted*

Brachioradialis *helps the biceps bend the arm by pulling the lower arm bones upwards*

Biceps *contracts to bend the arm*

Triceps *relaxed and stretched*

Muscles that raise and lower the forearm

OPPOSING MUSCLES

When a muscle contracts it shortens, pulling its insertion towards its origin. To produce movement in the opposite direction, there must be a separate antagonistic, or opposing, muscle. In the arm, for example, the biceps muscle pulls the forearm upwards towards the shoulder to bend the arm. Its antagonist, the triceps, pulls the forearm downwards to straighten the arm. The brachioradialis acts as a synergist to the biceps, helping it to bend the arm.

MUSCLES, BONES, AND LEVERS

Muscles and bones interact to move the body using lever systems. A lever is a bar that moves on a fixed point, the fulcrum, when a force is applied to one part of it to move a weight on another. In the body, bones are levers, a joint is a fulcrum, and muscle contraction provides the force to move a body part (weight). Levers fall into three classes according to the relative positions of the force, weight, and fulcrum.

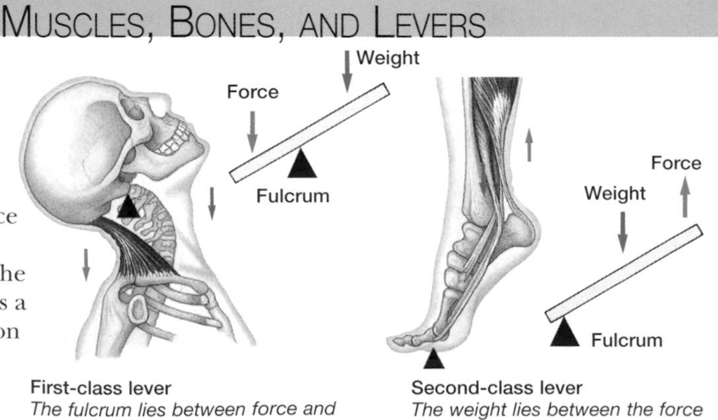

First-class lever
The fulcrum lies between force and weight, like a see-saw. Neck muscles pulling the back of the skull to tilt the head backwards produce a similiar action.

Second-class lever
The weight lies between the force and the fulcrum, like a wheelbarrow – for example, raising of the heel (and body weight) by the calf muscles.

Force
Weight
Fulcrum

Third-class lever
The most common type of lever in the body involves force applied between fulcrum and weight, like tweezers – for example, bending the elbow.

HAND MOVEMENTS

The hand owes its incredible precision and versatility to the muscles that move its flexible framework of 27 bones. These include the slender forearm muscles that taper into long tendons which reach into the hand. Extensor muscles in the outer forearm straighten the wrist and fingers, while opposing flexors in the inner forearm bend them.

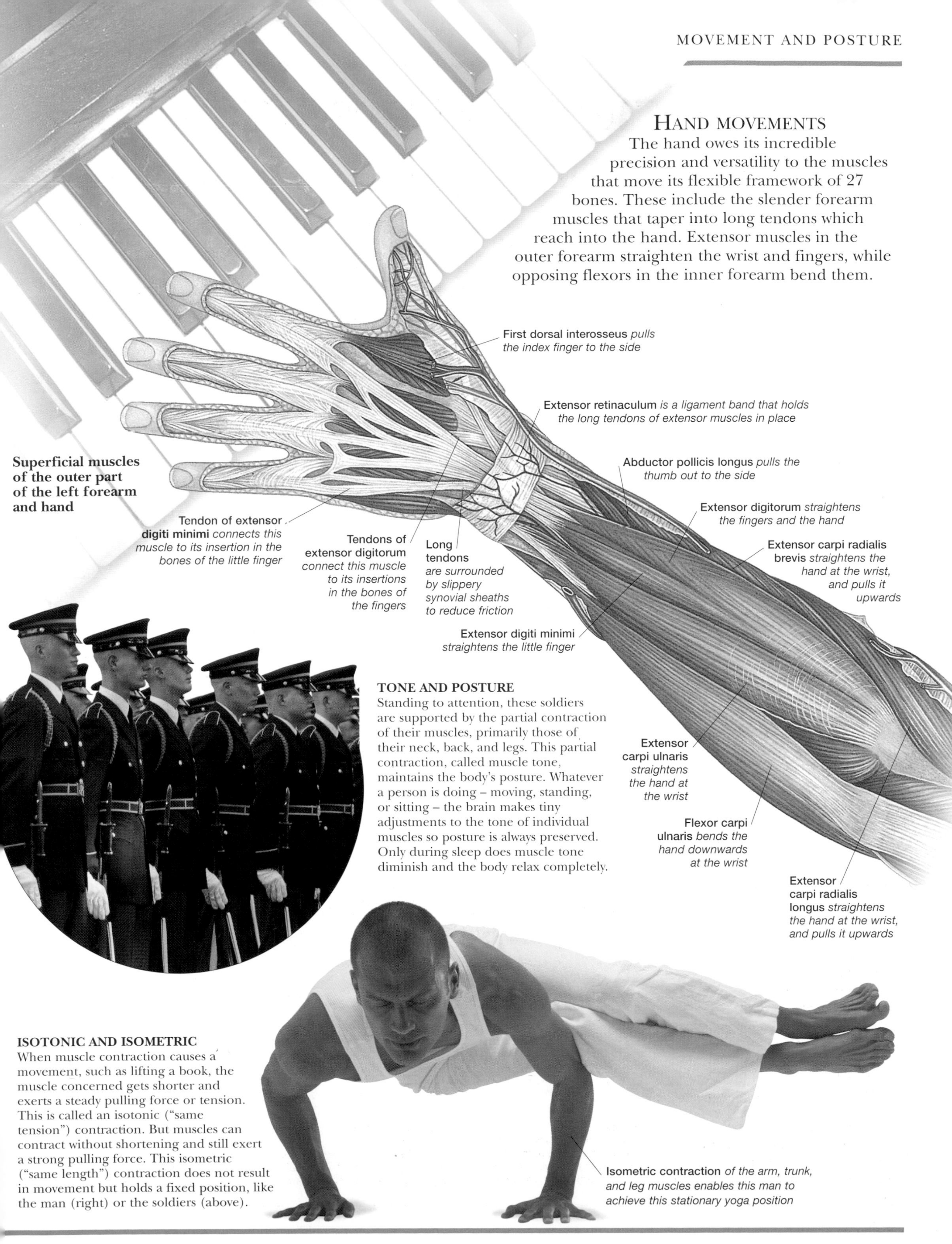
First dorsal interosseus pulls the index finger to the side
Extensor retinaculum is a ligament band that holds the long tendons of extensor muscles in place
Abductor pollicis longus pulls the thumb out to the side
Extensor digitorum straightens the fingers and the hand
Extensor carpi radialis brevis straightens the hand at the wrist, and pulls it upwards
Tendon of extensor digiti minimi connects this muscle to its insertion in the bones of the little finger
Tendons of extensor digitorum connect this muscle to its insertions in the bones of the fingers
Long tendons are surrounded by slippery synovial sheaths to reduce friction
Extensor digiti minimi straightens the little finger
Extensor carpi ulnaris straightens the hand at the wrist
Flexor carpi ulnaris bends the hand downwards at the wrist
Extensor carpi radialis longus straightens the hand at the wrist, and pulls it upwards

Superficial muscles of the outer part of the left forearm and hand

TONE AND POSTURE

Standing to attention, these soldiers are supported by the partial contraction of their muscles, primarily those of their neck, back, and legs. This partial contraction, called muscle tone, maintains the body's posture. Whatever a person is doing – moving, standing, or sitting – the brain makes tiny adjustments to the tone of individual muscles so posture is always preserved. Only during sleep does muscle tone diminish and the body relax completely.

ISOTONIC AND ISOMETRIC

When muscle contraction causes a movement, such as lifting a book, the muscle concerned gets shorter and exerts a steady pulling force or tension. This is called an isotonic ("same tension") contraction. But muscles can contract without shortening and still exert a strong pulling force. This isometric ("same length") contraction does not result in movement but holds a fixed position, like the man (right) or the soldiers (above).

Isometric contraction *of the arm, trunk, and leg muscles enables this man to achieve this stationary yoga position*

Muscles and exercise

THE BODY IS capable of responding to all manner of changes, both inside and outside itself. One of the most obvious responses is to increased activity or exercise. The heart, lungs, and muscles undergo changes – all carefully regulated by the brain – that ensure that muscle fibres obtain sufficient energy to contract more rapidly and more strongly. The ability of the body to react in this way depends on its fitness. A fit body is one that can carry out everyday activities, such as running for a bus or climbing stairs, without breathlessness or excessive tiredness. Fitness has three elements – stamina, strength, and flexibility – all of which can be increased by regular exercise. Some exercises focus on just one of these fitness elements, while others, such as swimming, improve all three.

Muscles *work harder during exercise and demand more energy*

STRENGTH

The amount of force muscles exert when performing an action such as lifting or jumping, and their ability to hold the body upright without tiring, are determined by their strength. Exercises such as weightlifting improve strength by increasing the size of muscle fibres. These exercises are anaerobic. That is, they are intensive exercises that last for a very short time, and use energy released without the need for oxygen. They do not improve stamina.

Fibres *in the arm's muscles increase in size, which improves strength*

Weightlifting is an anaerobic exercise that increases strength

STAMINA

Also called cardiovascular fitness or endurance, stamina reflects the ability of the heart and blood vessels to deliver oxygen and nutrients to the body's tissues, including muscles. Stamina is enhanced by aerobic exercises, such as running, brisk walking, or dancing, which use oxygen to release energy from "fuels" such as glucose. Performed for 20 minutes or more, at least three times a week, and demanding enough to produce sweating and slight breathlessness, aerobic activities improve stamina by increasing the strength and efficiency of the heart.

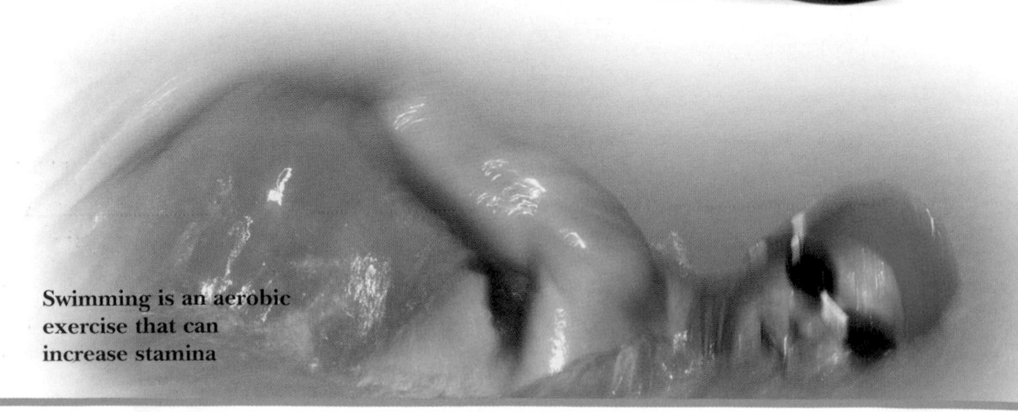

Swimming is an aerobic exercise that can increase stamina

ON THE MOVE

When exercise begins, several changes occur automatically to meet the demands for extra oxygen and nutrients needed by skeletal muscles that are working harder than normal. Blood flow to muscle fibres is dramatically increased during vigorous exercise from 1 to 12 litres (1.8 to 21.1 pints) per minute. This is achieved by widening blood vessels that serve muscle fibres, and by making the heart pump faster and more strongly so that blood flow rate increases from 5 to 20 litres (8.8 to 35.2 pints) per minute. At the same time, the volume of air taken in by the lungs goes from 6 up to 100 litres (0.2 to 3.5 cu ft) per minute by increasing the rate and depth of breathing.

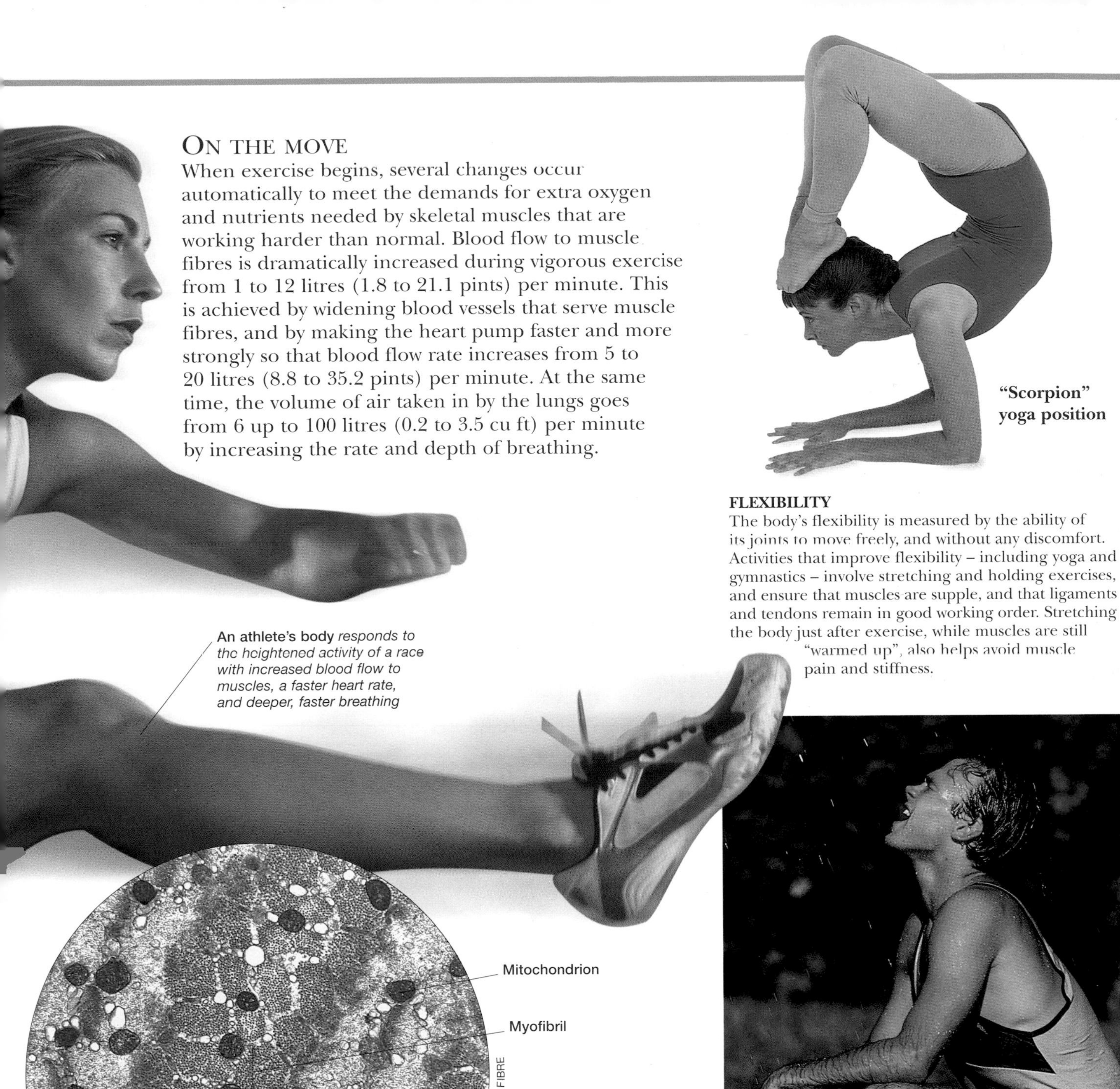

"Scorpion" yoga position

An athlete's body *responds to the heightened activity of a race with increased blood flow to muscles, a faster heart rate, and deeper, faster breathing*

FLEXIBILITY

The body's flexibility is measured by the ability of its joints to move freely, and without any discomfort. Activities that improve flexibility – including yoga and gymnastics – involve stretching and holding exercises, and ensure that muscles are supple, and that ligaments and tendons remain in good working order. Stretching the body just after exercise, while muscles are still "warmed up", also helps avoid muscle pain and stiffness.

EXERCISE BENEFITS

Regular exercise which combines elements that increase stamina, strength, and flexibility has numerous benefits. The heart pumps, and the lungs take in oxygen, more efficiently. Capillaries supplying muscle fibres with nutrients and oxygen increase in number. More mitochondria (organelles that release energy) and glycogen granules (energy stores) appear in muscle fibres, and there is more myoglobin (the substance that carries oxygen). Muscle fibres increase in size and work more efficiently, giving muscles greater strength and resistance to tiredness.

OXYGEN DEBT

An athlete pants after a race because his body needs to get extra oxygen to its cells – over and above their resting oxygen consumption – in order to "pay off" an "oxygen debt". This "debt" is generated during hard exercise by anaerobic respiration, when muscle fibres obtain energy without using oxygen. The waste product of this process – lactic acid – must be disposed of by aerobic respiration, a process that requires extra oxygen, before muscles can work normally again.

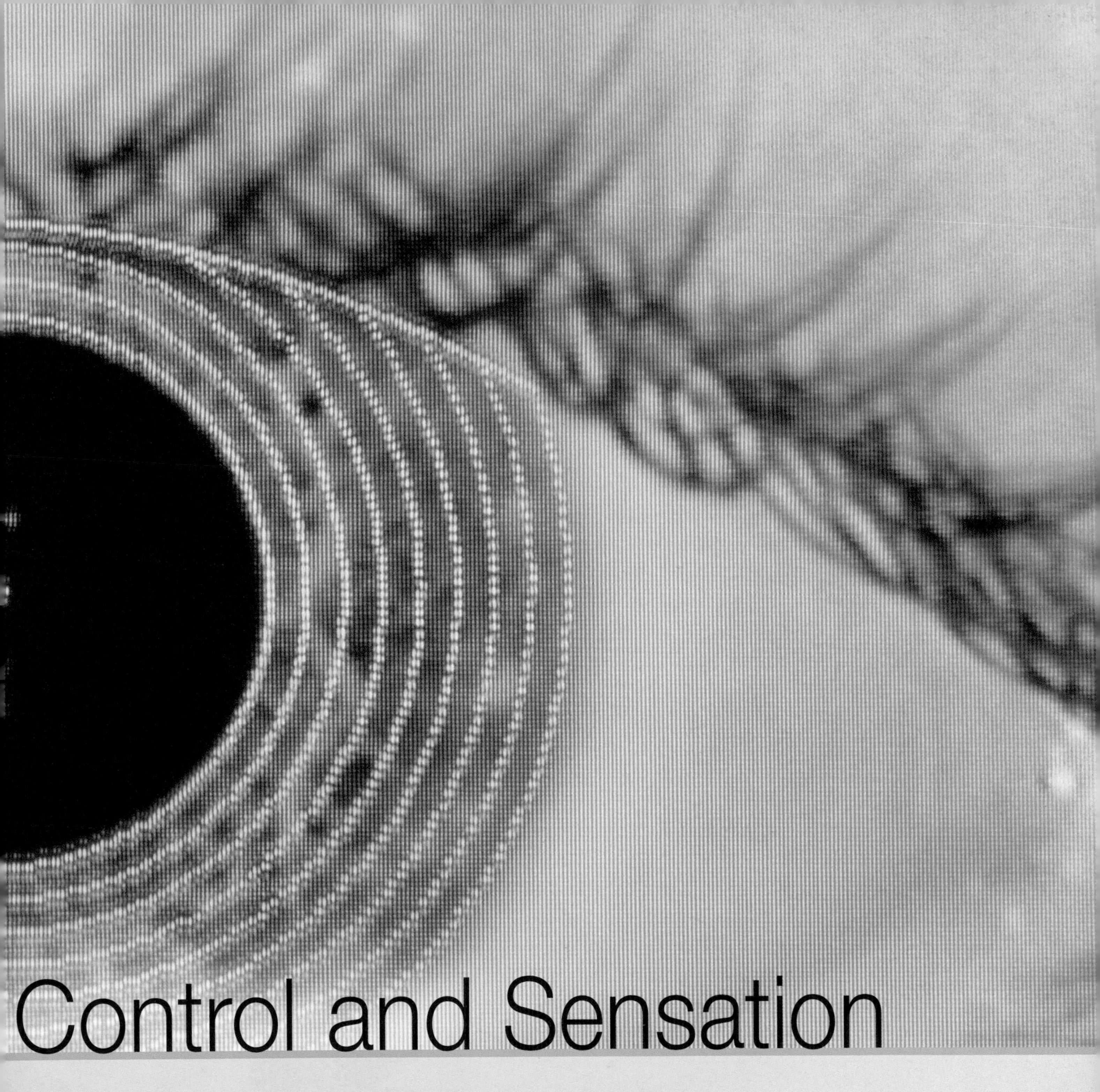

Control and Sensation

THE HUMAN BRAIN is a remarkable organ. It can simultaneously order leg muscles to run and the heart to beat faster. It gives humans their creativity, memory, intelligence, emotions, and personality. It turns messages received from sensors into the sensations that allow humans to see, hear, taste, smell, and touch. Together with its nerves and sensors, the brain forms the nervous system. A second, linked control system – the endocrine system – regulates growth, reproduction and some other processes.

NERVOUS system

OF ALL THE HUMAN body's systems, the nervous system is the most complex. It is on duty every second of every day, gathering information about the body and its surroundings, and issuing instructions that make the body react. Together with the endocrine system (see p. 118), it controls everything that the body does, and its speed and processing power mean that it can cope with an incredible range of tasks at the same time. It works through specialized cells called neurons, which carry signals in the form of tiny bursts of electricity. Some neurons carry signals to, or from, particular parts of the body, but most are packed into the nervous system's headquarters – the brain. This living computer allows us to think and to remember, and makes us who we are.

Computer artwork of the nervous system

NEURONS

Neurons, or nerve cells, are the basic units of the nervous system. They come in many shapes and sizes, but they all have slender fibres that can carry electrical impulses. These impulses flash along the cell, and "jump" from one neuron to another at chemical junctions called synapses. Some neurons have synapses with one or two neurons, but others can have hundreds of these connections.

SEM OF NEURONS

COMMUNICATION NETWORK

The nervous system reaches almost every part of the body, from muscles and sense organs to the inside of teeth and bones. Nerves are the system's main communication cables, fanning out from the spinal cord, and also from the brain. There are more than 80 major nerves, and each one may contain more than a million neurons. In the above diagram, the central nervous system, comprising the brain and spinal cord, is coloured blue-white; the spine, shown around the spinal cord, is coloured magenta; and the spinal nerves are orange.

Thigh muscle *contracts to straighten leg when stimulated by nerve impulses from the brain carried by motor neurons*

Pressure receptors *in skin detect the force with which the foot pushes down on the ground*

NERVOUS SYSTEM FUNCTIONS

Sensory	*Senses changes (stimuli) inside and outside the body, in conjunction with receptors or sense organs. The changes include a wide range of physical factors, such as light, pressure, or the concentration of dissolved chemicals.*
Integrative	*Analyses sensory information, and makes decisions on appropriate responses. Triggered or modified by information that is stored in and retrieved from memory.*
Motor	*Triggers responses by muscles or glands. The nervous system can either stimulate muscles and glands into action, or it can inhibit them.*

Brain *analyses incoming information from sensors, and sends out instructions to muscles to move the body and maintain its balance*
Optic nerve *from the eye sends a stream of signals to the brain, allowing it to track the ball*
Receptors *in the inner ear produce signals used to control balance*
Pressure receptors *in the hand tell the brain when the ball has been caught*
Motor neurons *of the autonomic nervous system stimulate the heart to beat faster to increase blood flow to the muscles*
Nerve impulses *carried by motor neurons make muscles straighten the arm to intercept the ball*
Stretch receptors *in the muscles detect muscle tone or tension and inform the brain so it can maintain balance*

Rapid response

On the sports field, split-second reactions are essential. The nervous system is ideally equipped for this, because it can flash signals along neurons at up to 100 m (328 ft) per second. This gets sensory signals in record time to the brain or spinal cord, where they are processed. Motor signals are then flashed to selected muscles, making the body respond. With something as complicated as catching a football, millions of signals race through the nervous system, ensuring that the body does the right thing at exactly the right time.

How the Nervous System is Organized

The nervous system is divided into two main parts: the central nervous system (CNS) and the peripheral nervous system (PNS). The CNS consists of the brain and spinal cord, and is the command centre of the nervous system. The PNS consists of the nerves (bundles of neurons) that extend from the brain and spinal cord, and which relay messages between the CNS and the rest of the body.

Central nervous system *The CNS co-ordinates the activities of the entire body. It receives and analyses incoming information from sense organs and other receptors, and sends out instructions that are based on past experience to muscles and glands.*

Peripheral nervous system *The PNS has three divisions: one relaying information to the CNS, and two carrying instructions from the CNS.*

Sensory division *gathers information from sensors to update the CNS about events occurring inside and outside the body.*

Somatic division *carries instructions to skeletal muscles to make them contract, enabling the body to respond under conscious control to outside events.*

Autonomic division *carries instructions to the body's internal organs to control their activities, thereby automatically regulating internal processes.*

Neurons

FOUND ONLY IN THE nervous system, neurons are among the body's most specialized cells. They carry electrical signals, or impulses, and pass them on through fibres called axons, which can be up to 1 m (3.3 ft) long. Neurons behave partly like wires and partly like batteries, because they charge themselves up. When something triggers a neuron to "fire", the charge reverses, creating a rapid burst of electricity that rushes along the cell. Some neurons are triggered by things that can be sensed. Others process this information, or carry signals that make the body react. Unlike most of the body's cells, neurons cannot divide once they are mature, and if they are badly damaged they cannot be replaced.

Axon terminals *The end of the axon divides to produce a collection of axon terminals which form synapses with other neurons, or with muscle or gland cells*

Schwann cell *is a glial cell that wraps itself around an axon*

NEURON STRUCTURE

A typical neuron is divided into a part that receives signals, and a long axon or fibre that carries the signals from one place to another. In a motor neuron, shown here, the receiving end has a swelling, the cell body. Attached to this are short filaments called dendrites, which connect with neighbouring neurons through chemical junctions called synapses. If one of these synapses is stimulated, an impulse flashes towards the cell body. It then travels along the neuron's axon, so that it can be passed on.

Axon or nerve fibre *conducts nervous impulses away from the cell body, so that they can be passed on to other neurons, or to muscles or glands*

Myelin sheath *insulates the axon and is formed in a Schwann cell*

Association neuron

Association neuron cell body

Synapse

Dendrite of motor neuron

Cell body of motor neuron

Synapse

Axon terminal

Sensory neuron axon

Axon of motor neuron

Sensory neuron cell body

Sensory neuron

Motor neuron

Sensory neuron dendrite

Myelin sheath *insulates nerve fibre*

Neuromuscular junction *(synapse between neuron and muscle)*

Path of nerve signal

Muscle fibres *receive nerve signal*

Touch sensor *in skin generates the nerve signal*

TYPES OF NEURON

There are three main types of neuron. Sensory neurons are triggered by physical stimuli, such as light. The strength of the stimulus affects the rate at which the neuron fires. Association, or intermediate, neurons are triggered by sensory neurons. They process the information from sensory neurons and issue outgoing commands. These commands are passed on to motor neurons, which in turn make parts of the body respond.

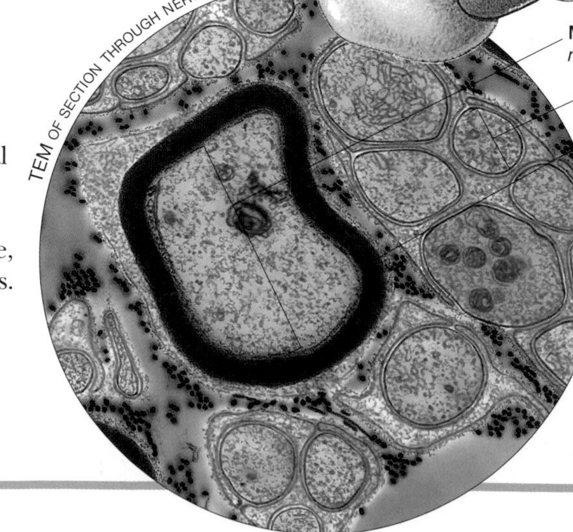

TEM OF SECTION THROUGH NERVE FIBRES

Myelinated axon *surrounded by narrower non-myelinated axons*

Non-myelinated axon

Myelin sheath

MYELIN SHEATH

Bare axons do not carry impulses quickly, because their electrical charge leaks away. Many sensory and motor neurons are insulated by a fatty substance called myelin. Myelin is produced by Schwann cells, which wrap themselves around axons, forming layers like those in a Swiss roll. The Schwann cells are separated by small gaps called nodes, and nerve impulses move by jumping from one node to the next.

IMMUNOFLUORESCENT LM OF ASTROCYTES

SUPPORTING CELLS

Neurons make up only about one in ten of the nervous system's cells. The remaining nine-tenths are known as glial cells. They support the neurons and keep them alive. Glial cells called astrocytes, which are shown here, help to supply neurons with nutrients, while other glial cells mop up invading bacteria. Another type – Schwann cells – wrap themselves around neurons, providing protection and insulation. Unlike neurons, glial cells can replace themselves if they are damaged.

High-speed impulses

The nervous system works very quickly, as is illustrated by the split-second reactions of racing car drivers. Neurons can fire up to 2,500 times a second and, in myelinated neurons, impulses sometimes travel at speeds of 350 kmh (218 mph). When a neuron is at rest, it uses energy to pump electrically charged particles, or ions, across its cell membrane. When the neuron is triggered by a synapse, the ions rush back across the cell membrane, and the charge is reversed. This reversal sweeps along the cell, and the result is a nerve impulse.

Direction of impulses

Node of Ranvier *is a gap between adjacent Schwann cells that allows nervous impulses to leapfrog along the axon*

Nucleus

Cell body *contains most of the neuron's cytoplasm, as well as mitochondria and other organelles*

Branching dendrites *enable a single neuron to make contact with dozens, or even hundreds, of its neighbours*

Golgi's Stain

Because they are so slender, neurons are difficult to see – even when magnified several hundred times. Until the late 19th century, little was known about neurons. But in 1873, Italian histologist Camillo Golgi (1844–1926), who identified different neuron types, found that they could be stained black by silver nitrate, making their fine structure easier to see.

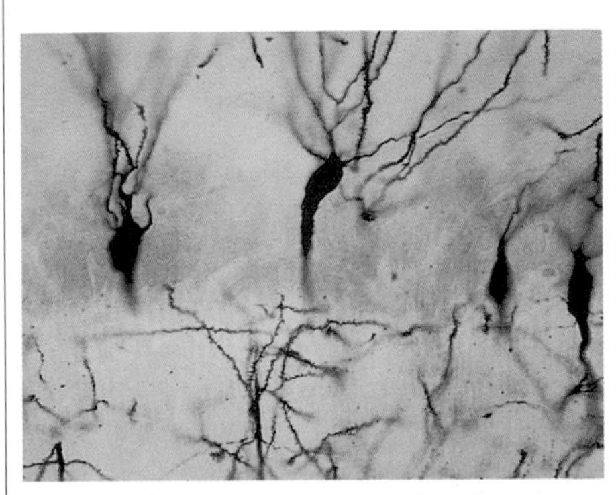

Section through the brain, showing neurons stained using Golgi's stain

Synapses

IN THE NERVOUS SYSTEM – just like in a computer – information is only useful if it can be communicated, or passed on. Neurons do this through microscopic junctions called synapses. At a synapse, a slender terminal fibre from a neuron reaches out to make contact with another cell. If a nerve impulse flashes along the fibre, it makes the synapse release a chemical called a neurotransmitter. In less than one-thousandth of a second, this chemical travels across a tiny gap between the two cells, and triggers the second cell to respond. A single neuron can have several hundred of these minute connections, and the total number in the nervous system runs into many trillions. Together, they create a vast array of circuits, which constantly interact with each other to control the body.

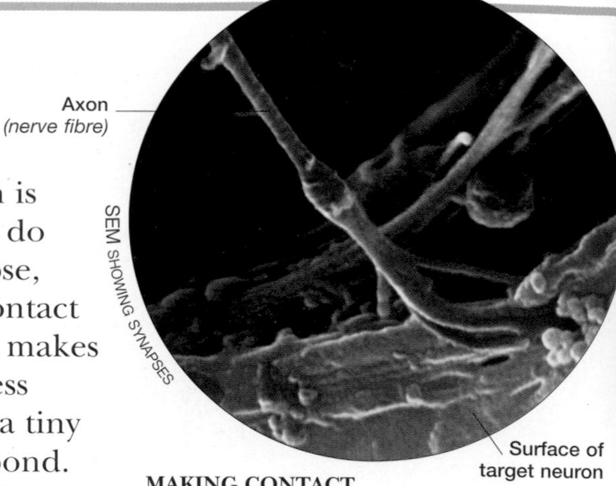

MAKING CONTACT
In this electron microscope view of a nerve cell, an axon can be seen connecting one neuron with another. The axon (coloured purple) divides into smaller fibres that end in small swellings, or synaptic bulbs. These produce neurotransmitters, and release them when the axon "fires". Most synapses are between neurons, though some synapses also connect neurons with other kinds of cell.

SYNAPSE CONNECTIONS

Nervous impulses travel throughout the body via synapses. Each synapse is like a one-way switch, because it cannot work in reverse. Some neurons have just a handful of synapses, but association neurons in the brain – like the ones shown on the left – often form synapses with hundreds of neighbouring cells. Synapses can change as time goes by. Extra synapses form in circuits that are used frequently, making it easier for nerve impulses to follow these particular paths. This explains why many things – such as playing a musical instrument or driving a car – get easier with practice.

Myelinated axon, *or nerve fibre, is surrounded by Schwann cells*

Neuron cell body

Synaptic bulb *releases neurotransmitter*

Dendrite *of target neuron receives stimuli from synapses*

Axon terminal

Cell body nucleus

DRUGS AND SYNAPSES
Many drugs work by interfering with the way that synapses work. For example, stimulants such as caffeine reduce the amount of neurotransmitter that is needed to make neurons "fire". After several cups of coffee, the body's neurons are triggered more frequently than normal, producing a feeling of being more awake and alert. Depressants, including alcohol, have the opposite effect.

Coffee beans

NEUROTOXINS
Some of the most lethal substances in nature are ones that switch synapses permanently on or off. Known as neurotoxins, these include chemicals produced by some bacteria, and also secretions produced by black widow spiders and poison dart frogs. Neurotoxins that switch on synapses cause paralysis, by making all the body's muscles contract at once. Those that switch off synapses stop muscles contracting, which means that the victim cannot breathe. The fatal dose can be less than one-millionth of a gram.

Poison dart frog

Stains and Synapses

Spanish doctor and medical researcher Santiago Ramón y Cajal (1852–1934) studied the human nervous system using the staining techniques developed by Camillo Golgi (see p. 79). However, unlike Golgi, he believed that neurons met at synapses, without actually merging. He could not prove this with the microscopes of the time, but after the development of the electron microscope, his theory was shown to be correct. Golgi and Ramón y Cajal shared the Nobel Prize in 1906.

CROSSING THE GAP
In the bulbous part of a synapse, neurotransmitters are stored in bubble-like packages called vesicles. When a nerve impulse arrives at the synapse, some of the vesicles migrate to the edge of the bulb and spill their neurotransmitter outside. The neurotransmitter crosses the gap to the target cell, where it activates special receptors. If the target cell is a neuron, the receptors trigger it to fire. The neurotransmitter is then broken down by enzymes.

Axon terminal

Synaptic bulb *lies close to the target neuron's cell body or dendrites*

Synaptic vesicles *store molecules of the neurotransmitter*

Dendrite *of target neuron*

Mitochondrion

Vesicle *discharges neurotransmitters into the synaptic cleft*

Cell membrane

Synaptic cleft or gap

Target cell membrane

Receptor sites *combine with the neurotransmitter on the target cell membrane, which then produces a nerve impulse in the target neuron*

Membrane channel *allows sodium ions (blue) in to trigger nerve impulse*

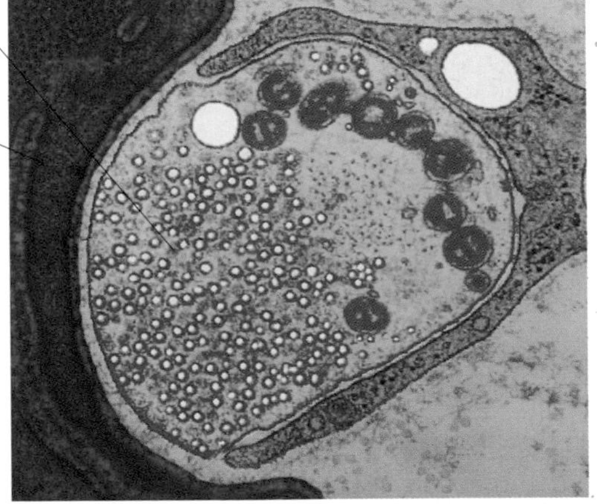

TEM OF SECTION THROUGH NEUROMUSCULAR JUNCTION

NEUROMUSCULAR JUNCTION
The cross-section above shows a special kind of synapse between neurons and skeletal muscle fibres that makes voluntary muscles contract. In the centre is the synaptic bulb (blue), while the muscle cell that it triggers is to the left (red). The bulb is packed with vesicles of neurotransmitter, waiting to be discharged. Smooth and cardiac muscles are triggered by synapses of a different kind, which pass electrical signals directly from one muscle cell to another.

Nerves

REGARDED AS THE INFORMATION highways of the nervous system, nerves can contain millions of individual neurons that fan out from the brain and spinal cord to reach every part of the body. Most carry two-way traffic, with sensory neurons flashing signals inwards to the central nervous system, and motor neurons transmitting signals in the opposite direction. Nerves do not recover well from injury, and the majority are buried deep in the body, with only small branches reaching upwards to make contact with the skin. The ulnar nerve, at the elbow, is one of the few that does sit close to the surface. If given a sudden knock against the humerus, or "funny bone", it sends a tingling sensation shooting down the arm.

Brain

Cervical plexus *provides nerves that supply the neck, shoulders and diaphragm*

Phrenic nerve *supplies the diaphragm, the chief muscle causing breathing movements*

Vagus nerve *helps control heart rate*

Musculocutaneous nerve *controls muscles that bend the elbow*

Intercostal nerves *control the muscles between the ribs*

Spinal cord

Lumbar plexus *provides nerves that supply abdominal wall and leg muscles*

Femoral nerve *controls the muscles that straighten the knee*

Ulnar nerve *controls muscles that bend the wrist and fingers*

Sacral plexus *provides nerves that supply the buttocks and leg*

Sciatic nerve, *the thickest and longest in the body, controls the thigh muscles that bend the leg*

Tibial nerve *supplies the posterior calf muscles that bend the foot downwards*

Common peroneal nerve *controls the lower leg muscles that lift the foot upwards*

Cross-section through nerve

Epineurium *surrounds the entire nerve*

Blood vessels *supply oxygen and nutrients to the nerve*

Individual axon

Myelin sheath *insulates and protects axons*

Fascicle *is a collection of axons surrounded by perineurium*

Fat-containing cells *act as a cushion against shocks*

Perineurium

INSIDE A NERVE

Nerves resemble telephone cables, although their individual "wires", or axons, are too small to be seen with the naked eye. In a nerve, groups of axons called fascicles are wrapped in tough but elastic tissue called perineurium. An outer layer – epineurium – encloses several fascicles, together with blood vessels and fat-containing cells. This protects the axons from damage, but allows the nerve to bend. In most nerves, the cell bodies of neurons are kept separate in swellings called ganglia, which are situated near the brain or spinal cord.

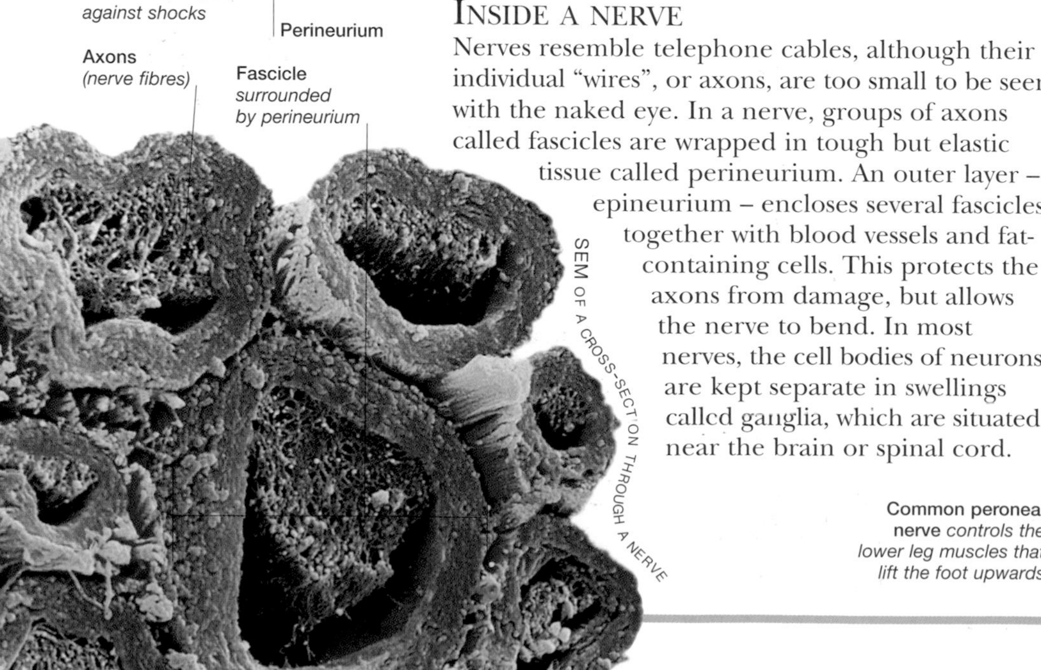

SEM OF A CROSS-SECTION THROUGH A NERVE

Radial nerve *controls muscles that straighten the elbow, wrist, and fingers*

Median nerve *controls muscles that bend the wrist and fingers*

Brachial plexus *provides the nerves that supply the arm and hand*

MICROSURGERY

If nerves, even small ones, are cut, it is possible to reconnect them using a delicate technique known as microsurgery. Surgeons use special binocular microscopes in the operating theatre. These devices enable the surgeons to see even the finest nerves – and blood vessels – in depth so that they can accurately rejoin the disconnected ends. This technique means that severed fingers or limbs can be successfully re-attached to the body with full feeling and control restored once the wound has healed.

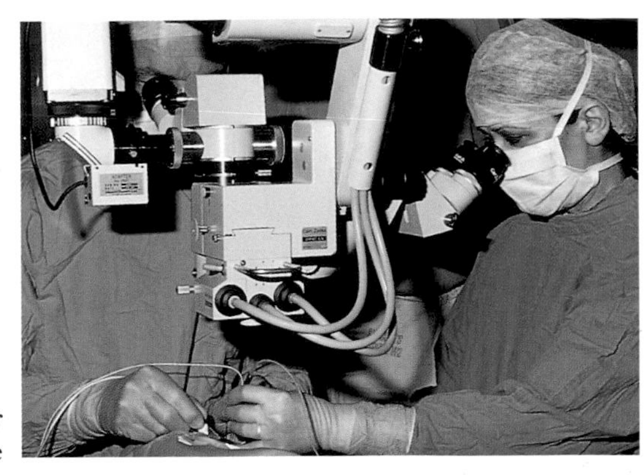

Surgeons using a binocular microscope as they operate

I **Olfactory nerve** *supplies the lining of the nose and relays signals from olfactory cells that are perceived as the sense of smell*

II **Optic nerve** *supplies the retina and relays signals from photoreceptors that are perceived as vision*

III **Oculomotor nerve** *controls movements of the eye and eyelid, and changes in shape of the pupil and lens*

IV **Trochlear nerve,** *in conjunction with the oculomotor and abducens nerves, controls movements of the eyeball*

VI **Abducens nerve** *controls movements of the eyeball*

V **Trigeminal nerve** *controls muscles involved in chewing and relays sensory information from the eye, teeth, and side of the face*

View of brain from underneath (front at top)

VII **Facial nerve** *controls muscles used in facial expressions and the salivary and tear glands. Relays sensory information from the taste buds*

VIII **Vestibulocochlear nerve** *relays sensory signals from the inner ear, which are perceived as sound and allow balance*

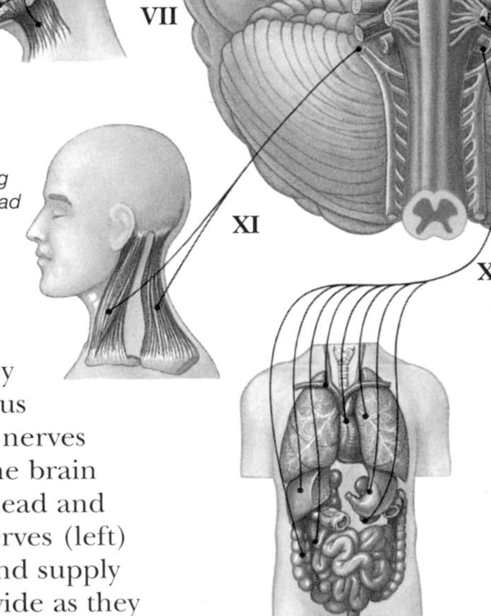

XI **Accessory nerve** *controls muscles involved in swallowing and in moving the head*

IX **Glossopharyngeal nerve** *controls the salivary glands and relays sensory signals from the tongue and pharynx*

XII **Hypoglossal nerve** *controls the movement of the tongue*

X **Vagus nerve** *controls muscles and glands in many internal organs, including the heart, lungs, and stomach*

MAJOR NERVES

Nerves are divided into two types, depending on where they connect with the central nervous system. The 12 pairs of cranial nerves (right) connect directly with the brain or brain stem and supply the head and neck. The 31 pairs of spinal nerves (left) connect with the spinal cord and supply the rest of the body. Nerves divide as they spread throughout the body, but some first converge in clusters called plexuses, which allow fine control of parts of the body such as the hand.

CRANIAL NERVES

The 12 pairs of cranial nerves fan out from the underside of the brain. Other than the vagus nerve (X), cranial nerves control muscles in the head and neck region, or carry nerve impulses from sense organs, such as the eyes, to the brain. Cranial nerves are not only named but are also numbered I to XII, traditionally using Roman numerals. The main functions of each nerve are detailed above.

Spinal cord

TOGETHER WITH THE BRAIN, the spinal cord carries out the vital work of processing information and keeping the body coordinated. It starts at the base of the brain, and reaches most of the way down the spine, giving it an average length in adults of about 44 cm (17 in). Like the brain itself, it contains two kinds of nervous tissue – grey matter, which contains neuron cell bodies, and white matter, which contains axons that carry signals up or down the spine. The spinal cord has two main functions. It relays information between the spinal nerves and the brain, but it also controls many automatic reactions, or reflexes. When it triggers a reflex, it often works on its own, making the body react without waiting to consult the brain.

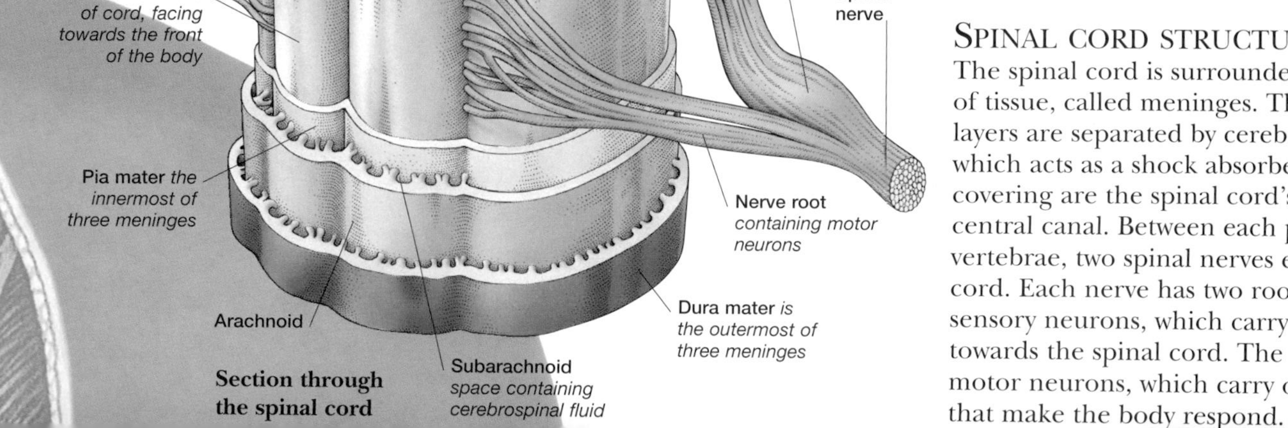

Section through the spinal cord

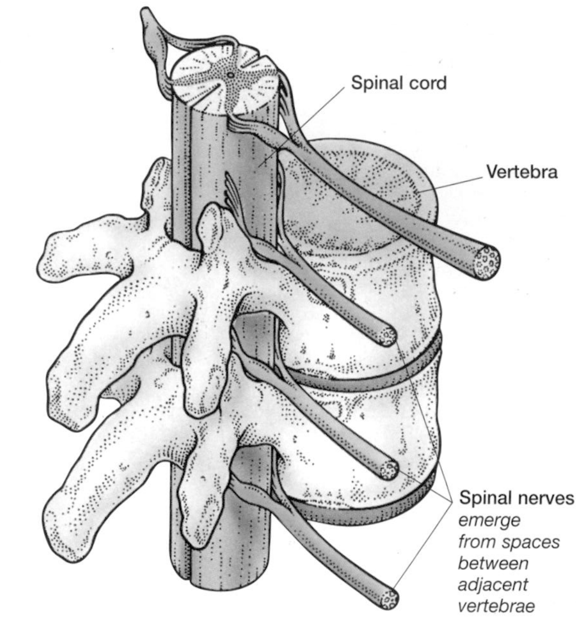

A PROTECTIVE TUNNEL

Although it is no thicker than a finger, the spinal cord is packed with millions of neurons, and it needs special protection to avoid being damaged. It gets this from the bony arches of the vertebrae, which form an open tunnel around it. The spinal cord is anchored at the base of the spine to the base of the brain, and to the vertebrae in between. This prevents it being displaced by sudden jolts.

SPINAL CORD STRUCTURE

The spinal cord is surrounded by three layers of tissue, called meninges. The two innermost layers are separated by cerebrospinal fluid, which acts as a shock absorber. Inside this triple covering are the spinal cord's nerve cells, and a central canal. Between each pair of neighbouring vertebrae, two spinal nerves emerge from the cord. Each nerve has two roots. One contains sensory neurons, which carry incoming signals towards the spinal cord. The other contains motor neurons, which carry outgoing signals that make the body respond.

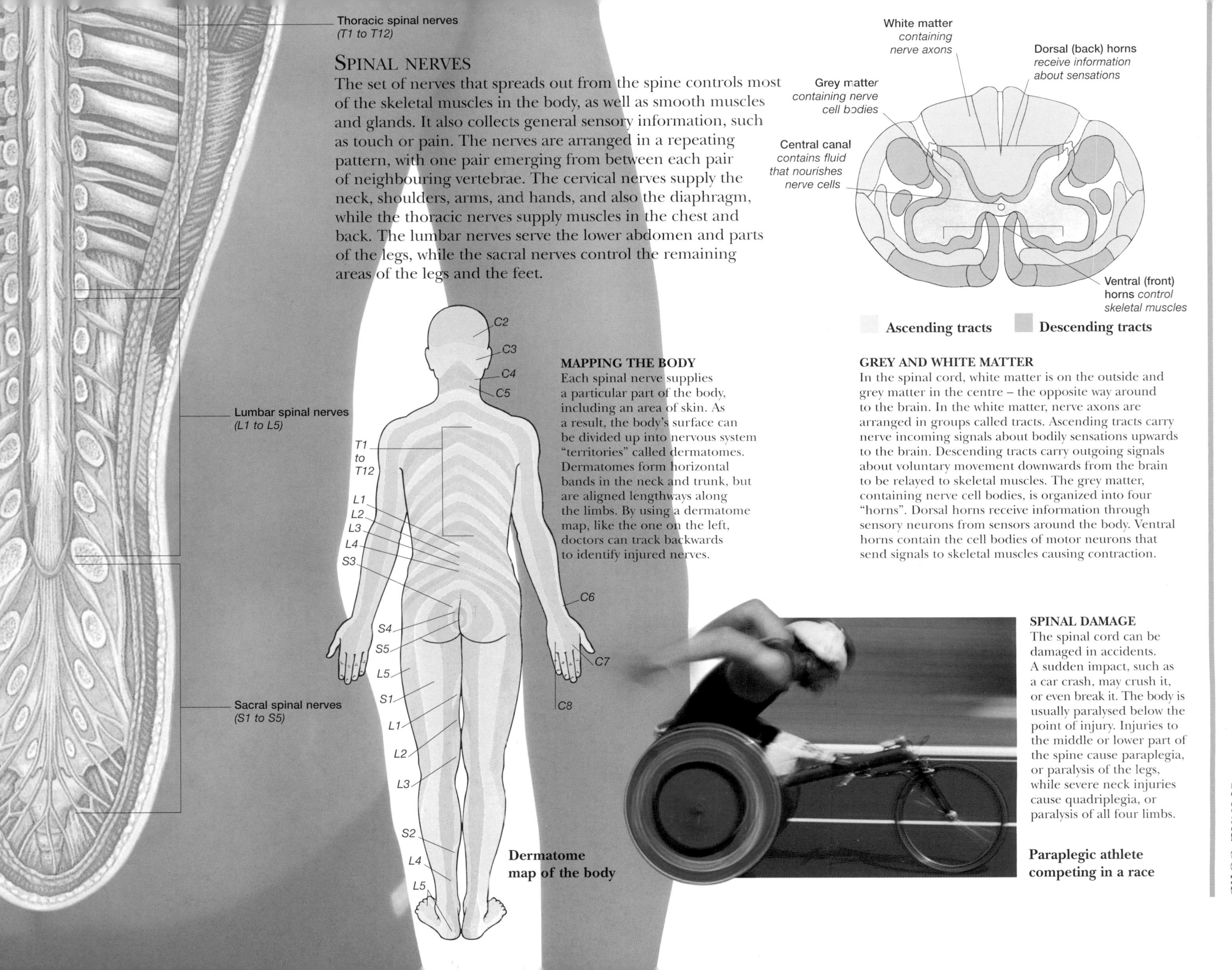

Spinal nerves

The set of nerves that spreads out from the spine controls most of the skeletal muscles in the body, as well as smooth muscles and glands. It also collects general sensory information, such as touch or pain. The nerves are arranged in a repeating pattern, with one pair emerging from between each pair of neighbouring vertebrae. The cervical nerves supply the neck, shoulders, arms, and hands, and also the diaphragm, while the thoracic nerves supply muscles in the chest and back. The lumbar nerves serve the lower abdomen and parts of the legs, while the sacral nerves control the remaining areas of the legs and the feet.

MAPPING THE BODY

Each spinal nerve supplies a particular part of the body, including an area of skin. As a result, the body's surface can be divided up into nervous system "territories" called dermatomes. Dermatomes form horizontal bands in the neck and trunk, but are aligned lengthways along the limbs. By using a dermatome map, like the one on the left, doctors can track backwards to identify injured nerves.

Dermatome map of the body

GREY AND WHITE MATTER

In the spinal cord, white matter is on the outside and grey matter in the centre – the opposite way around to the brain. In the white matter, nerve axons are arranged in groups called tracts. Ascending tracts carry nerve incoming signals about bodily sensations upwards to the brain. Descending tracts carry outgoing signals about voluntary movement downwards from the brain to be relayed to skeletal muscles. The grey matter, containing nerve cell bodies, is organized into four "horns". Dorsal horns receive information through sensory neurons from sensors around the body. Ventral horns contain the cell bodies of motor neurons that send signals to skeletal muscles causing contraction.

SPINAL DAMAGE

The spinal cord can be damaged in accidents. A sudden impact, such as a car crash, may crush it, or even break it. The body is usually paralysed below the point of injury. Injuries to the middle or lower part of the spine cause paraplegia, or paralysis of the legs, while severe neck injuries cause quadriplegia, or paralysis of all four limbs.

Paraplegic athlete competing in a race

Reflexes

ALTHOUGH HUMANS are all individuals, there are times when we behave in exactly the same way. For example, we all pull our hands away if we accidentally touch something sharp or hot, and we blink if anything looks as though it is heading for our eyes. These responses are known as reflexes. They happen automatically, and are a key part of protecting the body from danger. Unlike more complicated kinds of behaviour, reflexes are triggered by simple nervous pathways. A stimulus, such as pain, is picked up by a sensory nerve and then flashed to the spinal cord, or to the lower part of the brain. A motor signal travels back, making part of the body respond. As well as controlling emergency action, reflexes also manage many internal processes, such as movements involved in digesting food.

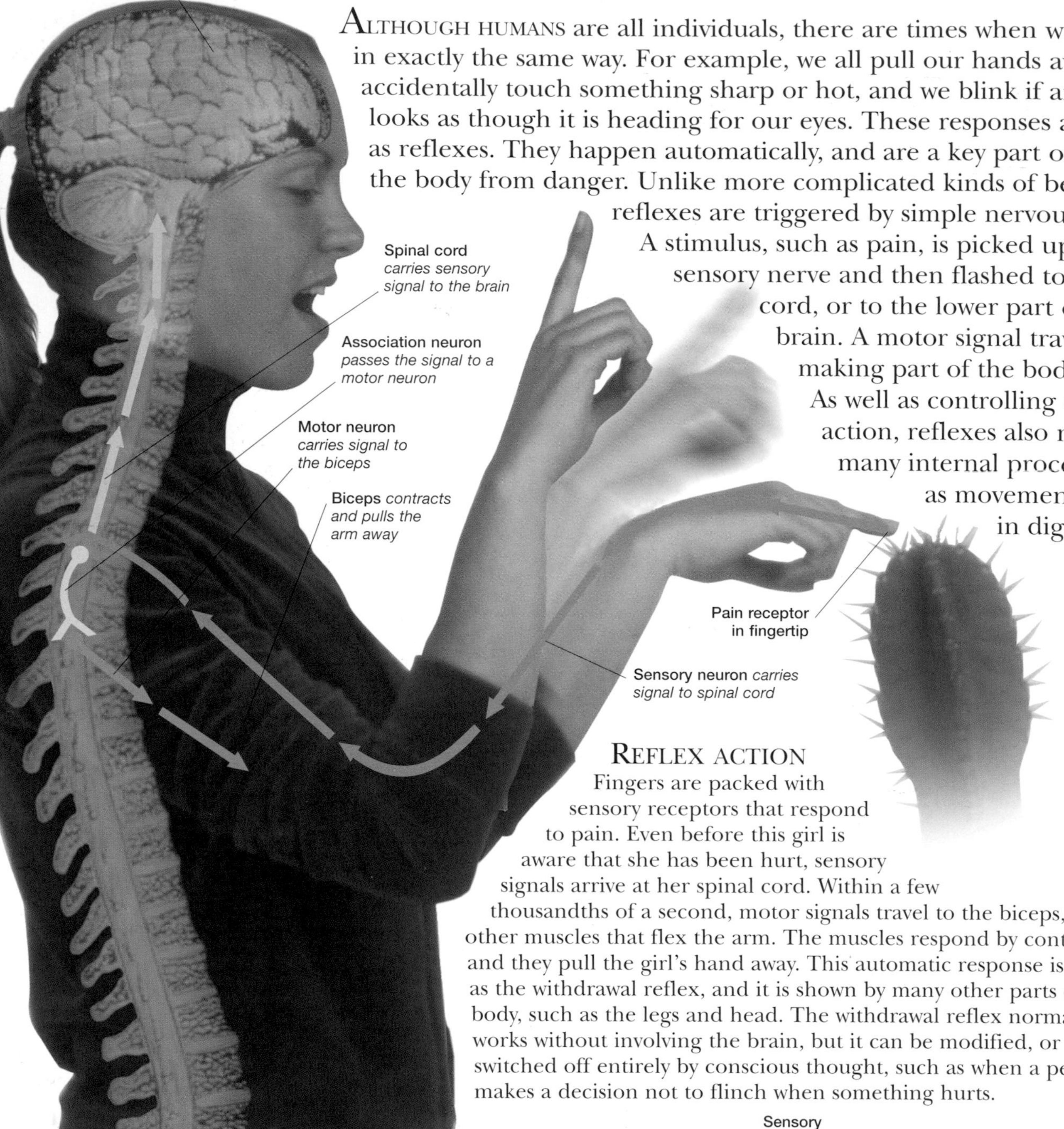

Brain *registers the painful stimulus, but after a reflex action has happened*

Spinal cord *carries sensory signal to the brain*

Association neuron *passes the signal to a motor neuron*

Motor neuron *carries signal to the biceps*

Biceps *contracts and pulls the arm away*

Pain receptor in fingertip

Sensory neuron *carries signal to spinal cord*

REFLEX ACTION

Fingers are packed with sensory receptors that respond to pain. Even before this girl is aware that she has been hurt, sensory signals arrive at her spinal cord. Within a few thousandths of a second, motor signals travel to the biceps, and to other muscles that flex the arm. The muscles respond by contracting, and they pull the girl's hand away. This automatic response is known as the withdrawal reflex, and it is shown by many other parts of the body, such as the legs and head. The withdrawal reflex normally works without involving the brain, but it can be modified, or even switched off entirely by conscious thought, such as when a person makes a decision not to flinch when something hurts.

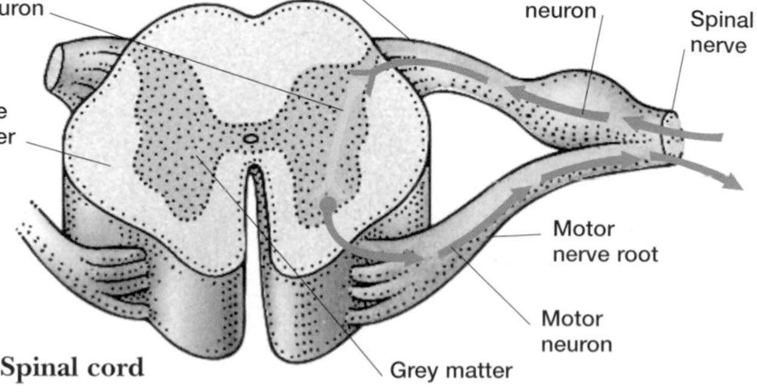

Association neuron

Sensory nerve root

Sensory neuron

Spinal nerve

White matter

Motor nerve root

Motor neuron

Grey matter

Spinal cord

REFLEX ARC

The withdrawal reflex, and reflexes like it, happen quickly because they typically involve just three sets of neurons. A sensory neuron carries a signal to the spinal cord, or to the lower part of the brain. An association neuron then passes the signal on to one or more motor neurons, which make muscles contract and the body react. This pathway is called a reflex arc.

Newborn reflexes

Babies are born with many reflexes that help them to survive. For example, they instinctively suckle at their mother's breast, and they grasp anything that is put in the palm of their hands. If they are underwater, they hold their breath and make swimming movements – even if they have had no experience of water before. These newborn reflexes disappear with age. The grasping reflex is one of the briefest reflexes, fading away after about three months.

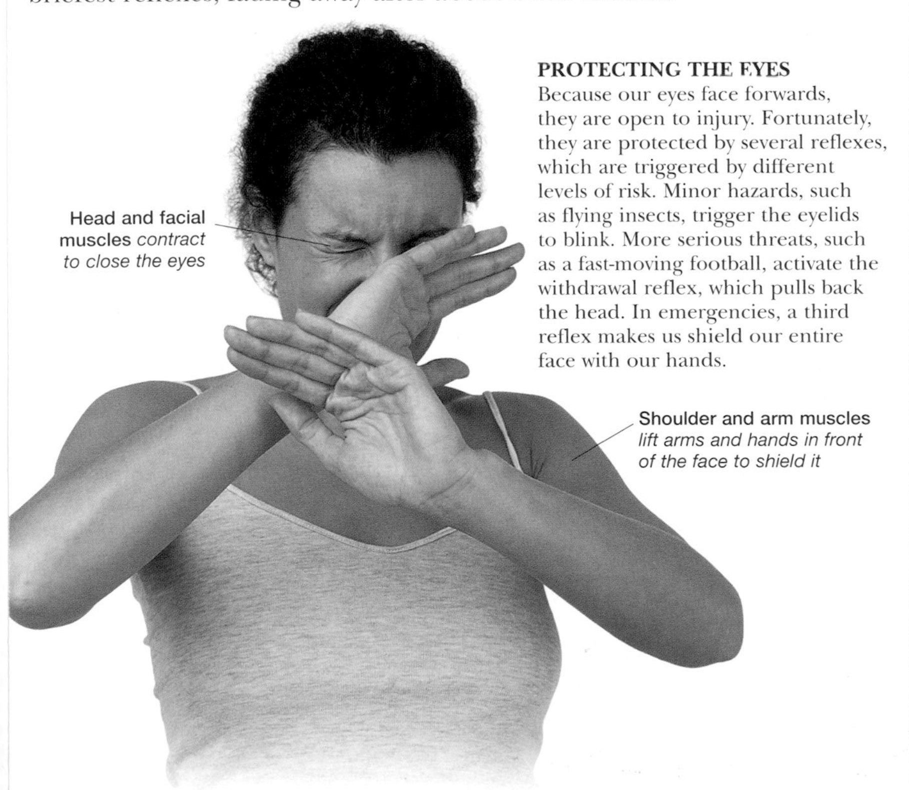
Head and facial muscles *contract to close the eyes*
Shoulder and arm muscles *lift arms and hands in front of the face to shield it*

PROTECTING THE EYES
Because our eyes face forwards, they are open to injury. Fortunately, they are protected by several reflexes, which are triggered by different levels of risk. Minor hazards, such as flying insects, trigger the eyelids to blink. More serious threats, such as a fast-moving football, activate the withdrawal reflex, which pulls back the head. In emergencies, a third reflex makes us shield our entire face with our hands.

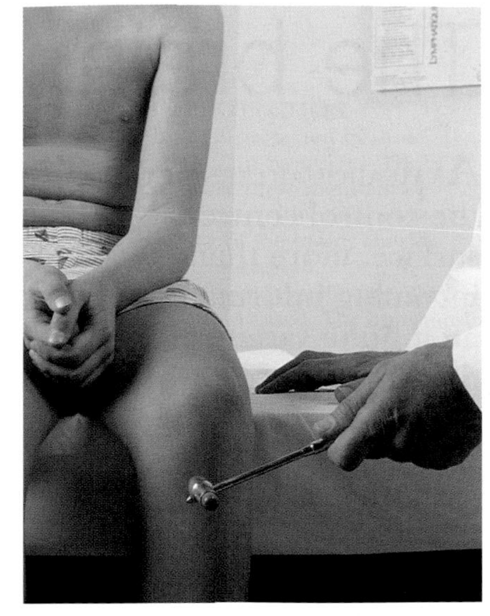

TELL-TALE TAP
In a healthy person, the stretch reflex maintains the body's posture by adjusting the tone, or tension, of skeletal muscles. It is activated by receptors deep inside muscles, and it makes muscles tighten if they stretch too far. Doctors test the stretch reflex by tapping a tendon just below the kneecap. The tapping stretches a muscle in the thigh, and the reflex makes the thigh muscle suddenly tighten, kicking the knee.

Animal Electricity

Although nerves have been known about for centuries, how they work remained a mystery until recently. The first breakthrough was made by the Italian anatomist Luigi Galvani (1737–98), who found that frogs' legs would twitch when they were pinned on a metal frame. He thought that the twitching was caused by "animal electricity" – an idea that turned out to be partly right.

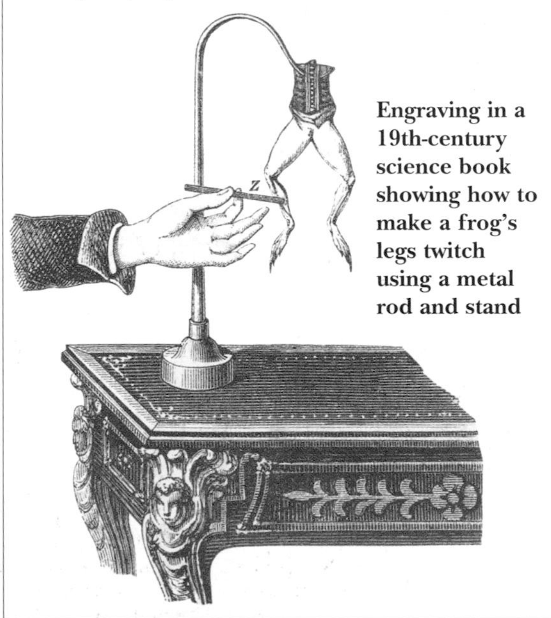

Engraving in a 19th-century science book showing how to make a frog's legs twitch using a metal rod and stand

Mapping the Brain

Anyone reading a book that describes the human brain – as this encyclopedia does – is likely to come across a brain "map" (see p. 92). Just as a street map directs a person through a city and its sights, a brain map provides a guide not just to the brain's parts but also their roles, whether it be thinking thoughts, feeling pain, moving a hand, or distinguishing a cat from a caterpillar. The first proof that different regions of the brain have different functions came in the second half of the 19th century, when the pioneers of brain mapping started their investigations. Early clues came from an accident that occurred in the USA.

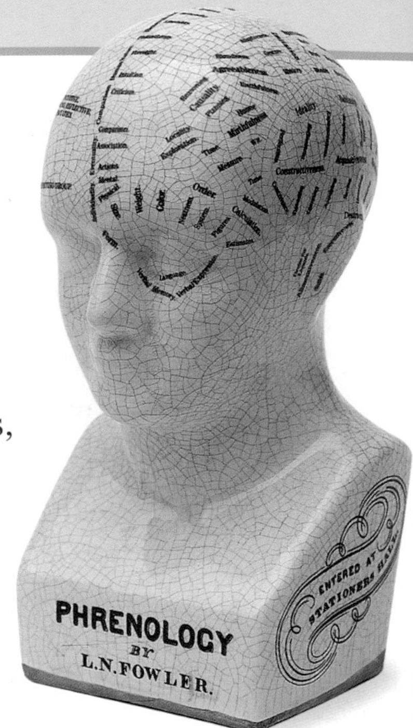

A mapped phrenology head

PHINEAS GAGE
A computerized reconstruction of Phineas Gage's skull, showing how the rod was driven upwards through his cheek and eye, and out through the front of his brain.

Changed personality

Phineas Gage was an American railroad construction worker. In 1848, as Gage used an iron tamping rod to push blasting powder into a hole, a spark set off an explosion that projected the rod up through Gage's left cheek and out through his forehead, taking with it part of the front of his brain. Remarkably, Gage recovered – but once friendly and reliable, he was now an anti-social drifter. Gage's misfortune did reveal, however, that although the front of the brain may not be vital for life, it does control personality. But the story of brain mapping began even earlier with the fanciful ideas of an Austrian doctor.

Reading the bumps

Franz Josef Gall (1758–1828) claimed that the brain had separate parts that controlled 32 different aspects of personality, such as cleverness, humour, or aggression. Gall suggested that the stronger the characteristic, the bigger the brain part that controlled it, so the more it "pushed" on the skull to form a bump. By feeling, or "reading", a person's skull bumps, Gall believed he could analyse their personality. Given the name phrenology, Gall's nonsensical theory became very popular, and people flocked to have their bumps read.

Lost for words

Phrenology met its match in the French doctor Pierre Paul Broca (1824–80). He was interested in which part of the brain controlled speech, and in 1861 examined a patient nicknamed "Tan", because that was the only sound he could utter. When Tan died, six days after the consultation, Broca found that part of his brain's left side was damaged, and concluded that this brain area was responsible for producing speech. Thirteen years later, Austrian doctor Karl Wernicke (1848–1905) was working with a patient who spoke fluently but talked gibberish. Wernicke later

SPEECH AREA
French physician Pierre Paul Broca was a pioneer of brain mapping who discovered that a small area on the left side of the brain controls the production of speech. This region of the left cerebral cortex was named Broca's area in his honour.

discovered that he also had brain damage on the left side, but further back. He concluded that this region – called Wernicke's area – dealt with choosing the right words to speak.

INVASIVE TECHNIQUES

Some researchers took a more direct approach. In 1870, when German army surgeon Eduard Hitzig (1838–1907) was treating wounded soldiers whose brains were exposed, he applied electrical currents to different parts of their brains and noted what happened. In the 1950s, Canadian brain surgeon Wilder Penfield (1891–1976) carried out similar, but more sophisticated, experiments. During surgery, when a patient's brain was exposed – but he or she was conscious – Penfield electrically stimulated different parts of the brain and carefully recorded which regions produced sensations, caused movements, or stored memories.

HAND CONTROL
These two MEG scans of the left side of the brain show brain–hand control in real time. On the left, milliseconds (msecs) before the person moves their right index finger, neurons in the motor cortex "light up" as they send instructions to the finger-moving muscles. On the right, just 40 msecs later, the sensory cortex "lights up" as it receives a message from the muscles that the finger is moving.

SCANNING THE BRAIN

Today's brain mappers use non-invasive methods to investigate brain function. PET and MRI scans (see pp. 30–31) produce indirect images of activity in living brains by measuring oxygen uptake or blood flow in brain regions that are active. MEG (magnetoencephalography) scans measure brain activity directly and in real time – as it happens – by monitoring the electrical activity of brain cells themselves. What all these techniques have shown is that the brain is made up of interactive, not isolated, units.

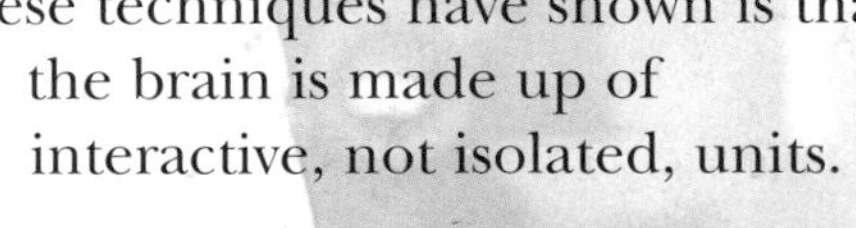

VICTIMS OF WAR
War inevitably produces head wounds, as shown by these soldiers from the First World War. Some wounded soldiers became subjects for researchers who were interested in using their exposed brains to find out which area controlled which action.

Soldiers wounded in the Somme Offensive, 1916

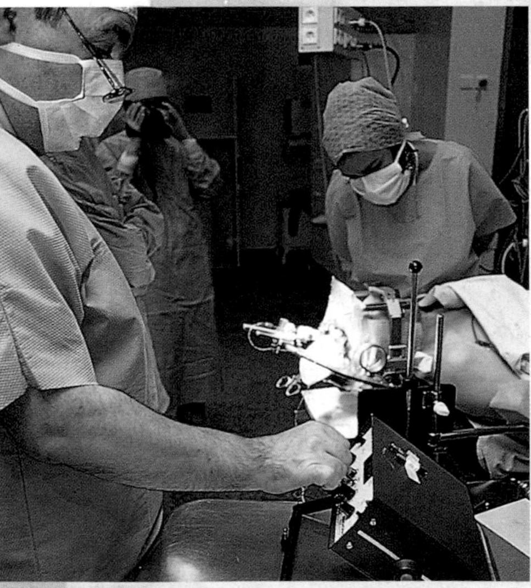

BRAIN PROBE
Surgeons use a frame fixed to a patient's head to provide support and guidance for probes that will go into specific parts of the brain through holes drilled in the skull. These probes are used to remove diseased tissue, but have also been used to find out more about brain function.

Consciousness and sleep

EVERY DAY, THE BRAIN switches between two quite different states – being awake and fully conscious, and being asleep. When the brain is awake, it is aware of its surroundings and is able to think in a purposeful way, so that it can meet the demands of daily life. During sleep, its activity is reduced, and its thought patterns are largely disconnected from the outside world. Sleep allows the body to rest, but just as importantly, it gives the brain time to sort and store the information it has accumulated during the day. Consciousness and sleep are both triggered and maintained by neurotransmitters. These chemicals are made and released by the reticular activating system – a regulator located in the brain stem at the base of the brain.

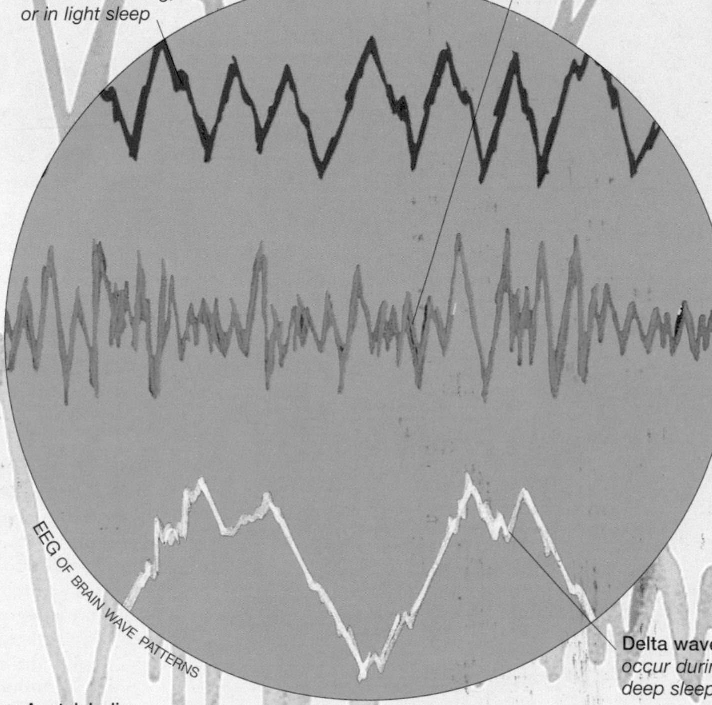

Alpha waves *occur when someone is awake but resting, or in light sleep*

Beta waves *occur when someone is awake and mentally alert*

EEG OF BRAIN WAVE PATTERNS

Delta waves *occur during deep sleep*

BRAIN WAVES
The invention of the electroencephalograph in the 1920s revealed the existence of brain waves – patterns of electrical activity created by the combined output of millions of neurons. Alpha waves are produced when the brain is awake but relaxed, while beta waves occur during times of intense activity. Widely spaced delta waves occur during deep sleep, when the brain is at its least active. The reading produced by the instrument is caused an electroencephalogram (EEG).

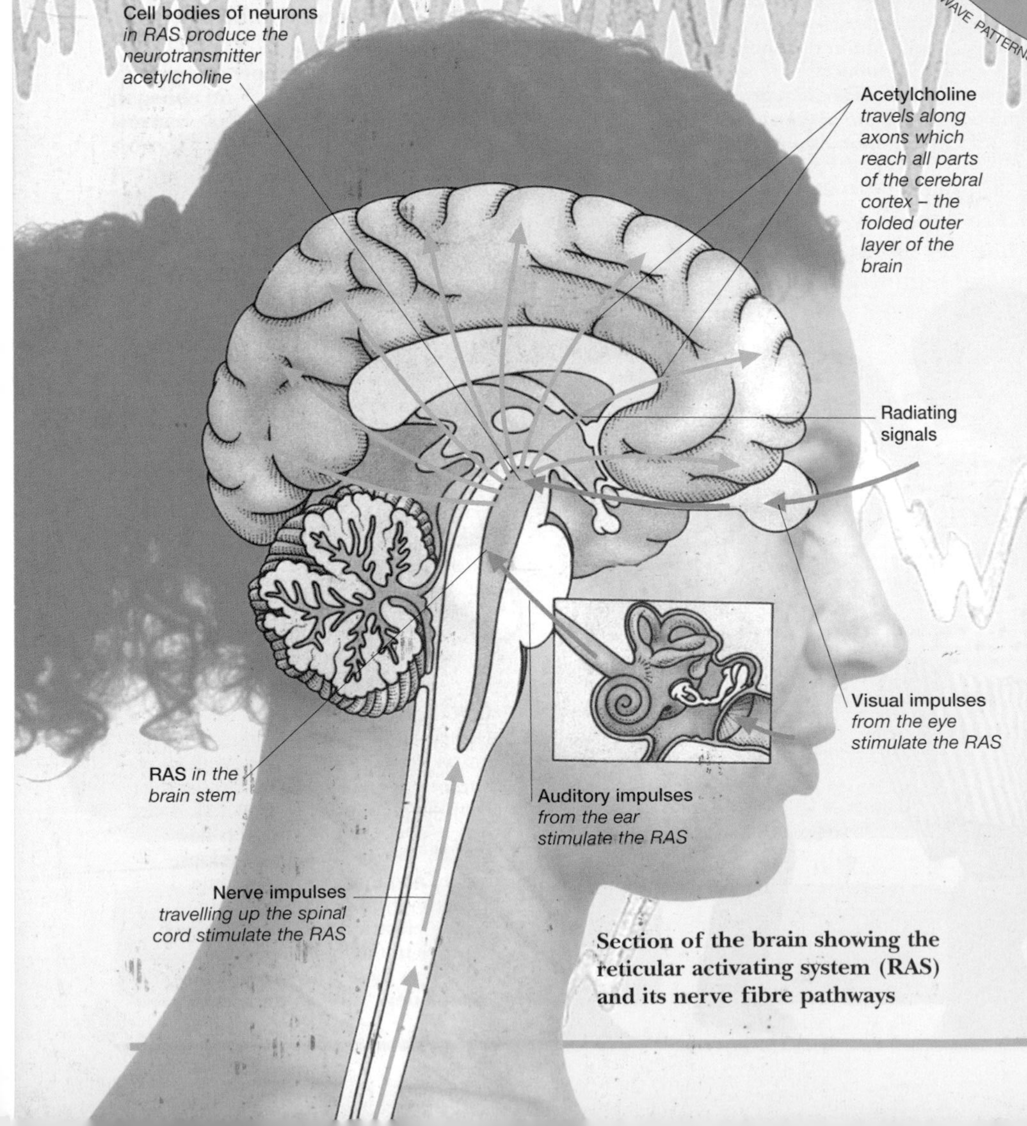

Cell bodies of neurons *in RAS produce the neurotransmitter acetylcholine*

Acetylcholine *travels along axons which reach all parts of the cerebral cortex – the folded outer layer of the brain*

Radiating signals

Visual impulses *from the eye stimulate the RAS*

RAS *in the brain stem*

Auditory impulses *from the ear stimulate the RAS*

Nerve impulses *travelling up the spinal cord stimulate the RAS*

Section of the brain showing the reticular activating system (RAS) and its nerve fibre pathways

STAYING ALERT

When a person is awake, their brain deals with a torrent of sensory information from inside and outside the body. To do this, its association neurons have to work with maximum efficiency. This level of activity is triggered by neurons in the reticular activating system (RAS). These reach throughout the cerebral cortex – the "thinking" part of the brain – where they release acetylcholine, a neurotransmitter that aids alertness and concentration. The brain uses many other neurotransmitters as well. One of them, called serotonin, triggers sleep and also affects mood. Another, called dopamine, helps to regulate movements. People suffering from Parkinson's disease have low levels of dopamine, causing weakness and muscle tremors.

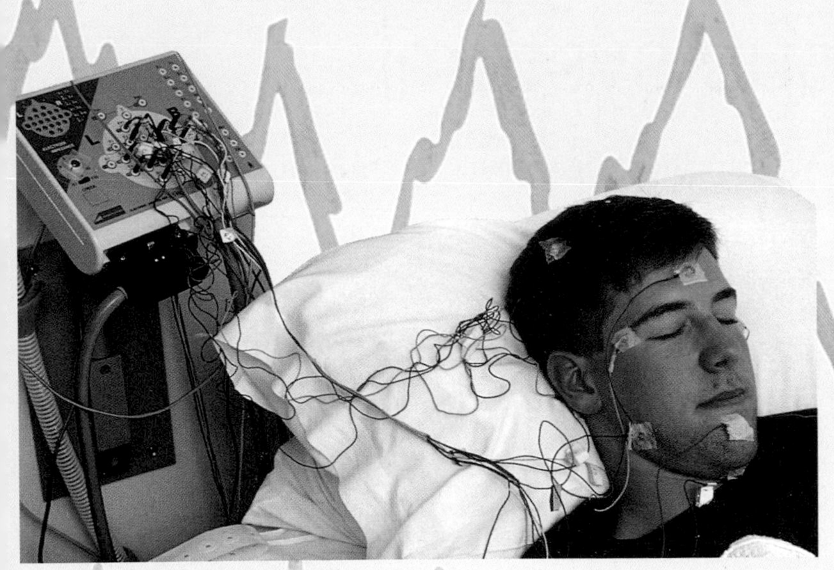

A circadian, or daily, rhythm experiment is conducted on a sleeping patient

SLEEPING

The amount of sleep that people need varies greatly from one person to another. Most adults sleep between seven and eight hours a day even though their bodies need less than half this time, while infants and young children need much more. Scientists do not know exactly why the brain needs sleep, but the most likely explanation is that freedom from external distractions allows it to carry out internal "housekeeping", which makes memory storage more efficient. During sleep, the brain is still partly alert, and an unexpected sound or movement will quickly rouse it.

DREAMS

Almost everyone dreams – although they may forget their dreams before they wake up – but exactly why we dream is still unknown. One possibility is that dreams are simply the results of nerve cells firing at random; another is that dreams play some part in the brain's memory storage, with old experiences being retrieved, and new ones filed away. Repeated dreams may reveal hidden anxieties, and some psychiatrists believe that they can be used as a window into the human subconscious – an idea proposed by the founder of psychoanalysis, Austrian doctor Sigmund Freud (1856–1939).

SLEEP PATTERNS

During a typical night, two kinds of sleep follow each other in succession. In the first type, called NREM (non-rapid eye movement sleep) the sleeper often moves about, but their brain activity drops to a low level. By contrast, in REM (rapid eye movement sleep) the body becomes immobile, but the eyes dart about. Most dreams occur during REM sleep.

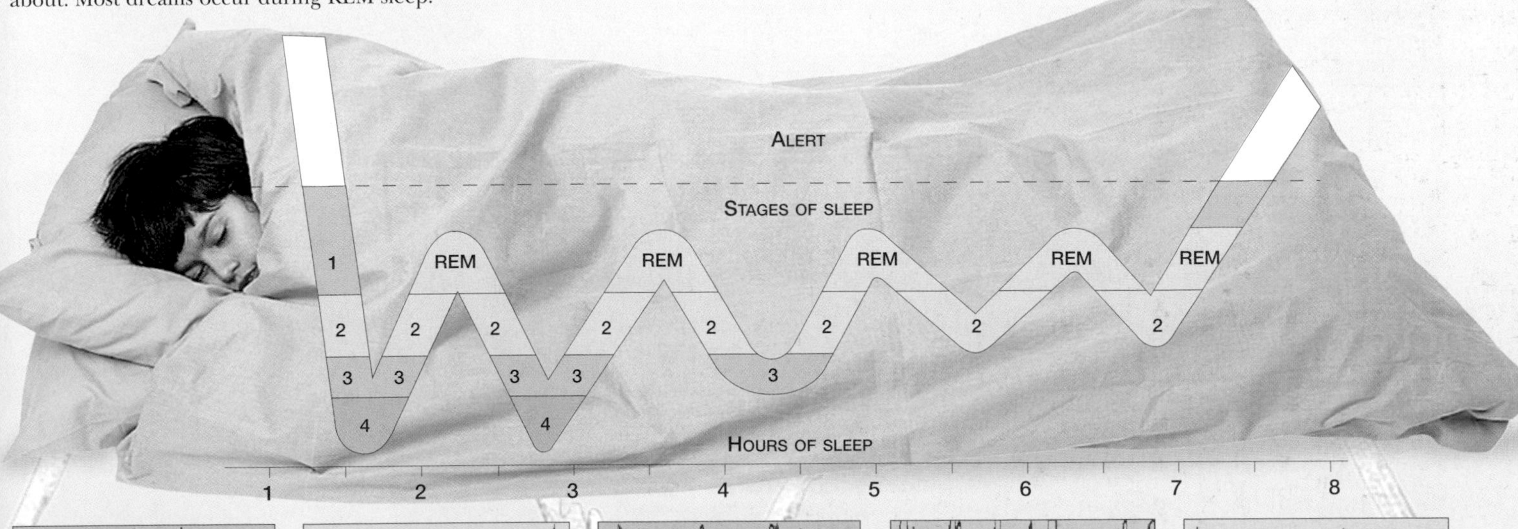

NREM SLEEP: STAGE 1
EEG shows alpha waves. The body is relaxed but the person wakes immediately if disturbed.

NREM SLEEP: STAGE 2
The EEG pattern becomes more irregular. It becomes more difficult to wake the sleeper.

NREM SLEEP: STAGE 3
Delta waves appear in the EEG. Vital signs – breathing, heart rate, and body temperature – decrease.

NREM SLEEP: STAGE 4
In deep sleep, delta waves dominate the EEG. Vital signs are at their lowest. Arousal is difficult.

REM SLEEP
Alpha waves appear. Vital signs increase, while skeletal muscles are inhibited. Dreaming occurs.

Autonomic nervous system

EVERY DAY, THE NERVOUS SYSTEM issues streams of instructions that enable the body to deal with the outside world. These are handled by the somatic branch, which is chiefly concerned with body movements. But the nervous system also helps with the body's internal "housekeeping" through its autonomic branch, which works without any conscious control. The autonomic nervous system (ANS) acts mainly on smooth and cardiac muscle – the kinds found in internal organs and the heart, respectively. It has two separate divisions, the sympathetic and parasympathetic, which often have opposite effects. By triggering one or the other in response to changing internal or external conditions, the brain can adjust all kinds of physical factors, from the size of the eyes' pupils, to the speed of digestion. By so doing, the ANS helps maintain homeostasis – stable conditions inside the body.

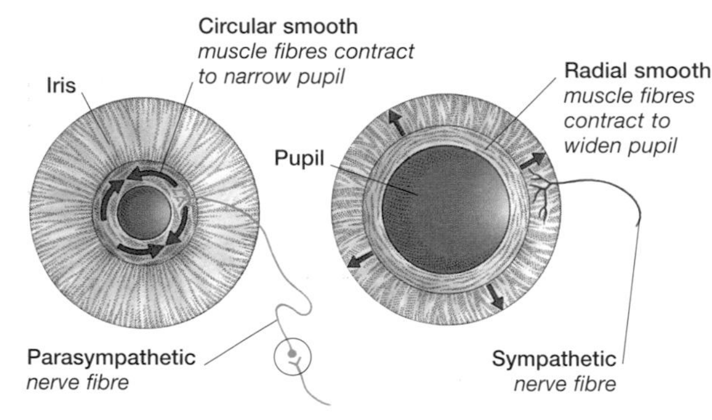
Circular smooth muscle fibres contract to narrow pupil
Iris
Radial smooth muscle fibres contract to widen pupil
Pupil
Parasympathetic nerve fibre
Sympathetic nerve fibre

CHANGING PUPIL SIZE
Pupil size is constantly adjusted by the iris under the control of the autonomic nervous system to change the amount of light entering the eyes. Pupil narrowing occurs when the iris's concentrically arranged smooth muscle fibres are stimulated by parasympathetic nerve fibres. Pupil widening occurs when radial smooth muscle fibres – arranged like the spokes of a wheel – are stimulated by sympathetic nerve fibres. This is one of the many autonomic reflexes of which people are usually unaware.

THE SYMPATHETIC DIVISION

This division of the ANS activates a wide range of organs and tissues. It also steps up processes that help the body to cope with stress. For example, it triggers the release of glucose into the blood, and increases the blood's oxygen level by speeding up the heart rate and widening the airways in the lungs. The nerves of this division all emerge from the central section of the spinal cord. Unlike motor nerves that drive skeletal muscles, they have two consecutive sets of neurons. One set runs from the spinal cord to swellings called ganglia, and another set carries signals from these ganglia to their final destination.

SYMPATHETIC GANGLION CHAIN
After leaving the spinal cord, nerves of the sympathetic division connect with a chain of ganglia. They then travel onwards through the body, sometimes connecting with more ganglia on the way. Ganglia are like junction boxes – they allow nerve signals to be passed on from one neuron to several, so that they reach a wide variety of organs.

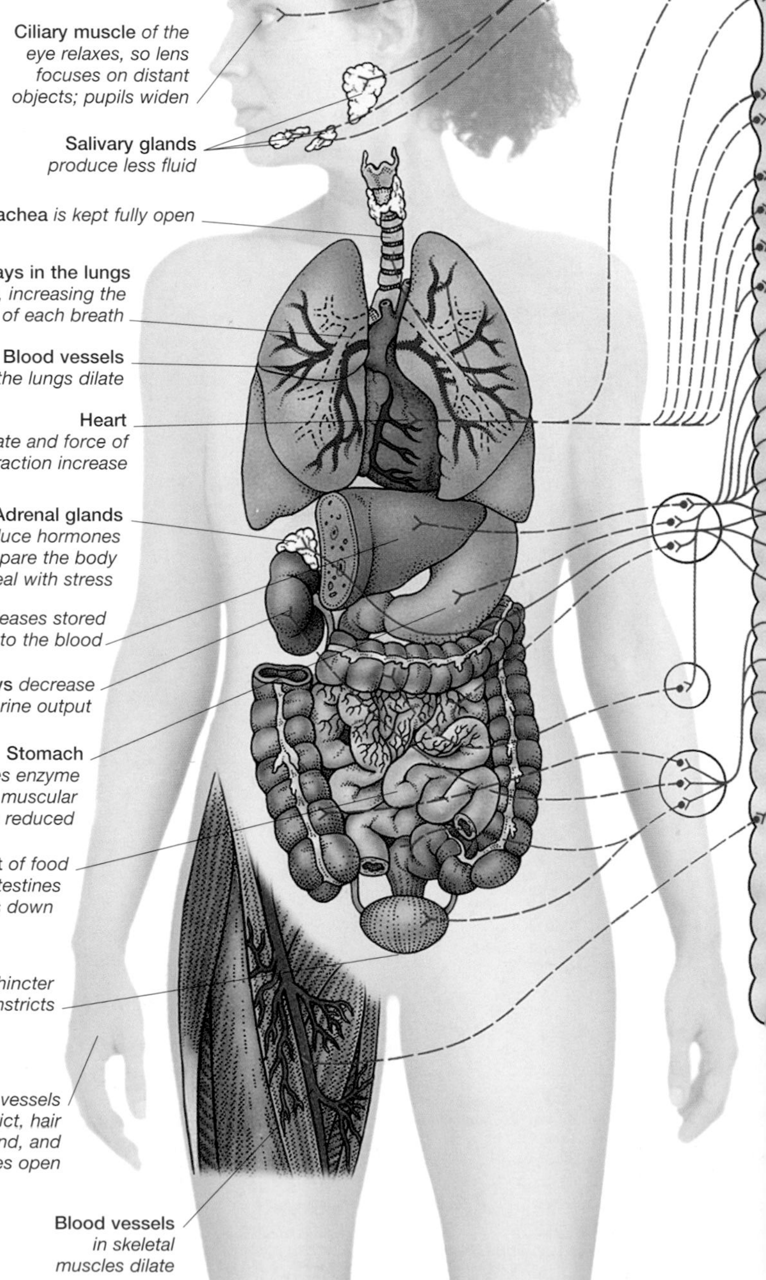
Ciliary muscle of the eye relaxes, so lens focuses on distant objects; pupils widen
Salivary glands produce less fluid
Trachea is kept fully open
Airways in the lungs dilate, increasing the volume of each breath
Blood vessels in the lungs dilate
Heart rate and force of contraction increase
Adrenal glands produce hormones that prepare the body to deal with stress
Liver releases stored glucose into the blood
Kidneys decrease urine output
Stomach decreases enzyme production, and muscular movements are reduced
Movement of food through the intestines slows down
Bladder sphincter muscle constricts
Skin: blood vessels constrict, hair stands on end, and sweat pores open
Blood vessels in skeletal muscles dilate

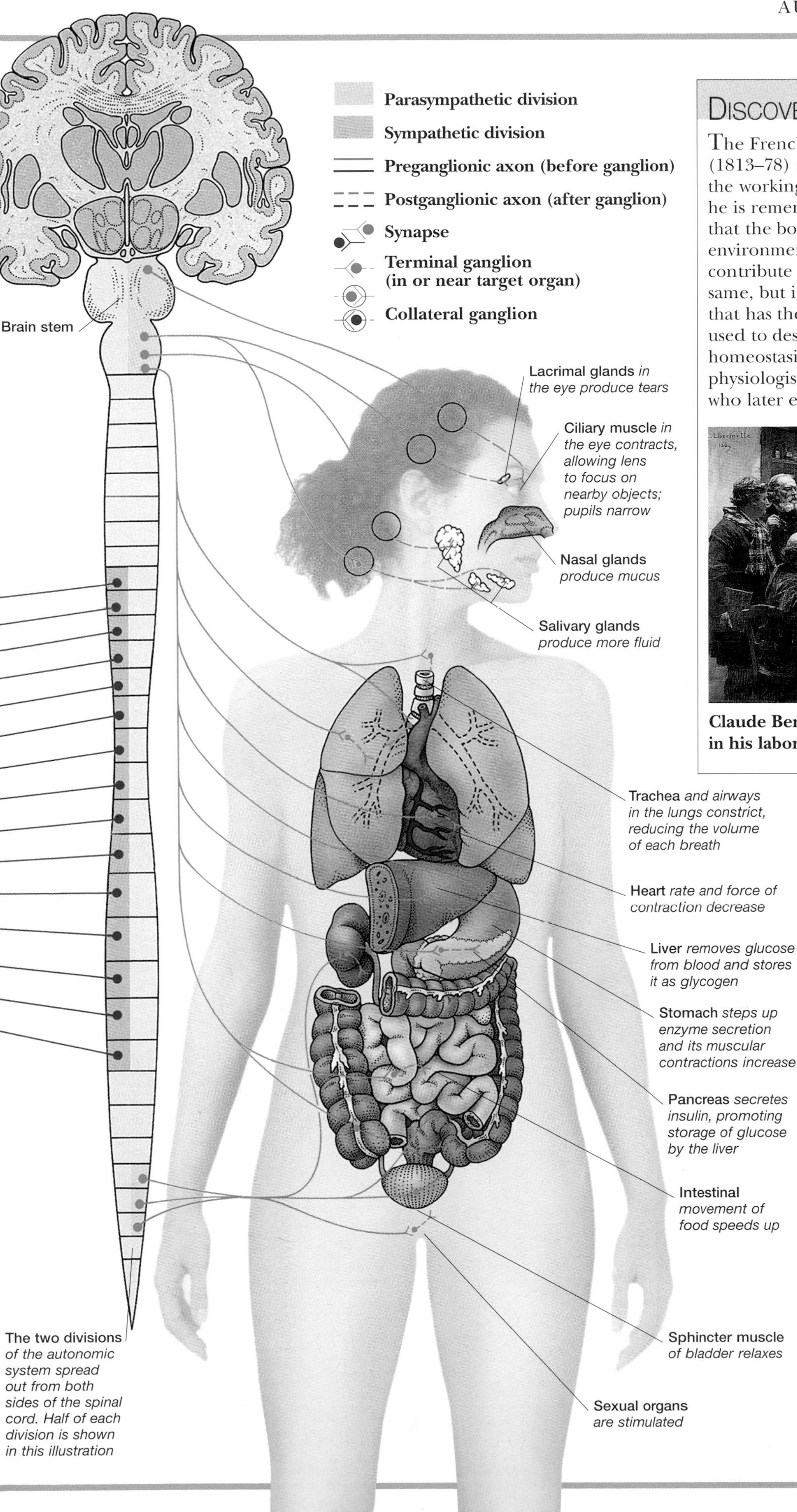

Parasympathetic division
Sympathetic division
Preganglionic axon (before ganglion)
Postganglionic axon (after ganglion)
Synapse
Terminal ganglion (in or near target organ)
Collateral ganglion
Brain stem
Lacrimal glands in the eye produce tears
Ciliary muscle in the eye contracts, allowing lens to focus on nearby objects; pupils narrow
Nasal glands produce mucus
Salivary glands produce more fluid
Trachea and airways in the lungs constrict, reducing the volume of each breath
Heart rate and force of contraction decrease
Liver removes glucose from blood and stores it as glycogen
Stomach steps up enzyme secretion and its muscular contractions increase
Pancreas secretes insulin, promoting storage of glucose by the liver
Intestinal movement of food speeds up
Sphincter muscle of bladder relaxes
Sexual organs are stimulated

The two divisions of the autonomic system spread out from both sides of the spinal cord. Half of each division is shown in this illustration

DISCOVERING HOMEOSTASIS

The French physiologist Claude Bernard (1813–78) made several discoveries about the workings of the human body. However, he is remembered particularly for the idea that the body's cells need a constant internal environment to survive. All body systems contribute to keeping internal conditions the same, but it is the autonomic nervous system that has the most important role. The word used to describe maintaining a stable state – homeostasis – was coined by the American physiologist Walter Cannon (1871–1945), who later expanded the concept.

Claude Bernard performing an experiment in his laboratory at the College de France

THE PARASYMPATHETIC DIVISION

This division of the ANS comes into play when the body needs to conserve its resources, rather than getting them ready for use. It gives priority to digestion, because this supplies the body with raw materials and energy, and it ensures that glucose is stored away in the liver for future use. The parasympathetic division is also involved in the regulation of sexual response – which is damped down when the body is under stress. The nerves of this division originate from the brain stem and base of the spine. Like the sympathetic division, they have two consecutive sets of neurons. In the parasympathetic division, however, the ganglia – called terminal ganglia – are near or inside the target organs.

Communication

HUMANS ARE INTENSELY social beings, and communication plays a central role in the way we live. We communicate not only to show how we feel, but also to pass on information. Today, people often communicate via machines, but the most natural way of expressing ourselves still involves being face to face. When people meet, they communicate in various ways. Facial expressions and body language convey a multitude of signals about a person's mood, but they relate only to the present time. Spoken language is far more powerful. It conveys an unlimited range of ideas, not only about what is actually happening, but also about the past and future. Together with our intelligence, it is one of the features that makes humans unique.

Frontalis *raises the eyebrow and wrinkles the forehead*

Levator labii superioris *raises the upper lip and makes it curl*

Corrugator supercilii *(under orbicularis oculi) pulls the eyebrow down and wrinkles it*

Orbicularis oculi *closes the eye*

Levator anguli oris *turns corner of mouth upwards*

Orbicularis oris *purses the lips and shapes them when speaking*

Mentalis *protrudes the lower lip and wrinkles the chin*

Zygomaticus major *pulls the corner of the mouth outwards and upwards when smiling*

Risorius *pulls the corner of the mouth outwards when smiling*

Depressor anguli oris *pulls the corner of the mouth downwards*

Depressor labii inferioris *pulls the lower lip downwards*

FACIAL EXPRESSIONS

Compared with most other animals, humans have extremely expressive faces. These expressions are produced by a set of more than 30 muscles, which pull small areas of facial skin when they contract. Most of these muscles work as pairs, but with a little practice, some – such as the ones that raise the eyebrows – can be used individually as well. Some facial expressions mean different things in different parts of the world, but many, such as smiling and crying, can be understood by anybody anywhere.

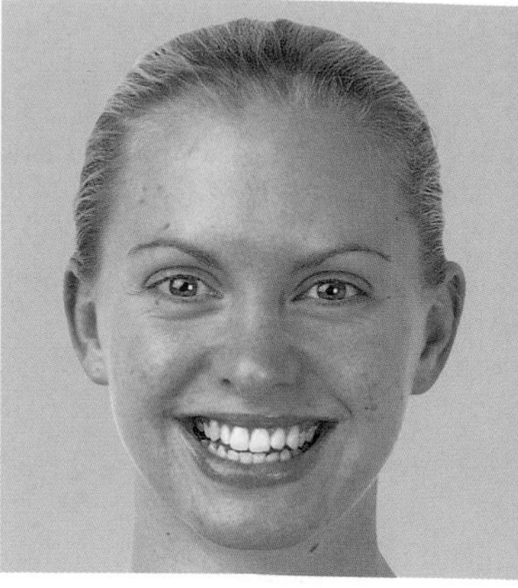

Smiling
This infectious expression is produced mainly by the risorius muscles. A really broad smile also involves the zygomaticus major, which stretches the mouth and lifts its corners up. These muscles – and more – are also used in laughing.

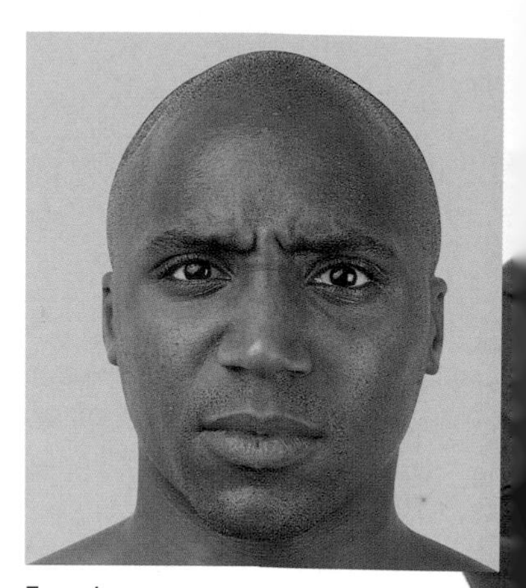

Frowning
Frowning involves the two corrugator supercilii muscles, which contract to pull the eyebrows down, making them wrinkle. Frowning conveys many emotions, from suspicion to deep thought. It also helps to shade the eyes from bright sunshine.

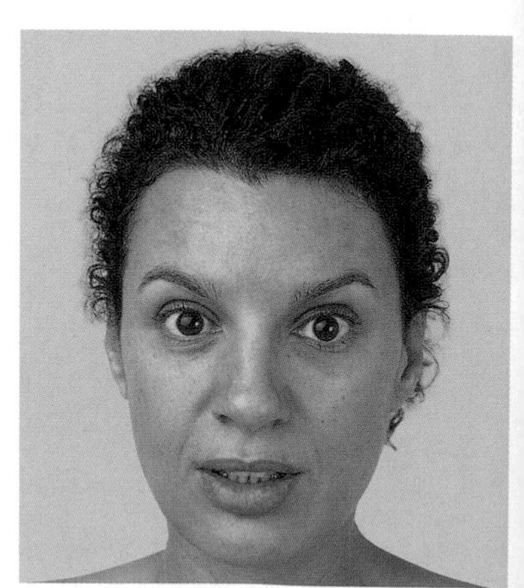

Surprise
When someone shows surprise, the frontalis muscle contracts, raising the eyebrows and making the forehead wrinkle. People use a variant of this expression, called the "eyebrow flash", to acknowledge people they know without stopping to talk.

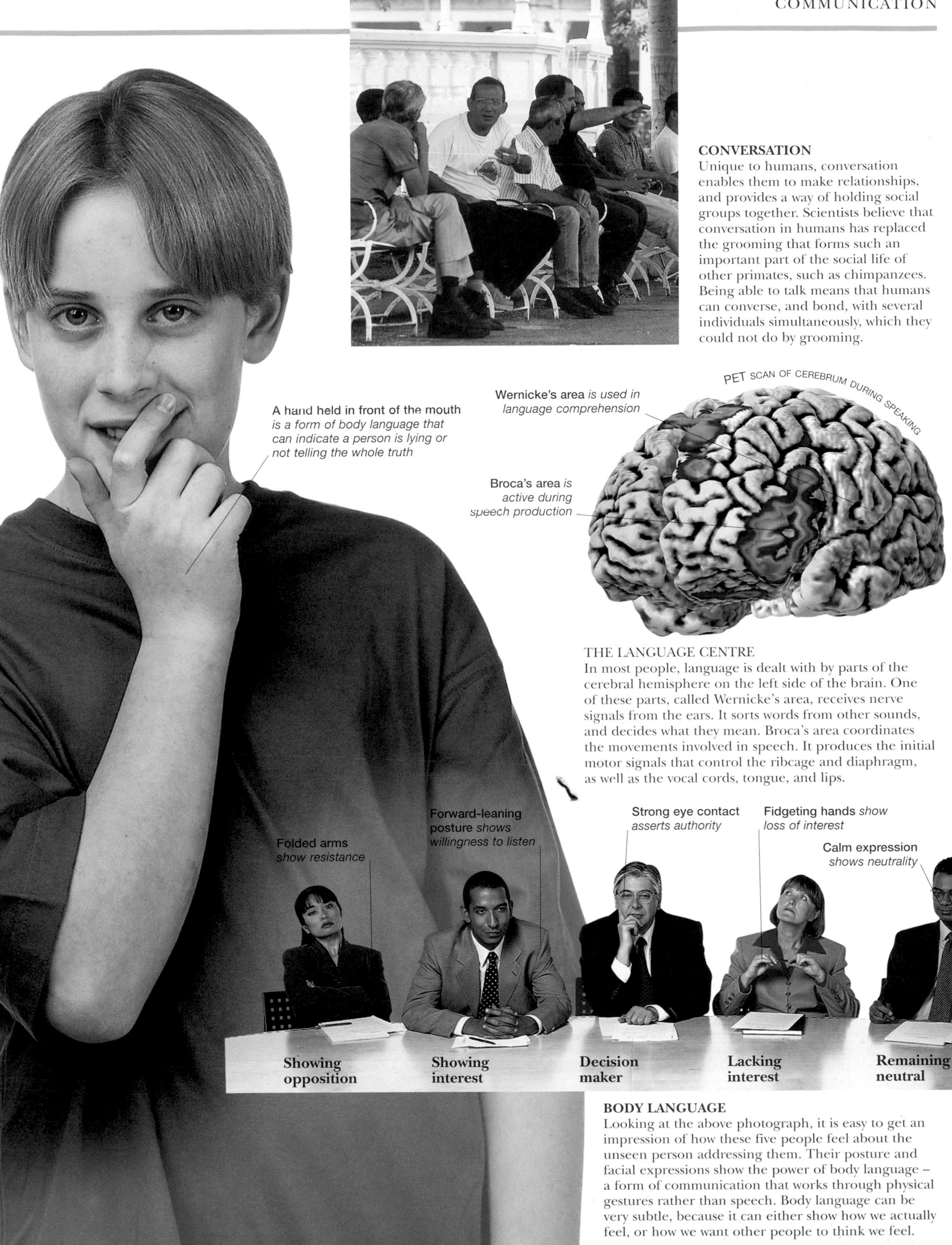
A hand held in front of the mouth *is a form of body language that can indicate a person is lying or not telling the whole truth*
PET SCAN OF CEREBRUM DURING SPEAKING
Wernicke's area *is used in language comprehension*
Broca's area *is active during speech production*
Forward-leaning posture *shows willingness to listen*
Folded arms *show resistance*
Strong eye contact *asserts authority*
Fidgeting hands *show loss of interest*
Calm expression *shows neutrality*
Showing opposition
Showing interest
Decision maker
Lacking interest
Remaining neutral

CONVERSATION
Unique to humans, conversation enables them to make relationships, and provides a way of holding social groups together. Scientists believe that conversation in humans has replaced the grooming that forms such an important part of the social life of other primates, such as chimpanzees. Being able to talk means that humans can converse, and bond, with several individuals simultaneously, which they could not do by grooming.

THE LANGUAGE CENTRE
In most people, language is dealt with by parts of the cerebral hemisphere on the left side of the brain. One of these parts, called Wernicke's area, receives nerve signals from the ears. It sorts words from other sounds, and decides what they mean. Broca's area coordinates the movements involved in speech. It produces the initial motor signals that control the ribcage and diaphragm, as well as the vocal cords, tongue, and lips.

BODY LANGUAGE
Looking at the above photograph, it is easy to get an impression of how these five people feel about the unseen person addressing them. Their posture and facial expressions show the power of body language – a form of communication that works through physical gestures rather than speech. Body language can be very subtle, because it can either show how we actually feel, or how we want other people to think we feel.

UNDERSTANDING THE MIND

FOR THOUSANDS of years, philosophers and scientists have tried to explain the mind, the intangible "thing" that gives a person his or her unique thoughts, feelings, behaviour, and sense of being. Until the 18th century, it was believed that the mind and brain were separate entities, but today it is widely accepted that the mind is a product of brain activity. One way of exploring the mind is through the study of mental illness. In the past, sheer ignorance made people react to mental illness in others with fear and hostility. But in the 19th century, a more scientific approach was taken to understanding the mind's problems.

RELEASING SPIRITS
This skull from around 2000 BC shows clearly the holes made by trepanning. This ancient practice of drilling, or cutting, holes in the skull was believed to "cure" mental illnesses, or milder conditions such as migraine, by releasing "evil spirits" from the head.

EVIL SPIRITS

What makes disorders of the mind differ from other diseases is that they affect how a person behaves. Ancient peoples believed mental illness was the result of someone being "possessed" by evil spirits. Treatment, if any, came in the form of magical amulets or charms, or, more drastically, by drilling holes into the head so those spirits could escape.

PINEL THE REFORMER
The reformer Philippe Pinel (centre left, with cane) is celebrated in heroic style as he orders the removal of chains from mental patients at the Bicêtre asylum, Paris, in 1795.

VIEWS OF MENTAL ILLNESS

In the Middle Ages, the Church saw mental illness as a sign of the devil at work, a belief that led to the persecution of people as witches. Doctors, however, viewed it as an imbalance in the four humours, the fluids that were believed to make up the body.

Most mentally-ill people were treated as less than human, with many languishing locked up in mad-houses. But by the end of the 18th century, reformers were suggesting a more humane approach. French doctor Philippe Pinel controversially insisted that his patients be released from their chains and allowed to receive medical attention. The work of Pinel helped establish the field of psychiatry, the branch of medicine that deals with mental illness.

SCIENCE IN MIND

In the 19th century, large mental hospitals were established which enabled doctors to observe patients and catalogue a wide range of disorders. The work of Broca and Wernicke (see pp. 90–1) proved that abnormal brain structure could alter behaviour, indicating that the mind really was part of the brain. In Paris, physician Jean-Martin Charcot (1825–93) drew together all that had been learned about mental illness and the brain to create a scientific approach to understanding the mind.

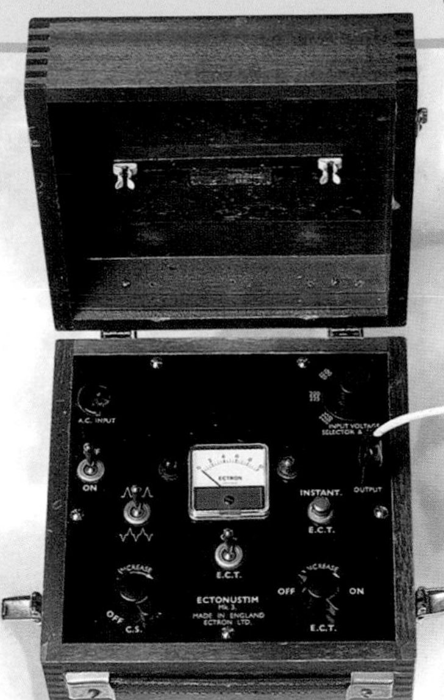

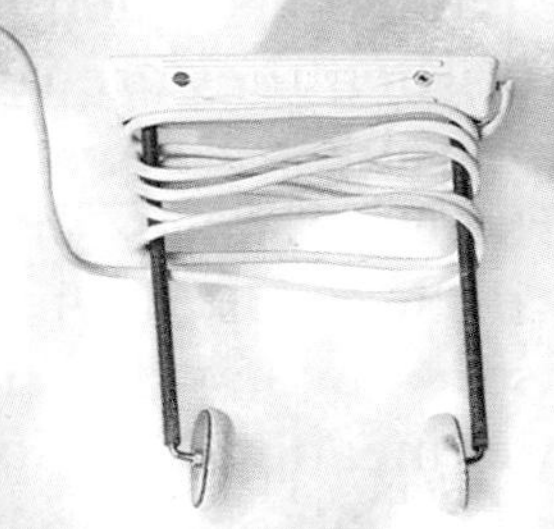

SHOCK TREATMENT
An electroconvulsive therapy (ECT) machine from 1950 is attached to a headset complete with paired electrodes. These padded electrodes were placed on the side of the head in order to deliver an electrical pulse – or shock – to the brain.

PRESSURE ON THE BRAIN
This MRI scan of a vertical section through the brain (orange) shows an abnormal growth, or tumour (blue), in the right hemisphere. Tumours like these may cause changes in behaviour and sensation, but can, in some cases, be removed by surgery.

PROBING THE UNCONSCIOUS

In 1885, an Austrian doctor called Sigmund Freud visited Charcot in Paris. On returning to Vienna, Freud evolved a method by which some patients could discuss their problems by free association – talking openly about their feelings and emotions to a person, called the analyst. Freud believed that problems were caused by subconscious fears, worries, and conflicts, often developed in childhood. His technique, called psychoanalysis, raises problems to a conscious level – by talking about them – so that the patient, with the help of the analyst, can deal with them.

THE MODERN APPROACH

But by the middle of the 20th century, there were still few treatments for mental illnesses, because there was little idea of their causes. Electroconvulsive therapy (ECT), for example, was used to treat severe depression, but its mode of action was unknown. But since then there have been considerable advances. Drugs have been developed that treat specific mental illnesses. Research has shown that in many cases these drugs work because they correct imbalances of neurotransmitters (see pp. 80–1) in the brain. Modern scanning techniques, such as MRI and PET scans, enable doctors and researchers to look for abnormal brain structures and to watch brain activity in action. Psychotherapy allows people to talk through and address their problems.

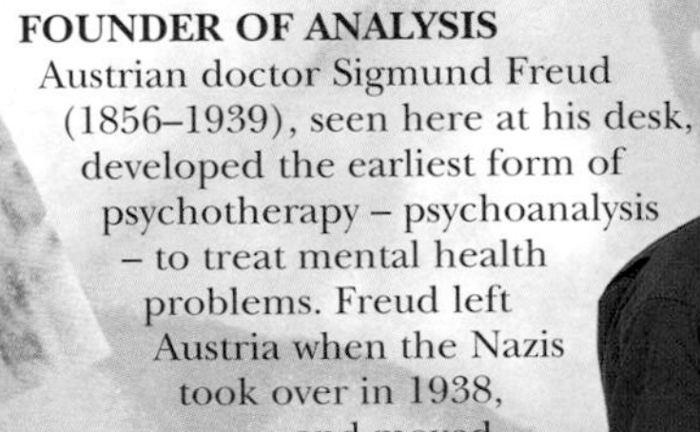

FOUNDER OF ANALYSIS
Austrian doctor Sigmund Freud (1856–1939), seen here at his desk, developed the earliest form of psychotherapy – psychoanalysis – to treat mental health problems. Freud left Austria when the Nazis took over in 1938, and moved to London.

PSYCHOTHERAPY
A person with a mental or emotional problem seeks help from a psychotherapist. The person is encouraged to talk about their symptoms and problems, while the therapist is trained to listen to and evaluate what has been said, and help the person to understand themselves more.

Smell and taste

COMPARED WITH THE OTHER special senses, taste and smell are close partners, and work in similar ways. Both are chemical senses: taste detects substances that are dissolved in saliva, while smell detects those present in the air. To do this, they use chemoreceptors, which are specialized cells that respond to specific molecules. Together, these two senses allow us to identify things that are good to eat or drink, but they also warn us about things that might be dangerous. The ability to taste and smell varies a great deal from person to person. This explains why some people make good wine tasters or perfumiers, while others have trouble perceiving some flavours or odours at all.

CHEMICAL DETECTORS

Taste receptors are found on the tongue, while smell receptors are located in the roof of the nasal cavity. Both are triggered by contact with chemicals, but their sensitivity is quite different. Taste receptors can detect only four overall tastes – sweet, salty, sour, and bitter – but smell receptors can distinguish between more than 10,000 different odours. Smell receptors are also much better at detecting faint chemical traces. They can pick up some particularly smelly substances in concentrations of just a few parts per billion, which is why a skunk's scent can be smelt some distance away.

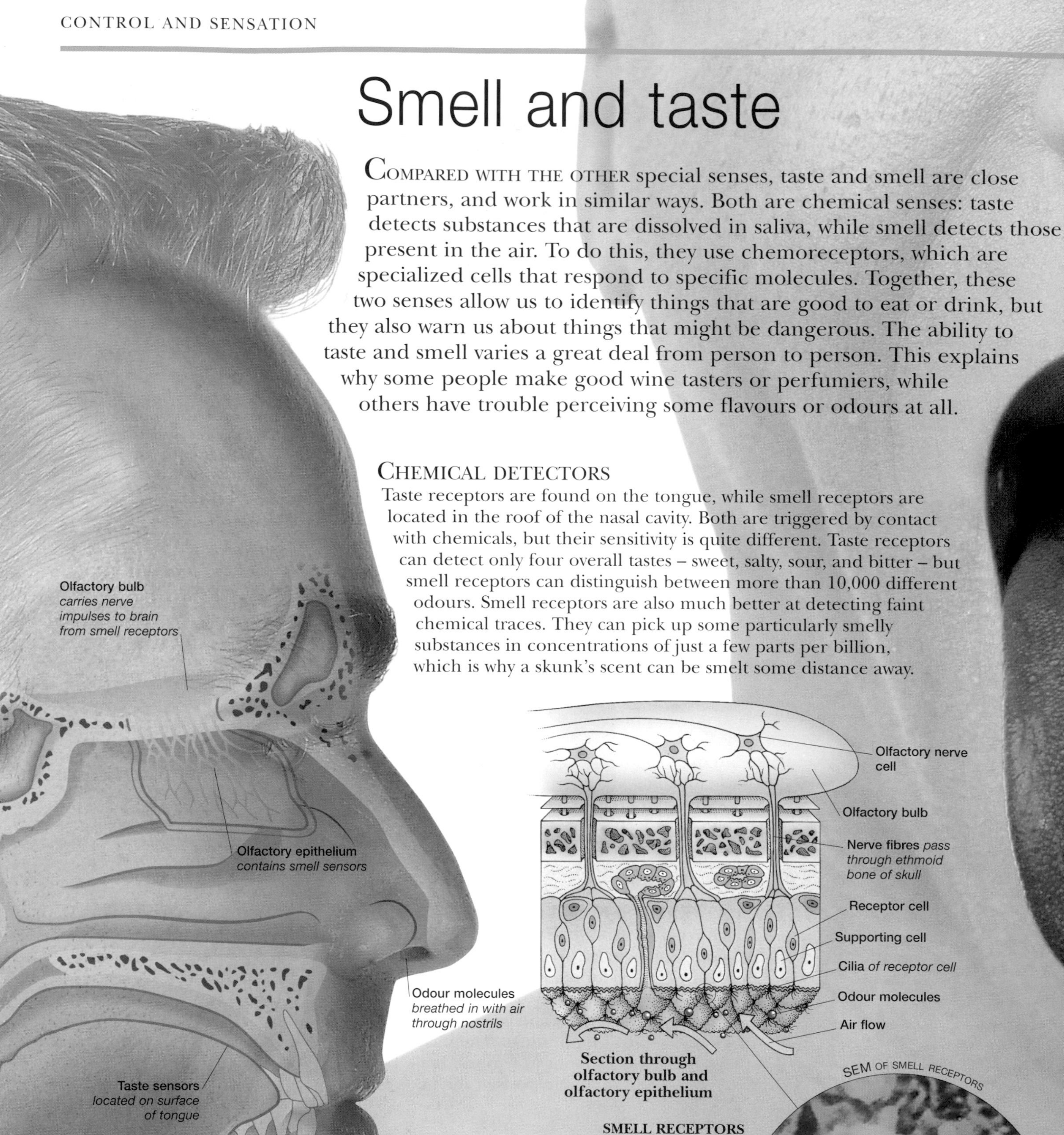

Section through olfactory bulb and olfactory epithelium

SMELL RECEPTORS

Located deep inside the nasal cavity are smell receptors, or olfactory cells, in a thumbnail-sized patch of tissue called the olfactory epithelium. One end of each receptor connects with the olfactory bulb, which is an extension of the brain. The other ends in a cluster of cilia, which look like microscopic hairs. These hairs project into the mucous film lining the nasal cavity, where they respond to dissolved molecules from the incoming air.

VANISHING FLAVOURS

The flavour of food or drink is more than just its taste, because smell and vision also play a part. A wine taster relies on all three of these senses to determine the flavour of a wine. If one of these senses is taken away, it is very difficult to identify foods by taste alone. This is easily demonstrated by blindfolding a person and pegging their nose. Flavours can also be confused by giving food unfamiliar colours, because what we see dictates what we expect to taste.

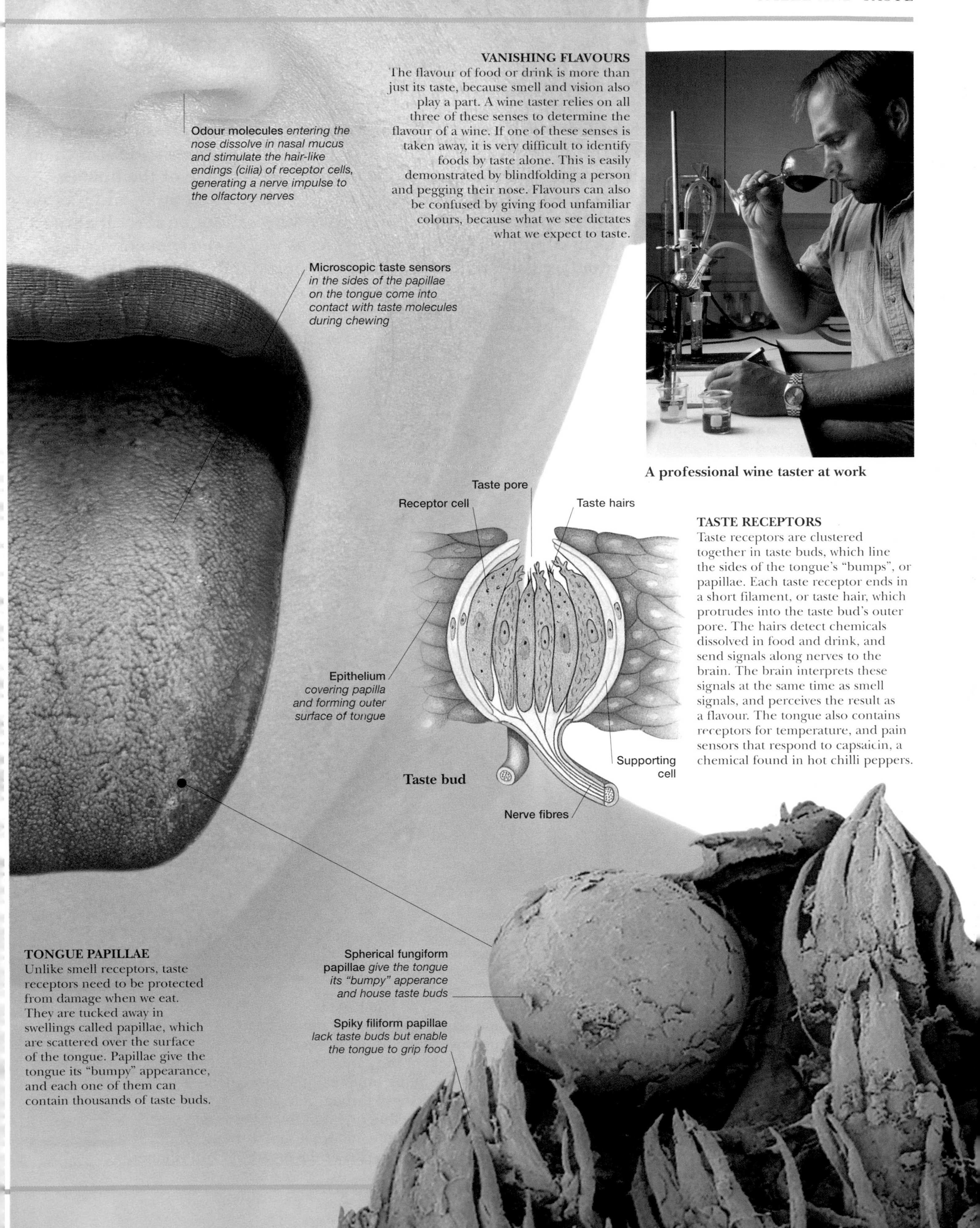
Odour molecules *entering the nose dissolve in nasal mucus and stimulate the hair-like endings (cilia) of receptor cells, generating a nerve impulse to the olfactory nerves*
Microscopic taste sensors *in the sides of the papillae on the tongue come into contact with taste molecules during chewing*
Taste pore
Receptor cell
Taste hairs
Epithelium *covering papilla and forming outer surface of tongue*
Supporting cell
Nerve fibres
Spherical fungiform papillae *give the tongue its "bumpy" apperance and house taste buds*
Spiky filiform papillae *lack taste buds but enable the tongue to grip food*

A professional wine taster at work

Taste bud

TASTE RECEPTORS

Taste receptors are clustered together in taste buds, which line the sides of the tongue's "bumps", or papillae. Each taste receptor ends in a short filament, or taste hair, which protrudes into the taste bud's outer pore. The hairs detect chemicals dissolved in food and drink, and send signals along nerves to the brain. The brain interprets these signals at the same time as smell signals, and perceives the result as a flavour. The tongue also contains receptors for temperature, and pain sensors that respond to capsaicin, a chemical found in hot chilli peppers.

TONGUE PAPILLAE

Unlike smell receptors, taste receptors need to be protected from damage when we eat. They are tucked away in swellings called papillae, which are scattered over the surface of the tongue. Papillae give the tongue its "bumpy" appearance, and each one of them can contain thousands of taste buds.

The eye

VISION IS THE FOREMOST of the special senses, and the one that dominates our impressions of the outside world. We use it every moment we are awake, and we dream with visual images when we sleep. Visual expressions crop up almost every time we speak, and when we think of ourselves or other people, it is almost always in a visual way. The organ responsible for triggering this imagery – the eye – is one of the most complex in the body. Set in a protective socket inside the skull, it is always on the move, and is quick to adapt to changing light conditions. It automatically focuses on anything that comes into view, collecting light and converting it into a stream of billions of nervous impulses. Once they have arrived in the brain, these signals have to be analyzed – an awesomely complex process that makes sense of what we see.

Eyebrows *direct sweat away from the eye and help to keep out some light*
Eyelashes *stop too much light entering the eye, as well as protecting it from foreign particles*
Tear ducts *empty tears onto the surface of the eye*
Eyelids *protect the eye from bright light and foreign objects; when we blink, they wipe the eyes and keep them clear of dust*
The tear gland *is under the outer edge of the eyebrow and produces tears*
Tears *move down and across the eye, carrying debris with them*
Tears *drain away through two tear openings in the corner of the eye*
Iris *colour is determined by the amount of melanin present – brown eyes have the most melanin and blue eyes the least*
Pupil *is the hole that lets light into the eye*
Nasolacrimal duct *drains tears intc the nasal cavity*
Opening *into nasal cavity*
Nostril

EYE PROTECTION

Only the iris and pupil of each eyeball – about one sixth – can be seen from the outside. The rest lies inside the eye socket, or orbit, where it is cushioned by pads of fat. Most features surrounding the eye have a protective function – the eyebrows, eyelashes, and eyelids all help shield the eye from excessive light and dust. The front of the eye is kept moist by tears, which wash away specks of dust and help to prevent infection. Tears are washed across the eye by blinking – a reflex action that also helps to protect the eye if anything heads its way.

Contracted pupil in bright light

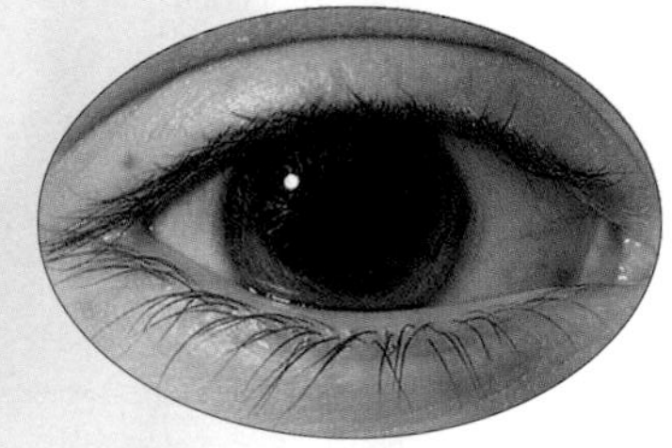

Dilated pupil in dim light

PUPIL SIZE

Human eyes have to cope with a huge range of light intensity. In bright conditions, muscles in the iris contract, narrowing the pupil and cutting down the amount of light entering the eye. If it is dim, the reverse happens. This reflex action takes about a fifth of a second. Pupils also narrow when the eyes look at something very close, because this increases the eye's depth of field.

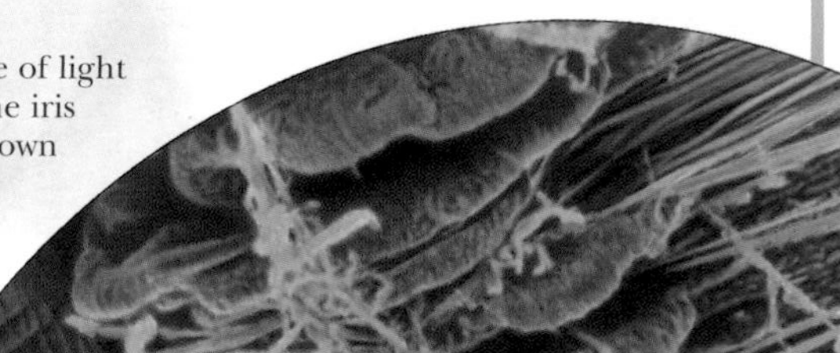

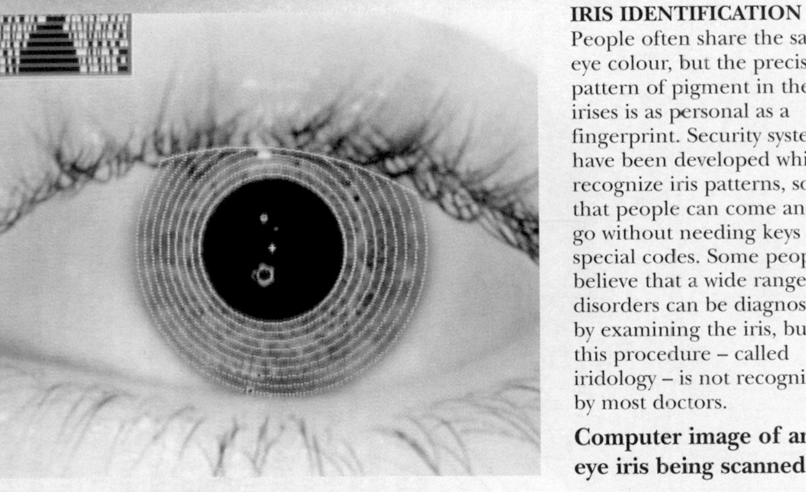

Computer image of an eye iris being scanned

IRIS IDENTIFICATION

People often share the same eye colour, but the precise pattern of pigment in their irises is as personal as a fingerprint. Security systems have been developed which recognize iris patterns, so that people can come and go without needing keys or special codes. Some people believe that a wide range of disorders can be diagnosed by examining the iris, but this procedure – called iridology – is not recognized by most doctors.

Ciliary processes *form part of the ciliary body whose muscles alter the shape of the lens during focusing*

Iris *surface folds mark the position of circular and radial muscle fibres that contract to alter pupil size*

Suspensory fibres *run from the ciliary processes to the lens (unseen, to the left of the SEM)*

SEM OF THE INNER SURFACE OF THE IRIS

INSIDE VIEW

This SEM shows an inside view of the eyeball, looking forwards from behind the iris (mauve). Visible here are the inner surface of the iris and the ciliary processes (red) that lie behind the iris and surround its base. These folded processes are the point of attachment of the suspensory fibres (yellow and green) that hold the lens in place. They also secrete the fluid that fills the front section of the eyeball.

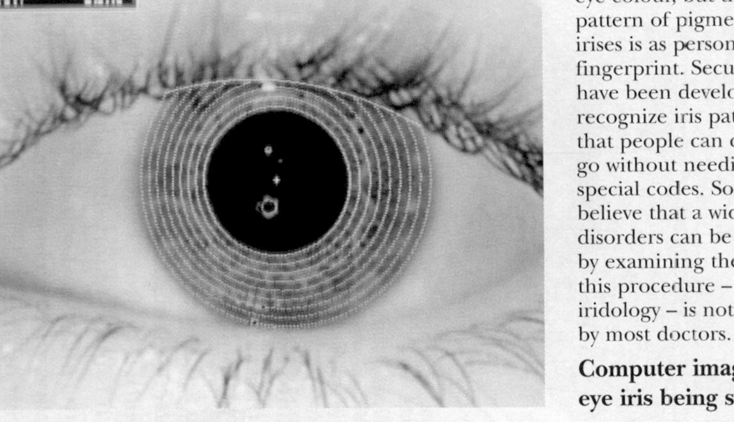
Superior rectus makes the eyeball look upwards
Superior oblique swivels the eye so that it looks downwards and outwards
Medial rectus pulls the eyeball in, so that it looks towards the nose
Optic nerve
Ciliary body controls the thickness of the lens so it focuses correctly
Suspensory ligament fibres attach the lens to the ciliary body
Lens changes shape to focus light from both near and far objects
Lateral rectus pulls the eyeball so that it looks out towards the side
Inferior oblique makes the eye look upwards and outwards
Inferior rectus makes the eye look downwards and inwards
Fat cells cushion eyeball inside bony eye socket

Side view of right eye

EYE MOVEMENT

The eyeball is moved by six small, strap-shaped muscles. Five of them are anchored to the back of the eye socket, although one of them – called the superior oblique – initially runs forwards, before doubling back through a "pulley" made of cartilage. The sixth, called the inferior oblique muscle, is anchored to the point nearest the nose. Compared to other voluntary muscles, all six can make extremely precise movements, allowing the eye to track objects on the move.

KEEPING ON THE MOVE

Eyes move smoothly when they follow a moving object, but they jump rapidly when they are looking at different parts of a scene, words on a page, or a face. These abrupt jumps – represented in the photograph on the left by green lines – are called saccades. Eyes also make much smaller movements called tremors. Tremors are essential to vision. If they are stopped – for example by an anaesthetic – signals from the eye fade away.

ENDOCRINE system

THE TWO CONTROL systems of the human body work in quite different ways. While the nervous system uses electrical signals to make cells respond, the endocrine system uses chemical messengers, called hormones, that are released into the blood. Hormones often take a longer time than nerves to react, and they can have important and long-lasting effects. Together, they regulate the speed of thousands of chemical processes, and they are also responsible for the physical changes that occur during puberty. Hormones are produced by endocrine glands, which empty directly into the bloodstream. Many of these glands influence each other, and some are also triggered by nerves. The result is an intricate web of controls – one that keeps the whole body in a stable state.

Testosterone, *the male sex hormone, produces male secondary sexual characteristics at puberty*

Growth hormone *from the pituitary gland controls the body's growth; with too little, a child will fail to grow at the normal rate*

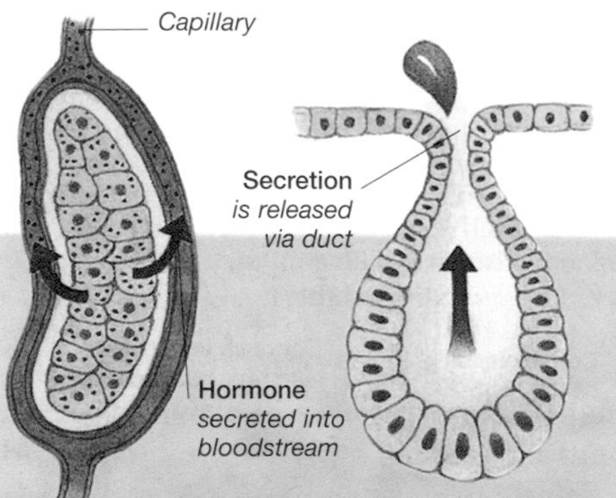

HORMONE-SECRETING GLANDS

Endocrine glands produce their secretions in tiny amounts, releasing the hormones directly into capillaries, so that they can be carried away by the bloodstream. Endocrine glands are found in two types of organ. Some of these organs are devoted exclusively to making hormones, while others – such as the pancreas and kidneys – carry out different functions as well. In contrast, exocrine glands such as salivary or sweat glands, release their secretions through a duct into body cavities or onto the skin.

ENDOCRINE SYSTEM FUNCTIONS

Function	Description
Homeostasis	*Stimulates or inhibits chemical processes in cells to keep the body in a stable state. These adjustments can affect just one type of target cell, or a wide range across the body.*
Reproduction	*Initiates and maintains production of sex cells. In women, it also initiates and controls the release of eggs cells, and prepares the body for possible pregnancy. Post fertilization, it maintains the uterus lining, prepares the mammary glands for milk production, and initiates birth.*
Development	*Initiates and governs physical changes that lead to sexual maturity and bring the body to adult size.*

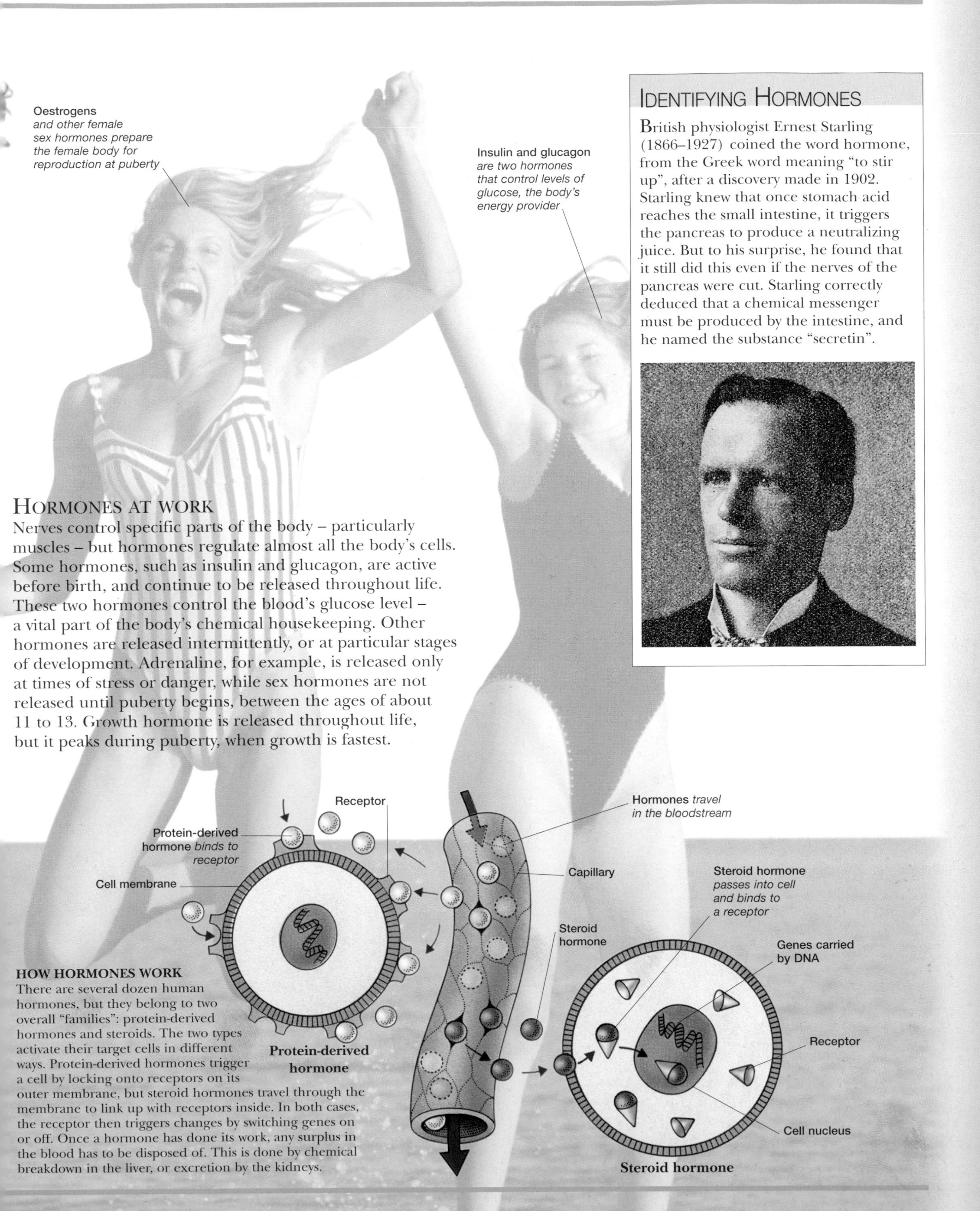

Oestrogens
and other female sex hormones prepare the female body for reproduction at puberty
Insulin and glucagon
are two hormones that control levels of glucose, the body's energy provider
Receptor
Protein-derived hormone binds to receptor
Cell membrane
Hormones travel in the bloodstream
Capillary
Steroid hormone passes into cell and binds to a receptor
Steroid hormone
Genes carried by DNA
Receptor
Cell nucleus
Protein-derived hormone
Steroid hormone

Identifying Hormones

British physiologist Ernest Starling (1866–1927) coined the word hormone, from the Greek word meaning "to stir up", after a discovery made in 1902. Starling knew that once stomach acid reaches the small intestine, it triggers the pancreas to produce a neutralizing juice. But to his surprise, he found that it still did this even if the nerves of the pancreas were cut. Starling correctly deduced that a chemical messenger must be produced by the intestine, and he named the substance "secretin".

Hormones at work

Nerves control specific parts of the body – particularly muscles – but hormones regulate almost all the body's cells. Some hormones, such as insulin and glucagon, are active before birth, and continue to be released throughout life. These two hormones control the blood's glucose level – a vital part of the body's chemical housekeeping. Other hormones are released intermittently, or at particular stages of development. Adrenaline, for example, is released only at times of stress or danger, while sex hormones are not released until puberty begins, between the ages of about 11 to 13. Growth hormone is released throughout life, but it peaks during puberty, when growth is fastest.

HOW HORMONES WORK
There are several dozen human hormones, but they belong to two overall "families": protein-derived hormones and steroids. The two types activate their target cells in different ways. Protein-derived hormones trigger a cell by locking onto receptors on its outer membrane, but steroid hormones travel through the membrane to link up with receptors inside. In both cases, the receptor then triggers changes by switching genes on or off. Once a hormone has done its work, any surplus in the blood has to be disposed of. This is done by chemical breakdown in the liver, or excretion by the kidneys.

Endocrine glands

UNLIKE THE NERVOUS system, the endocrine system consists of separate "outposts" scattered across the body. Three major endocrine glands – including the pituitary – are in the head, while the rest are in the neck and trunk. Together with some more diffuse regions of endocrine tissue, these glands produce all the hormones that keep the body under control. Each hormone has a specific list of target tissues, and in many cases these include other endocrine glands. This means that one hormone can stimulate the production of another, or counteract its effects. This kind of interaction is called feedback, and it lies behind almost all the subtle changes in hormone levels that keep the body on an even keel.

HORMONE PRODUCERS

Endocrine glands can be organs in their own right, or they can form part of organs that also carry out other tasks. The pituitary gland acts as the system's overall coordinator, and is directly linked to the hypothalamus, a hormone-producing region of the brain. Several glands, including the thyroid and pancreas, adjust the body's rate of energy use, while the adrenals produce an unusually fast-acting hormone that is released in moments of sudden stress. Sex hormones are released by the ovaries and the testes, but only once puberty has reached. Endocrine tissue is also found in some other organs, such as the kidneys and the heart.

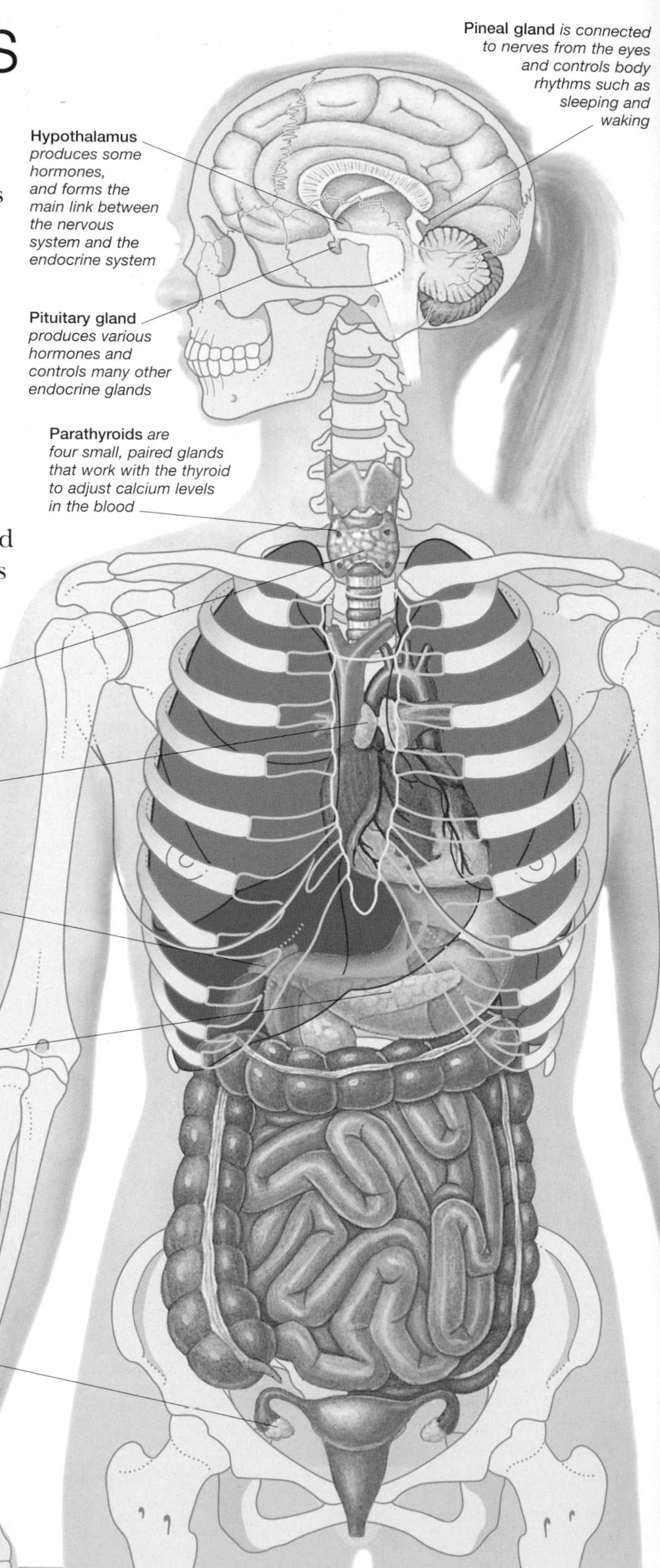
Pineal gland *is connected to nerves from the eyes and controls body rhythms such as sleeping and waking*
Hypothalamus *produces some hormones, and forms the main link between the nervous system and the endocrine system*
Pituitary gland *produces various hormones and controls many other endocrine glands*
Parathyroids *are four small, paired glands that work with the thyroid to adjust calcium levels in the blood*
Thyroid gland *produces thyroxine, which increases the body's overall metabolic rate*
Thymus gland *produces hormones needed for normal development of the immune system*
Adrenal glands *are paired glands that produce adrenaline, the hormone that prepares the body for action in emergencies*
Pancreas *releases insulin and glucagon – two opposing hormones that control the level of glucose in the blood*
Ovaries *are paired organs in women that form part of the reproductive system and endocrine system; their endocrine tissue releases female sex hormones*
Testes *are paired organs in men that form part of the reproductive system and endocrine system; their endocrine tissue releases male sex hormones*

FIGHT OR FLIGHT

One hormone – adrenaline – has an almost instantaneous effect. It is released by the adrenal glands whenever the body is put under sudden stress. This kind of stress includes any kind of alarming experience, from taking a ride at a fairground to being bitten by a dog. Adrenaline prepares the body to face danger, or to run away. It speeds up breathing and heart rate, and releases energy-rich glucose in the blood. At the same time, it sends more blood to the muscles and brain, and diverts it from non-essential areas, such as the digestive system. Adrenaline is used in medicine as an emergency treatment for a drastic drop in blood pressure – a condition known as shock.

Fear and excitement *experienced on the roller-coaster ride stimulate the release of adrenaline to enable the body to cope with the experience*

Adrenaline *speeds up heart rate and breathing*

FEEDBACK CONTROL

Hormone release is automatically controlled by feedback mechanisms to ensure that hormone levels remain stable. For example, production of thyroxine by the thyroid gland is controlled by the hypothalamus, which monitors thyroxine levels in the blood. The hypothalamus releases thyrotropin-releasing hormone (TRH). This triggers the pituitary gland to release thyroid-stimulating hormone (TSH), which activates the thyroid gland. If the level of thyroxine rises or falls, a corrective mechanism comes into play, as this feedback loop shows.

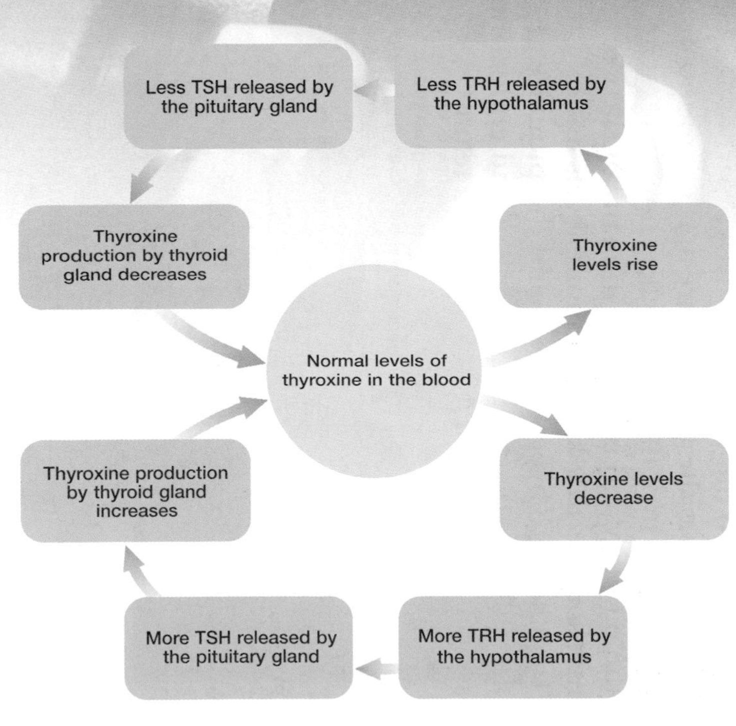

SEASONAL TIREDNESS
The body's internal clock is kept running by the pineal gland, which produces a hormone called melatonin at night. In the far north and south, where winter nights are very long, melatonin production hits an annual peak. Researchers believe this is why some people suffer seasonal affective disorder (SAD) – a condition that leaves them feeling tired and depressed. One way to treat it is by "phototherapy", or exposure to a bright light (above). Jet lag is another disorder caused by disruption of the body clock.

THE SEARCH FOR INSULIN

DIABETES MELLITUS IS A DISEASE that prevents sufferers from breaking down glucose in their cells. The glucose level in their blood rises rapidly and their body tries to remove it by increasing urine production, leading to uncontrollable thirst. Glucose is the major source of energy for the body but diabetes prevents cells from using it, so body fat is broken down instead and the body wastes away. Diabetes was always fatal until Canadian physiologists Frederick Banting and Charles Best discovered an effective treatment in 1922 – one of the greatest achievements in modern medicine.

THOMAS WILLIS
English physician Thomas Willis (1621–75) showed that sugar was excreted in diabetics' urine and re-discovered the ancient Chinese test of the sweet taste of urine indicating diabetes. "Taste thy patient's urine," he advised fellow doctors. "If it be sweet like honey, he will waste away, grow weak, fall into a sleep and die."

ISLETS OF LANGERHANS
The hormone insulin is produced in the pancreas, in distinctive groups of cells called islets of Langerhans. They are named after Paul Langerhans, who first described them in 1869, although it was left to others to discover their function.

EARLY DESCRIPTIONS OF THE SYMPTOMS

Symptoms of diabetes, including frequent urination, thirst, and weakness, were recognized by the ancient Egyptians, who left a description in the Ebers papyrus, dating from 1500 BC. Later, in the 2nd century AD, Aretaeus of Cappadocia referred to the way in which patients "never stop making water and the flow is incessant, like the opening of aqueducts". Aretaeus named the disease "diabetes", which comes from the Greek word meaning "siphon" and refers to a constant flow of water. By the 6th century, Indian physicians had recognized that sufferers have sweet urine, but the presence of sugar in their urine was not confirmed in Western medicine until the 17th century, when Thomas Willis showed that sugar crystals remain when urine is evaporated.

THE ROLE OF THE PANCREAS

The first conclusive proof that the pancreas played the central role in diabetes came in 1889, when German doctors Joseph von Mering (1849–1907) and Oskar Minkowski (1858–1931) showed that dogs developed diabetes if their pancreas was removed.

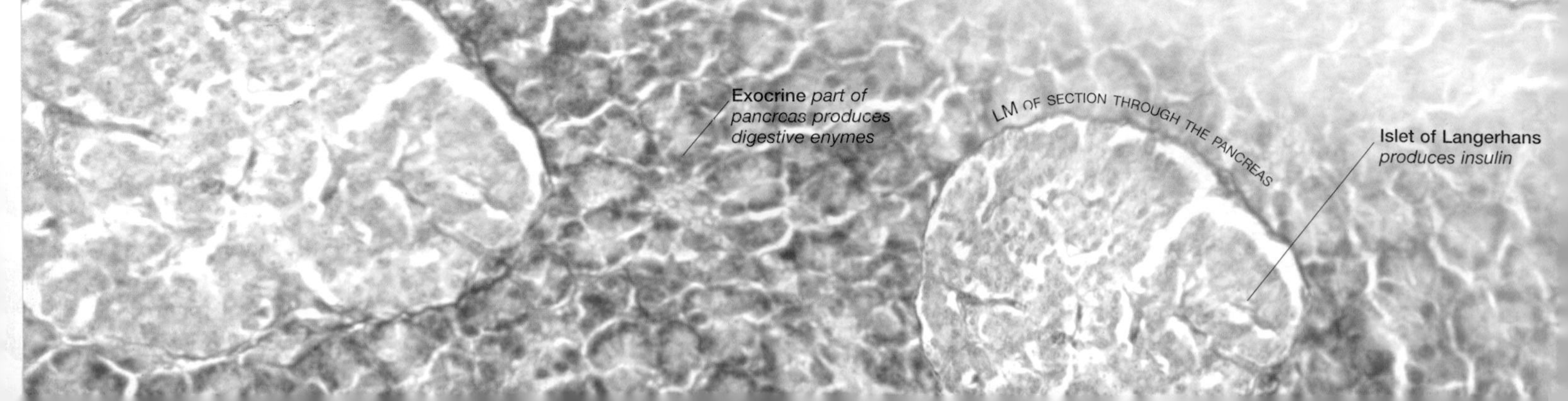

Exocrine *part of pancreas produces digestive enymes*

LM OF SECTION THROUGH THE PANCREAS

Islet of Langerhans *produces insulin*

BANTING AND BEST
Frederick Banting (1899–1978), far right, and Charles Best (1891–1941) isolated insulin. They injected it into Marjorie, a dying diabetic dog whose pancreas had been removed. She recovered. Banting won a Nobel Prize for his work but Best's achievement was overlooked, although Banting shared the prize money with him.

1922. The crucial test with a human patient came in 1923, when they treated Leonard Thompson, a 14-year-old diabetic who was close to death. The effect was miraculous. His blood sugar level fell, one day later he was on his feet, and he soon returned home to lead a normal life with the aid of insulin injections. The Eli Lilly pharmaceutical company immediately began large-scale production of insulin, extracted from pig pancreas. This method has now been superseded by the use of laboratory-produced human insulin. Diabetes is still an incurable disease, but diabetics, once condemned to a short and painful existence, can now live a full and active life.

INSULIN PEN
Some diabetics control their blood glucose by injecting insulin into their blood. The insulin pen gives a measured dose, according to the patient's varying blood glucose levels.

Twenty years earlier, a German scientist, Paul Langerhans (1847–88), had described unique groups of cells within the pancreas, later named islets of Langerhans. British scientist Edward Sharpey-Schafer (1850–1935) showed that a substance regulating glucose metabolism in the body was produced by these islets and called it "insuline" (later shortened to insulin), from the Latin *insula,* meaning island. English physiologist Ernest Starling (1866–1927) coined the name "hormone" in 1905 to describe substances secreted by endocrine glands, like the pancreas, that control metabolism. So now the race was on to isolate the pancreatic hormone, insulin.

Insulin *is a small protein made up of two amino acid chains*

Six insulin molecules *shown in green and purple are bound together here*

COMPUTER-GENERATED IMAGE OF INSULIN MOLECULE

ROLE OF INSULIN
Insulin molecules bind to the outer membrane of cells, triggering chemical processes inside that break down glucose molecules, so removing the sugar from the blood and generating energy. Insulin also inhibits breakdown of glycogen to glucose by the liver.

Insulin production sites *are stained orange*

MAKING INSULIN
Initially, insulin was laboriously extracted from pig pancreas, obtained from slaughterhouses. In some patients, the immune system reacts against pig insulin. This can be avoided by using human insulin, made in the laboratory using *E. coli* bacteria that have human insulin genes inserted into them.

Isolation of insulin

Attempts to treat diabetics by feeding them pancreas failed, so extracting the hormone and injecting it into their blood was the only option. After many unsuccessful attempts, Frederick Banting and Charles Best succeeded in isolating insulin in

TEM OF GENETICALLY ALTERED *E. COLI* BACTERIA

Escherichia coli bacteria *are genetically engineered to produce insulin*

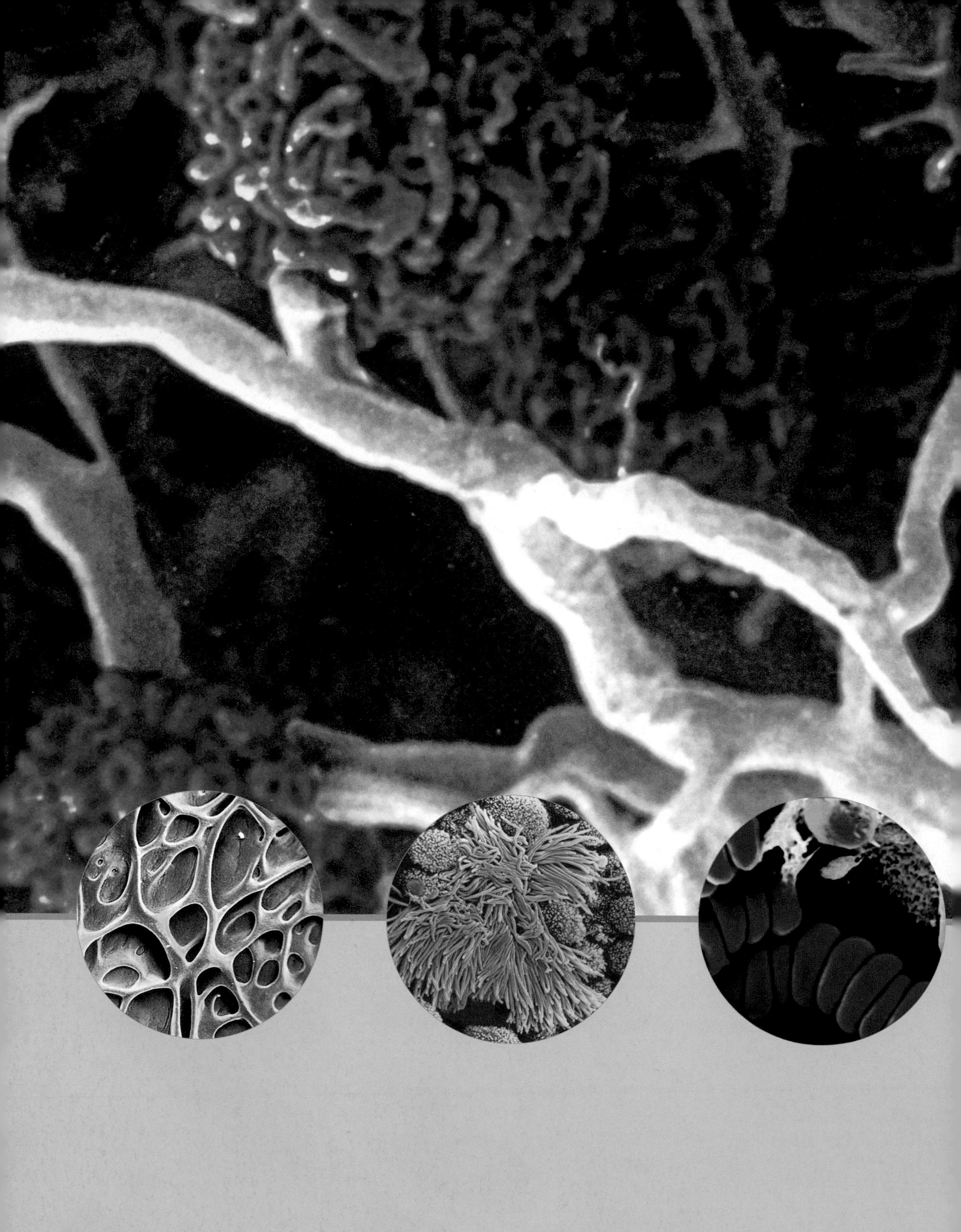

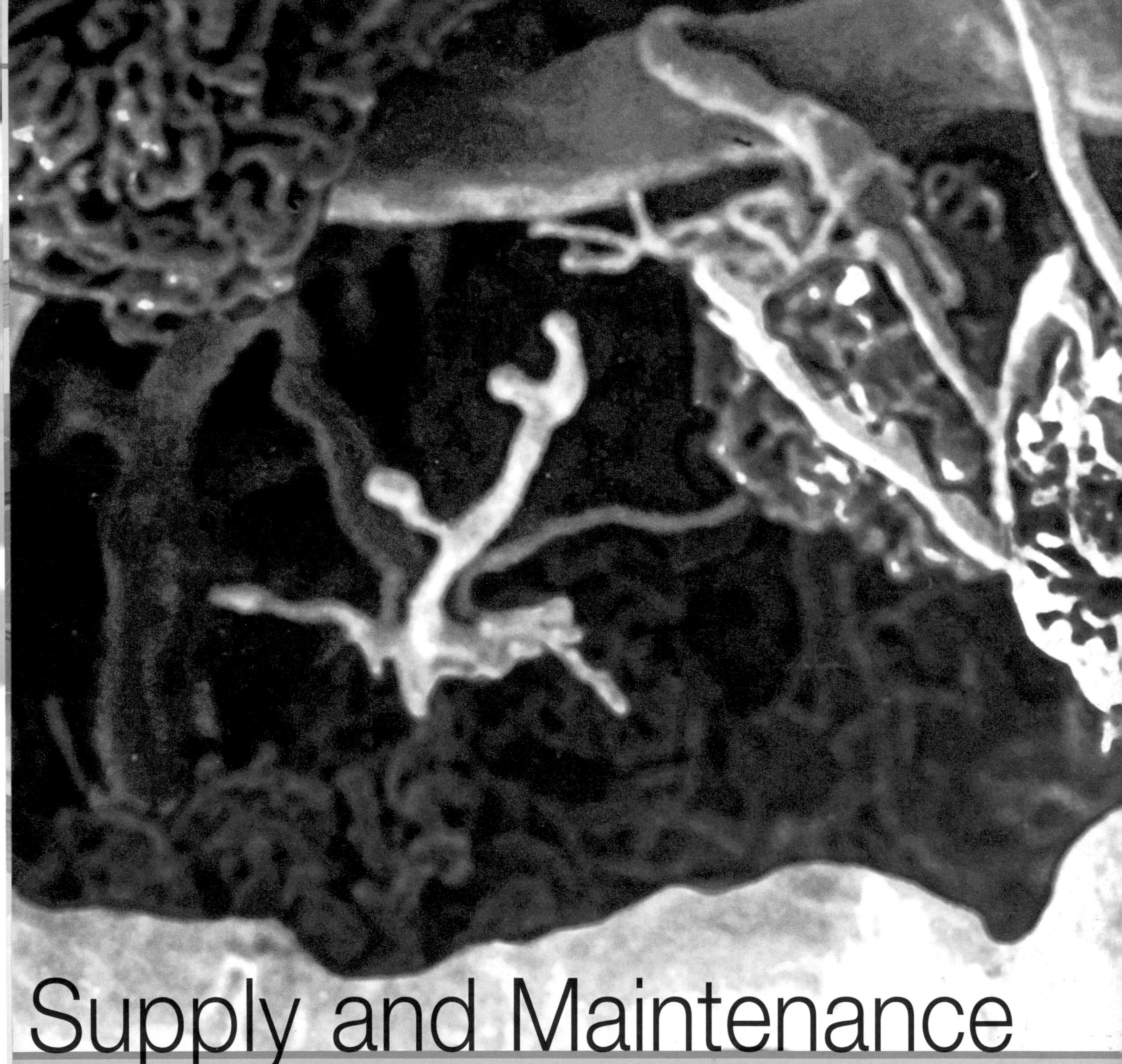

Supply and Maintenance

To work at their best, the body's trillions of cells must be kept in a stable environment, regardless of conditions outside the body. This stability is provided by the systems that supply and maintain body tissues. Together, they ensure that all cells have sufficient food and oxygen to meet their energy requirements, have their wastes removed, are protected from disease-causing pathogens, are kept at a constant, warm temperature of 37°C (98.6°F), and are bathed by a nurturing fluid, the composition of which fluctuates little.

Red blood cells

THE AVERAGE ADULT HAS around 25 trillion red blood cells in their blood at any one time. Also known as erythrocytes, these tiny cells make up 99 per cent of the cells in the blood and perform the vital function of carrying oxygen to all the tissues of the body. They are produced at a rate of over 2 million per second in the red marrow of certain bones. Their unique structure means they are able to squeeze through the tiniest of blood vessels and can pick up oxygen readily in the lungs and unload it in the tissues. In addition to transporting oxygen, they also carry some of the carbon dioxide produced as a waste product by the cells of the body.

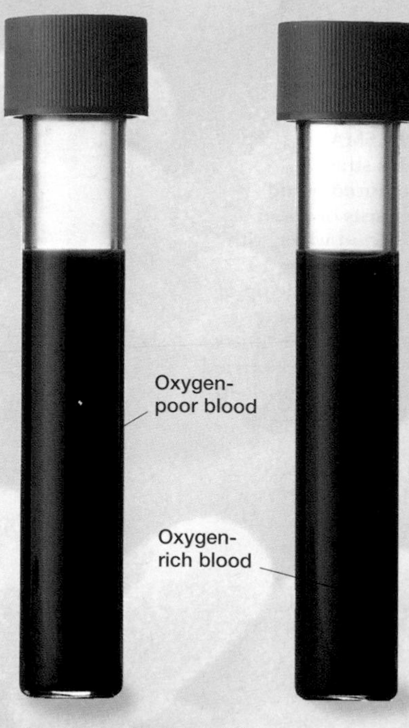

Oxygen-poor blood

Oxygen-rich blood

CHANGING COLOUR
The redness of blood alters as it travels around the body because haemoglobin changes its colour as it picks up or deposits oxygen. After haemoglobin picks up oxygen in the lungs, oxygen-rich blood travelling along the arteries to the tissues is bright red in colour. After haemoglobin unloads oxygen in the tissues, the oxygen-poor blood returning along veins to the heart is now a dull, dark red. This gives thin-walled veins just under the skin's surface a blue appearance.

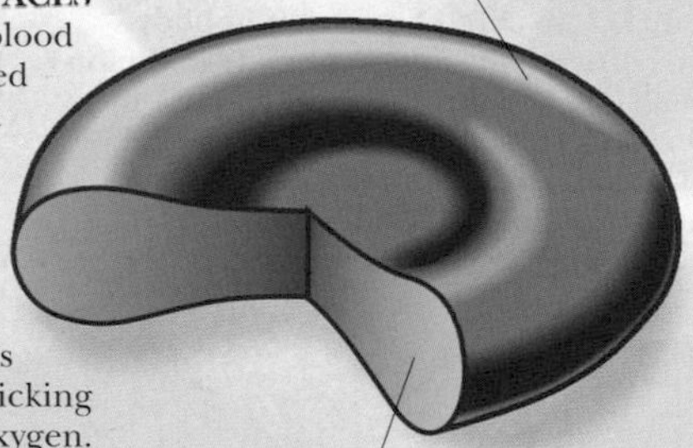

LARGE SURFACES
As this cutaway view shows, a red blood cell resembles a flattened, dimpled disc, a shape that gives the cell a large surface area in relation to its volume. It means that no oxygen-carrying haemoglobin molecule is ever far from the cell membrane, which explains why red blood cells are very efficient at both picking up and depositing oxygen.

Cell membrane

Haemoglobin *packs the interior of the cell*

Cut-away of red blood cell

OXYGEN CARRIERS

Red blood cells have a number of special features, including their dimpled shape and flexibility. They are also the only cells in the body that do not have a nucleus. Instead, they are packed with haemoglobin, a protein that is responsible for their red colour, and for the redness of the blood itself. Haemoglobin molecules pick up oxygen in the lungs and release it in the tissues. The formation and destruction of red blood cells takes place continuously. They begin their life as immature stem cells in the bone marrow, becoming fully developed and capable of transporting oxygen over a period of about five days. Each red blood cell lasts for about 120 days.

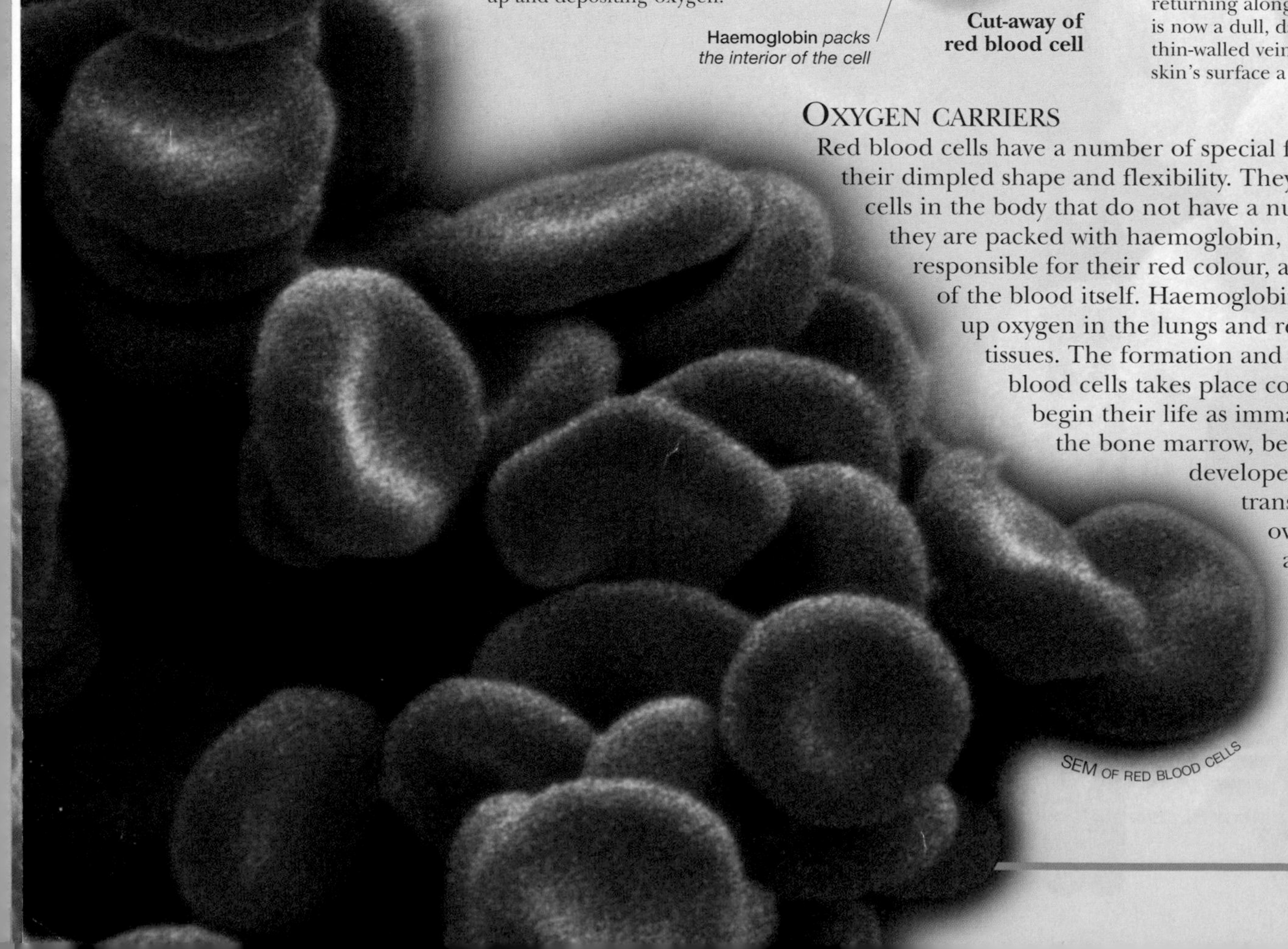

SEM OF RED BLOOD CELLS

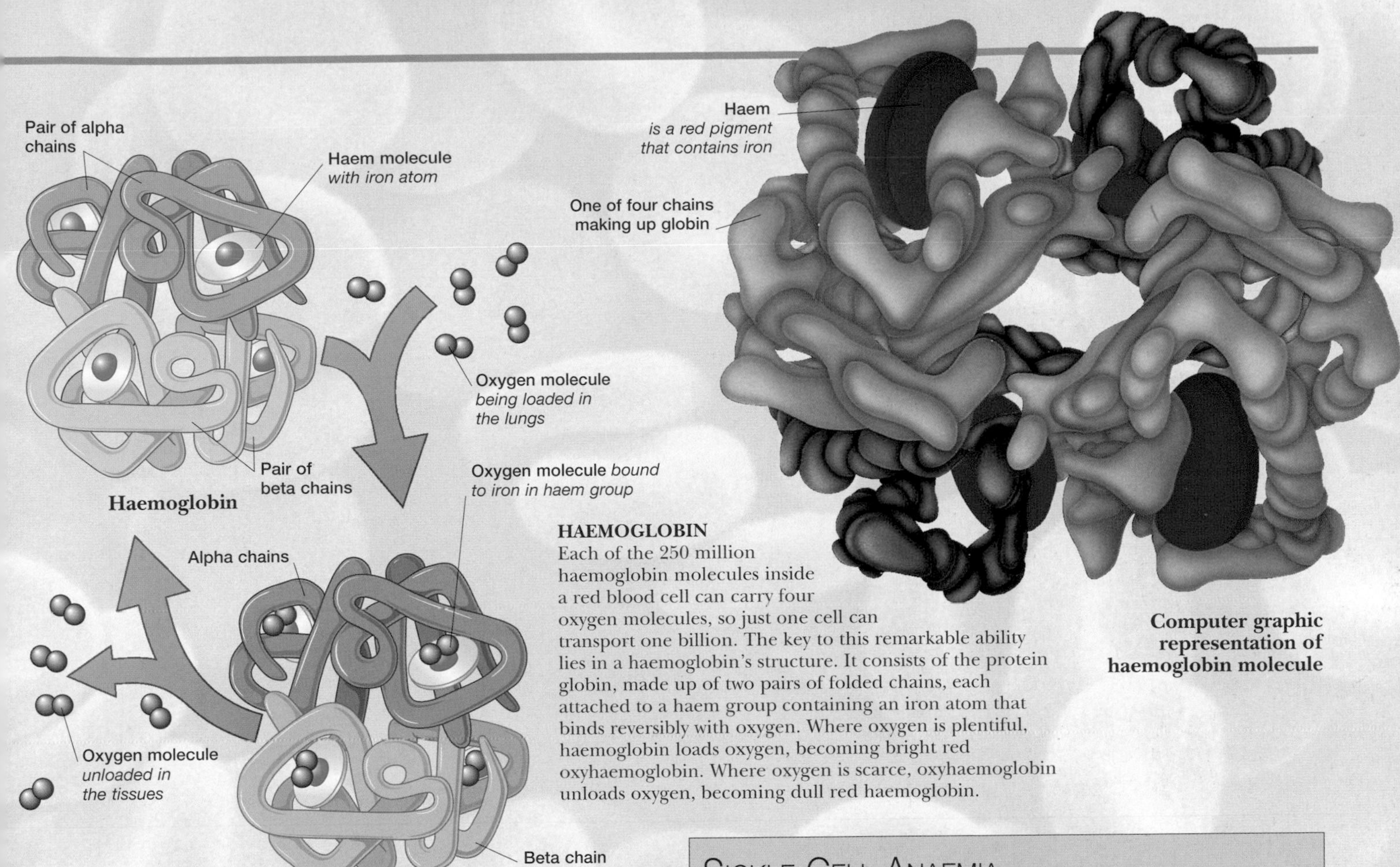

Computer graphic representation of haemoglobin molecule

HAEMOGLOBIN
Each of the 250 million haemoglobin molecules inside a red blood cell can carry four oxygen molecules, so just one cell can transport one billion. The key to this remarkable ability lies in a haemoglobin's structure. It consists of the protein globin, made up of two pairs of folded chains, each attached to a haem group containing an iron atom that binds reversibly with oxygen. Where oxygen is plentiful, haemoglobin loads oxygen, becoming bright red oxyhaemoglobin. Where oxygen is scarce, oxyhaemoglobin unloads oxygen, becoming dull red haemoglobin.

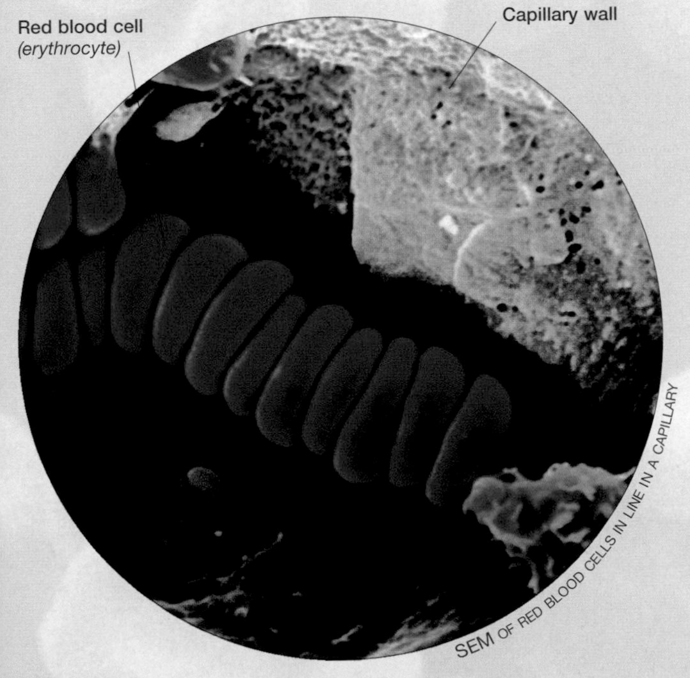

LINING UP
The cell membrane of red blood cells contains a protein called spectrin that makes them very flexible. This, together with their minute size, enables them to pass through tiny capillaries. Often, they must pass along blood vessels in single file, as many of the smallest capillaries are only just wider than one blood cell.

Sickle Cell Anaemia

Sickle cell anaemia is an inherited disorder. Caused by a defect of the gene responsible for producing haemoglobin, it results in an abnormality of the red blood cell structure. Whereas a normal red blood cell maintains its shape in oxygen-poor areas of the body, these cells become sickle-shaped, lose their flexibility, and obstruct narrow blood vessels, so blocking the blood supply and causing pain. People with sickle cell anaemia have inherited two copies of the abnormal gene, one from each parent. The ability of their blood to carry oxygen is reduced. Sickle cell trait is a milder form, which often does not cause symptoms; it is the result of one abnormal and one normal gene. For reasons that are unknown, sickle cell trait seems to give some protection against the infectious disease malaria.

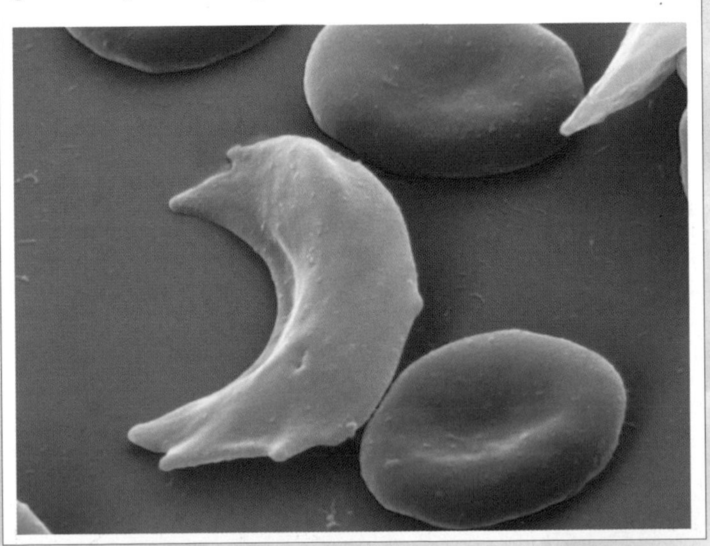

White blood cells

As BLOOD COURSES around the body, it carries with it a mobile defence force that is ready at all times to destroy invading pathogens before they can cause disease. Manning this force are the white blood cells, or leukocytes. White blood cells are outnumbered by oxygen-carrying red blood cells by about 700 to 1, but they are larger and have a nucleus, which their red partners lack. They can also change shape and slip out through capillary walls into the tissues in order to hunt down their prey. That prey includes not just foreign invaders such as disease-causing bacteria, viruses, protists, or fungi; white blood cells also tackle enemies from within, such as the rogue cells that cause cancer.

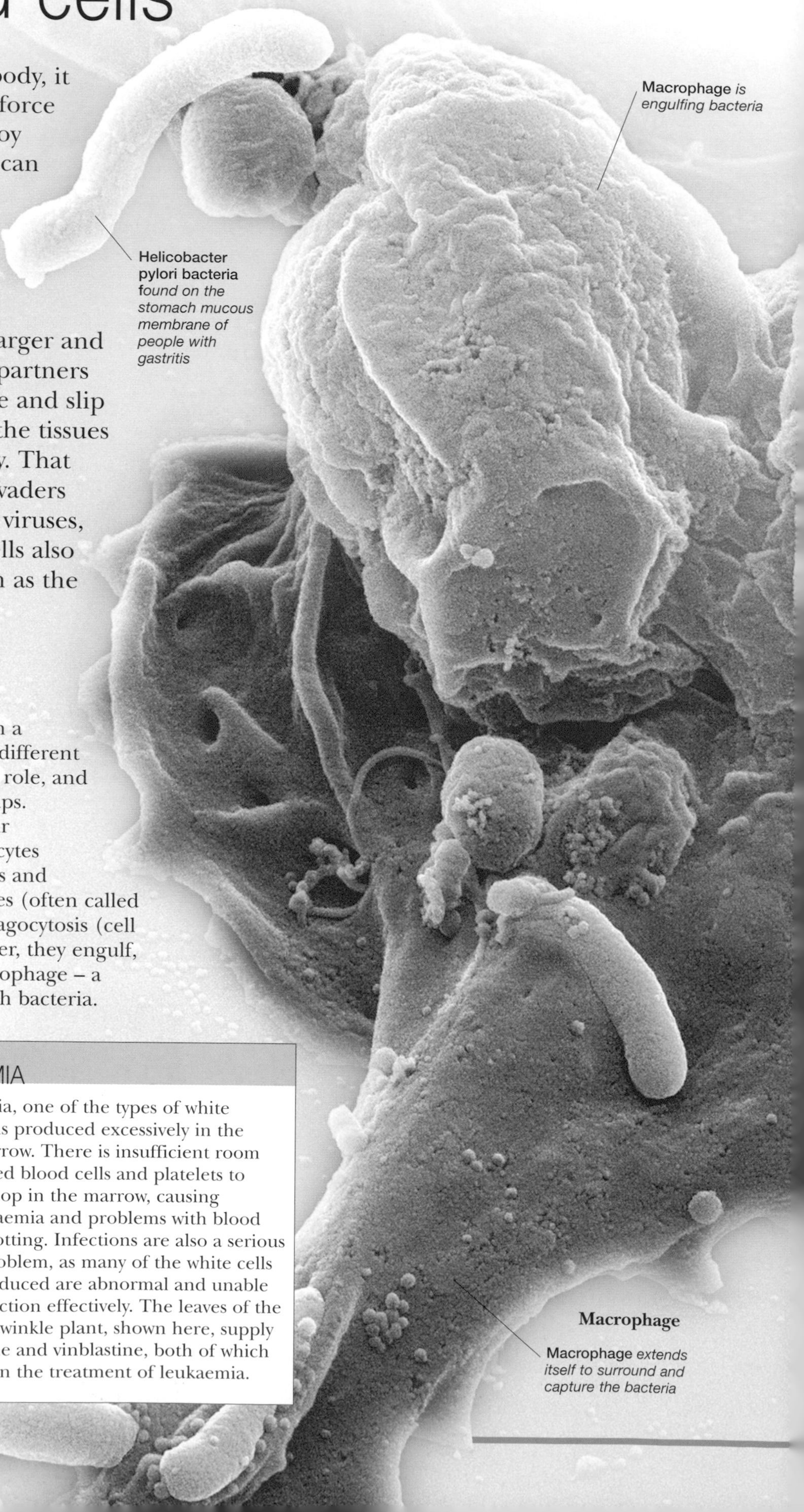

Macrophage *is engulfing bacteria*

Helicobacter pylori bacteria *found on the stomach mucous membrane of people with gastritis*

Body defenders

Among the 375,000 white blood cells in a single drop of blood, there are several different types, each with their own appearance, role, and life span, which fall into two main groups. Granulocytes are so called because their cytoplasm contains granules. Agranulocytes (without granules) include lymphocytes and monocytes. Granulocytes and monocytes (often called phagocytes) destroy their quarry by phagocytosis (cell eating). Having tracked down an invader, they engulf, then digest it. Here, a marauding macrophage – a monocyte – is doing just that to stomach bacteria.

Leukaemia

In leukaemia, one of the types of white blood cell is produced excessively in the bone marrow. There is insufficient room for the red blood cells and platelets to develop in the marrow, causing anaemia and problems with blood clotting. Infections are also a serious problem, as many of the white cells produced are abnormal and unable to function effectively. The leaves of the rosy periwinkle plant, shown here, supply vincristine and vinblastine, both of which are used in the treatment of leukaemia.

Macrophage

Macrophage *extends itself to surround and capture the bacteria*

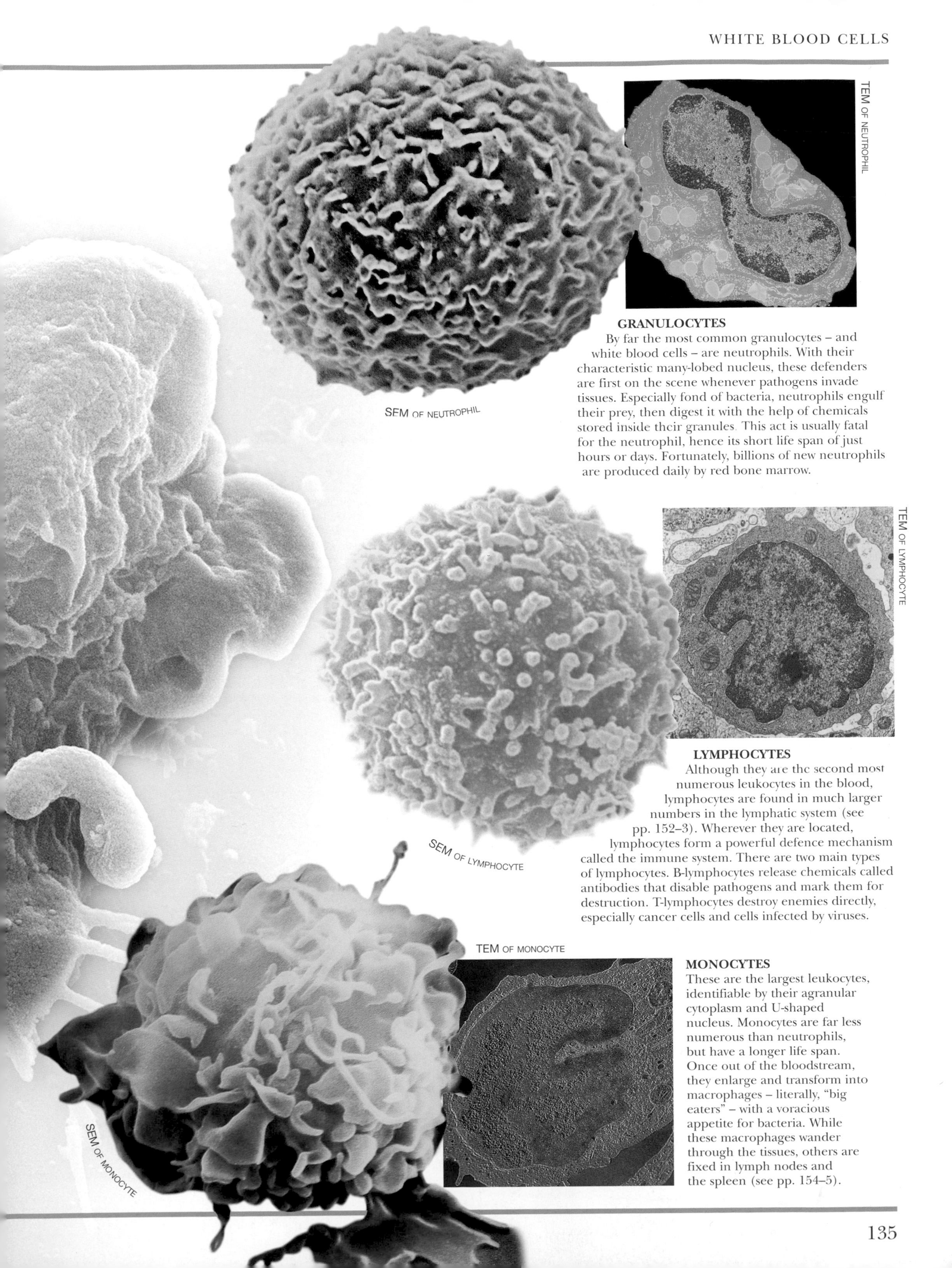

GRANULOCYTES

By far the most common granulocytes – and white blood cells – are neutrophils. With their characteristic many-lobed nucleus, these defenders are first on the scene whenever pathogens invade tissues. Especially fond of bacteria, neutrophils engulf their prey, then digest it with the help of chemicals stored inside their granules. This act is usually fatal for the neutrophil, hence its short life span of just hours or days. Fortunately, billions of new neutrophils are produced daily by red bone marrow.

LYMPHOCYTES

Although they are the second most numerous leukocytes in the blood, lymphocytes are found in much larger numbers in the lymphatic system (see pp. 152–3). Wherever they are located, lymphocytes form a powerful defence mechanism called the immune system. There are two main types of lymphocytes. B-lymphocytes release chemicals called antibodies that disable pathogens and mark them for destruction. T-lymphocytes destroy enemies directly, especially cancer cells and cells infected by viruses.

MONOCYTES

These are the largest leukocytes, identifiable by their agranular cytoplasm and U-shaped nucleus. Monocytes are far less numerous than neutrophils, but have a longer life span. Once out of the bloodstream, they enlarge and transform into macrophages – literally, "big eaters" – with a voracious appetite for bacteria. While these macrophages wander through the tissues, others are fixed in lymph nodes and the spleen (see pp. 154–5).

Platelets and clotting

IF A WATER PIPE develops a hole, water will leak out until the supply is turned off. But if a blood vessel is similarly damaged, the hole is rapidly sealed by the circulatory system's self-repair mechanism. This mechanism, called haemostasis ("blood halting"), leaps into action whenever a blood vessel breaks. Haemostasis involves three phases that happen in a rapid sequence. First, chemicals released by the damaged vessel make the vessel constrict (narrow) so that the flow of blood is immediately slowed down. Next, tiny platelets carried by the blood congregate around the damaged site and form a temporary plug to stop the leak. Finally, blood coagulates (clots) to form a more permanent seal where new tissue will grow to repair the hole in the blood vessel.

MEDICINAL LEECHES
When blood-sucking leeches like these bite through human skin, chemicals in their saliva stop blood from clotting and increase blood flow. Until the 19th century, doctors regularly used leeches to bleed their patients as a "cure" for all kinds of disorders. Today, leeches are still used in special cases, such as when severed body parts are reattached to the body, because they get blood flowing and drain any excess.

PLATELETS

Unlike blood cells, platelets are not complete cells, but are tiny cell fragments, round or oval in shape, each about one-third the size of a red blood cell. The 1,500 billion platelets circulating in the blood play a vital role in stopping bleeding, by first plugging the leak and then triggering the events that lead to the formation of a clot (see opposite). Under normal conditions, the lining of a blood vessel, or endothelium, is smooth, allowing blood to flow easily over it. But if the vessel is cut and its endothelium disrupted, blood platelets at the scene undergo a remarkable transformation. They swell, form spiky processes, and stick as if glued to the damaged site. These sticky, activated platelets release chemicals that attract more platelets. Within a few minutes, the mass of platelets forms a temporary plug to stop blood loss. Now the clotting process can begin.

Wound healing

Damage to blood vessels can happen deep inside the body or, as in this case, near the surface of the skin. Wherever the wound is, the body responds in the same way, except that a hard scab does not form inside the body. As soon as an injury occurs, platelets stick together to form a temporary plug in the wound. Once this plug is formed, the next stage of healing – coagulation – is triggered. A chain of events is set off that results in the formation of a clot, a more permanent structure than the plug, that remains in place until the damage has been repaired by cell division.

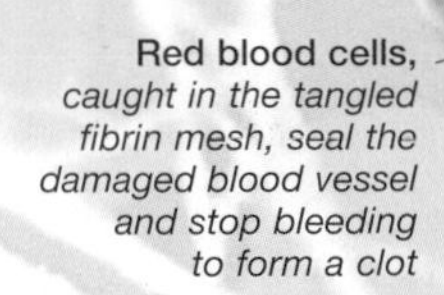

Red blood cells, *caught in the tangled fibrin mesh, seal the damaged blood vessel and stop bleeding to form a clot*

Threads of fibrin *form a mesh, like a fishing net, trapping red blood cells to form a clot*

SEM OF A BLOOD CLOT

CLOT FORMATION

Within minutes of an injury, a gel-like clot begins to form at the wound site as a result of coagulation of the blood. Platelets trigger the activation of chemicals called clotting factors that normally circulate in the blood in inactive form unless mobilized. Activation happens as a "cascade" sequence, with one factor activating the next until, finally, the soluble blood protein fibrinogen is converted into insoluble fibrin, which forms a mesh of tough threads which traps blood cells to make a clot.

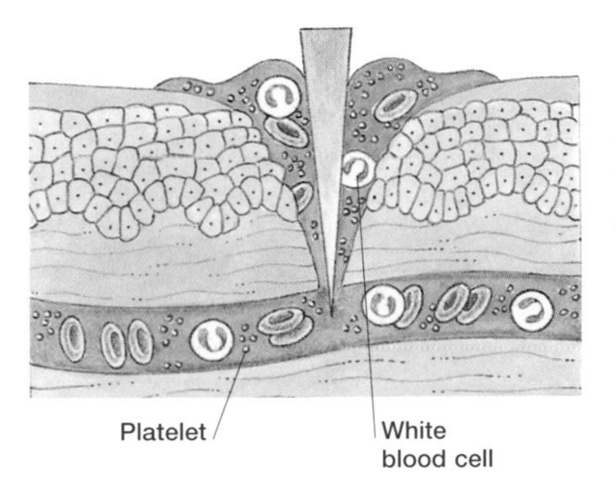

INJURY
A pin has punctured a small blood vessel just below the skin's surface. As blood oozes out, the damage to the blood vessel causes platelets to change and become "sticky". White blood cells track down any invading micro-organisms.

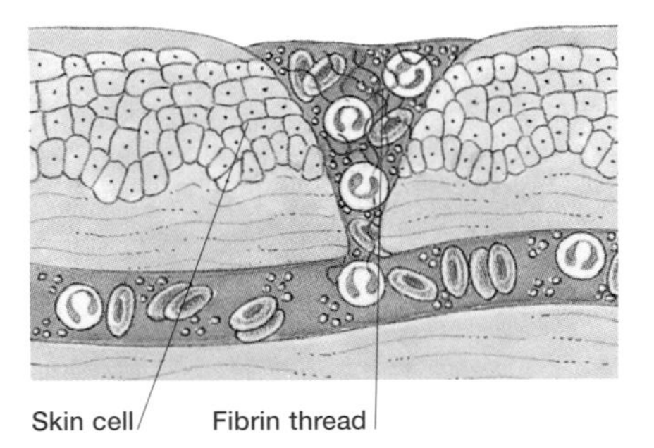

CLOT FORMATION
The "sticky" platelets clump together in the wound, forming a plug that stops blood leaking out. They also release chemicals that turn the soluble blood protein fibrinogen into fibrin threads. These trap blood cells and more platelets to make a clot.

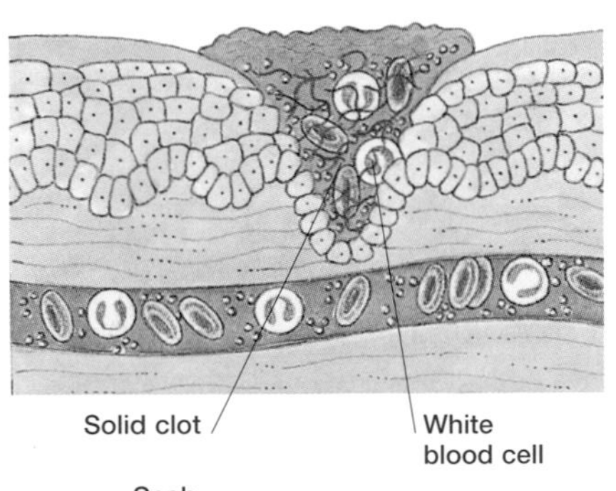

OLDER CLOT
In time, the clot becomes firmer and tightens, pulling the edges of the damaged blood vessel closer together so that blood loss is reduced even more. Cells in the skin and blood vessel wall divide, providing new cells to repair the damage.

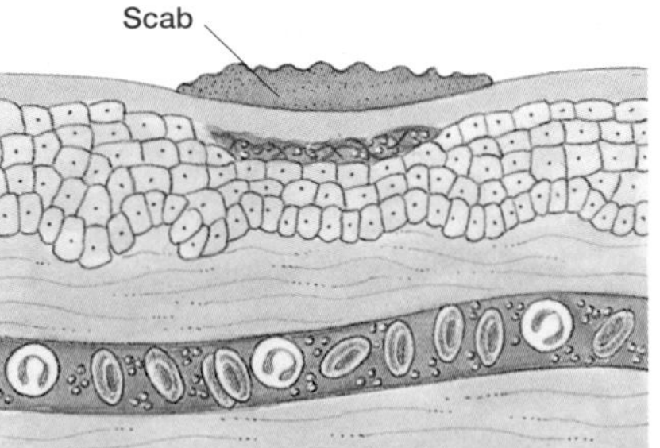

SCAB
On the skin's surface, the clot dries to form a hard scab. Beneath the scab, the clot shrinks as cell division continues to repair the skin and blood vessel. As wound repair reaches completion, the scab loosens and falls off, leaving fully healed tissue below.

Haemophilia

Affecting only males, haemophilia prevents blood from clotting properly, so excessive bleeding occurs. It is caused by an abnormal gene, located on the X chromosome, one of the sex chromosomes (see pp. 248–49). This stops production of one of the clotting factors described above, so that the clotting process is halted before a clot forms. Girls who inherit an abnormal gene are unaffected, but they can pass on the disease, because they have two X chromosomes, one of which contains a normal gene. The son of a female carrier has a 1 in 2 chance of inheriting haemophilia. Boys with a defective gene get the disease because they have only one X chromosome. The most famous occurrence of the condition was in the British royal family in the 19th century. Queen Victoria was a carrier, as were two of her daughters, but only one of her sons was affected.

Carrier of haemophilia **Sufferer of haemophilia** **Normal male or female**

BLOOD TRANSFUSIONS

LONDON PHYSICIAN William Harvey announced in 1628 that the heart is responsible for pumping blood around the body in arteries and veins, overthrowing the long-held belief that the body makes and consumes large amounts of blood daily. This discovery led to a new way of treating patients whose lives were threatened by blood loss or illness, by transfusing, or transferring, blood into their veins from a healthy donor. Early attempts at blood transfusions usually failed and were sometimes fatal. Routine, safe transfusions finally became possible in the early years of the 20th century, when scientists recognized that blood carries a complex defence system for destroying invading cells – including transfused blood cells.

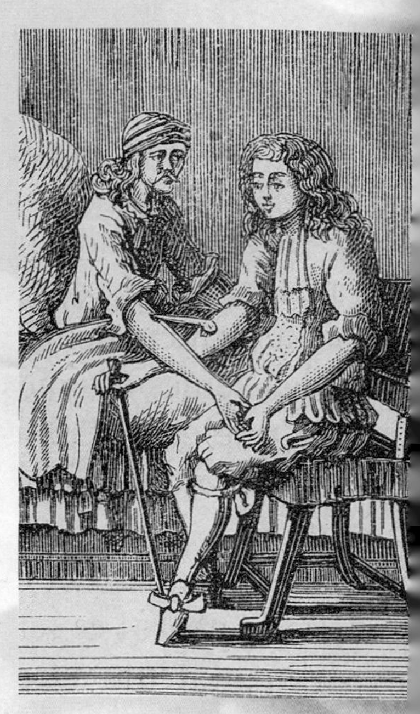

FIFTY-FIFTY SUCCESS RATE
Early person-to-person transfusions involved joining arteries and veins surgically, an awkward and painful method. Some surgeons used syringes to inject blood, but failure rates by both methods were high. Only 114 of the 243 patients transfused before 1873 fully recovered from the ordeal.

INITIAL ATTEMPTS

Some of the earliest blood transfusions were carried out between dogs by Jean-Baptiste Denis (1625–1704) in Paris, and by Richard Lower (1631–91) in Oxford. Lower is credited with performing the first blood transfusion on a patient in England, in November 1667. The donor was a sheep and the recipient a mentally ill student called Arthur Coga who, remarkably, survived. Similar experiments by Denis, however, using a calf and a human, led to the death of the patient. In 1818, British obstetrician James Blundell, of Guy's Hospital, London, performed the first successful transfusion of human blood to a human patient. Between 1825 and 1830, he performed a further ten documented transfusions, five of which were beneficial.

DISCOVERY OF BLOOD GROUPS

By the early 20th century, it was well known that blood from different individuals often thickened (coagulated) when it was mixed, blocking

KARL LANDSTEINER
Austrian-born American scientist Karl Landsteiner discovered the ABO blood group system in the early 1900s, the M, N, and P systems in 1927, and the rhesus system in 1940. Landsteiner was awarded the Nobel Prize in 1930.

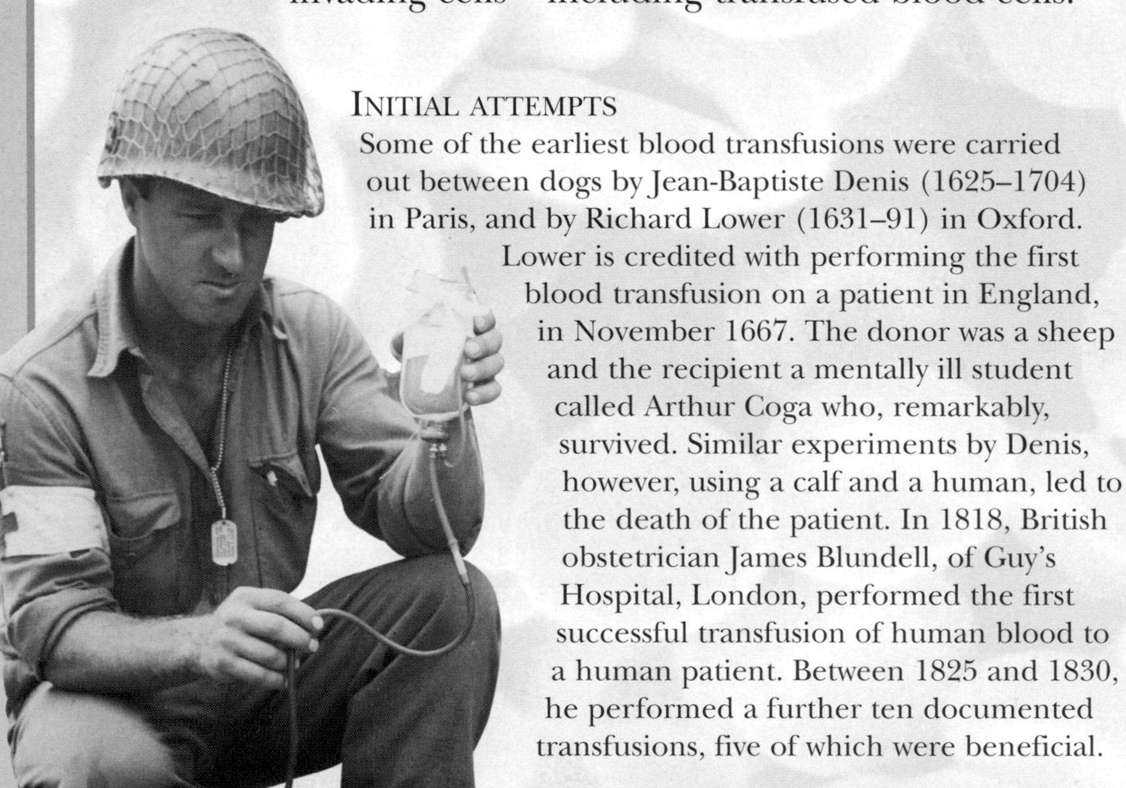

TRANSFUSIONS IN THE FIELD
Landsteiner's discovery of blood groups had an impact on the survival of wounded soldiers. From the First World War onwards, improved methods for storing blood made it possible to stockpile blood and transfuse casualties on the battlefield.

Blood Group Compatibility

Blood is always typed before transfusion to make sure a person receives blood that matches their own ABO and Rh blood groups to avoid adverse reactions. In an emergency, however, group O blood can be given to anyone, while someone who is group AB can receive blood from any donor. This chart shows compatible donors and recipients.

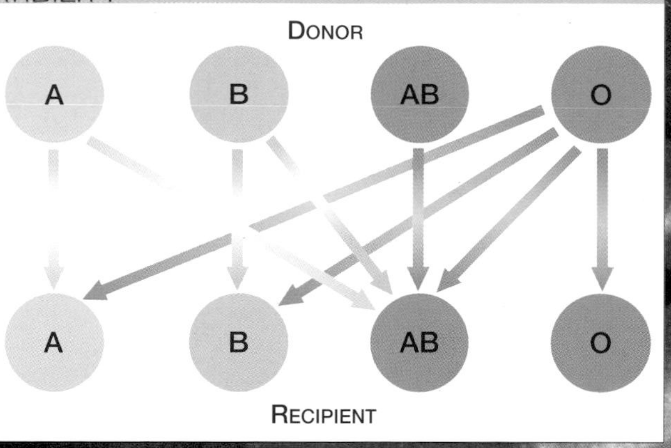

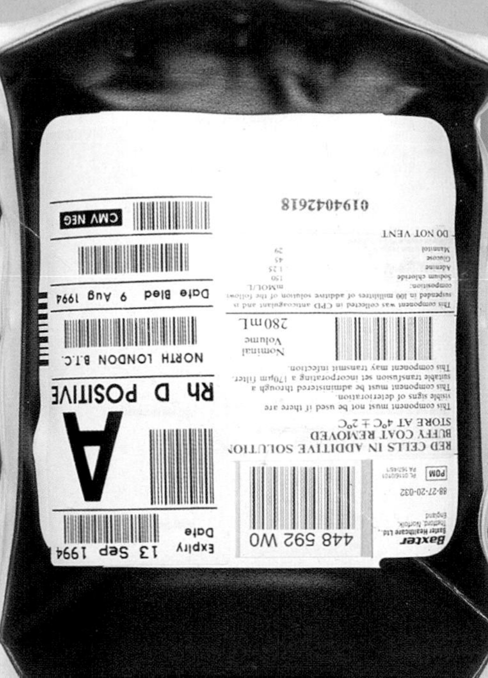

BLOOD BANK
Blood is tested for diseases and typed for its blood group. Chemicals are added to increase its shelf life, and it is stored at low temperature in hospitals until it is needed. It can be stored either as blood, or as one of its components, such as plasma.

blood vessels. The explanation for this was provided by Karl Landsteiner (1868–1943), who in 1901 discovered that within the human population there were different blood groups. Which of the four blood groups a person belonged to – A, B, AB, or O – depended on the presence or absence of two molecular markers called antigens – identified as A and B – carried on the surface of their red blood cells. Group A has the A antigen, group B the B antigen, group AB both antigens, and group O neither antigen. Also present in blood plasma are antibodies that act against antigens which are not present on a person's own red blood cells. Group A blood contains anti-B antibodies, group B blood contains anti-A antibodies, group O blood contains both antibodies, and group AB blood contains neither. This explains why, for example, if a group A person is given group B blood, their anti-B antibodies "recognize" the B antigens on the foreign red blood cells, make them stick together, and block blood vessels.

In 1940, Landsteiner recognized the rhesus antigen (Rh), first identified in rhesus monkeys. About 85 per cent of people are rhesus positive (Rh+) because they carry the antigen, while the remainder lack the antigen and are therefore rhesus negative (Rh-).

Safe transfusions

Once an individual's blood group could be identified by simple tests, safe blood transfusions became routine. But obtaining enough blood for emergency use is still a problem. Blood can be stored at low temperature in blood banks, but only for limited periods, so an important long-term goal is to develop artificial blood that can be given to a member of any blood group.

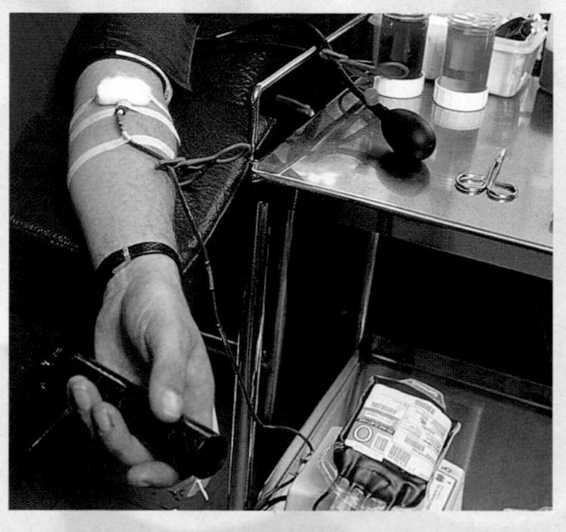

GIVING BLOOD
Healthy individuals can donate 500 ml (0.9 pint) of blood every 16 weeks. They can also donate parts of their blood, such as red cells, white cells, plasma, or platelets, more frequently. To give blood, the donor lies down and has a needle inserted into a vein near the elbow. Blood flows down a plastic tube and into a storage bag.

Heart

ONCE THOUGHT TO BE the source of feelings of love and emotion, the heart is actually the powerhouse of the circulatory system. Rhythmic contractions of this muscular pump push blood along the blood vessels to all parts of the body, even its far extremities, and back to the heart again. The beating heart ensures that every cell of the body has an uninterrupted supply of food, oxygen, and other essentials. So powerful is the heart that it can pump the body's entire blood volume of 5 litres (8.8 pints) around the body about once every minute. On average it beats, or pumps, 70 times a minute when the body is at rest, yet can increase this rate if the body is more active. Over a lifetime of 70 years, the heart beats some 2.5 billion times without tiring or stopping for a rest, thanks to the cardiac muscle in its walls.

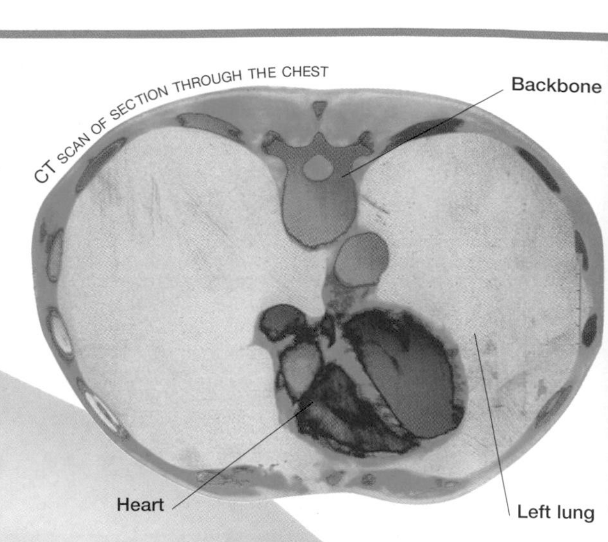

HEART AND LUNGS
This CT scan "slices" through the chest to show clearly how the heart is surrounded by right and left lungs. Being this close is important. It cuts to a minimum the distance blood has to travel from the heart to pick up oxygen in the lungs.

TWO SIDES
Divided into the left and right sides, the heart is two pumps in one. On each side, blood enters the atrium, then passes into the ventricle to be pumped on its onward journey. On the right side, oxygen-poor blood (blue) enters the right atrium, flows into the right ventricle, and is pumped to the lungs. On the left side, oxygen-rich blood (red) enters the left atrium, flows into the left ventricle, and is pumped to the rest of the body.

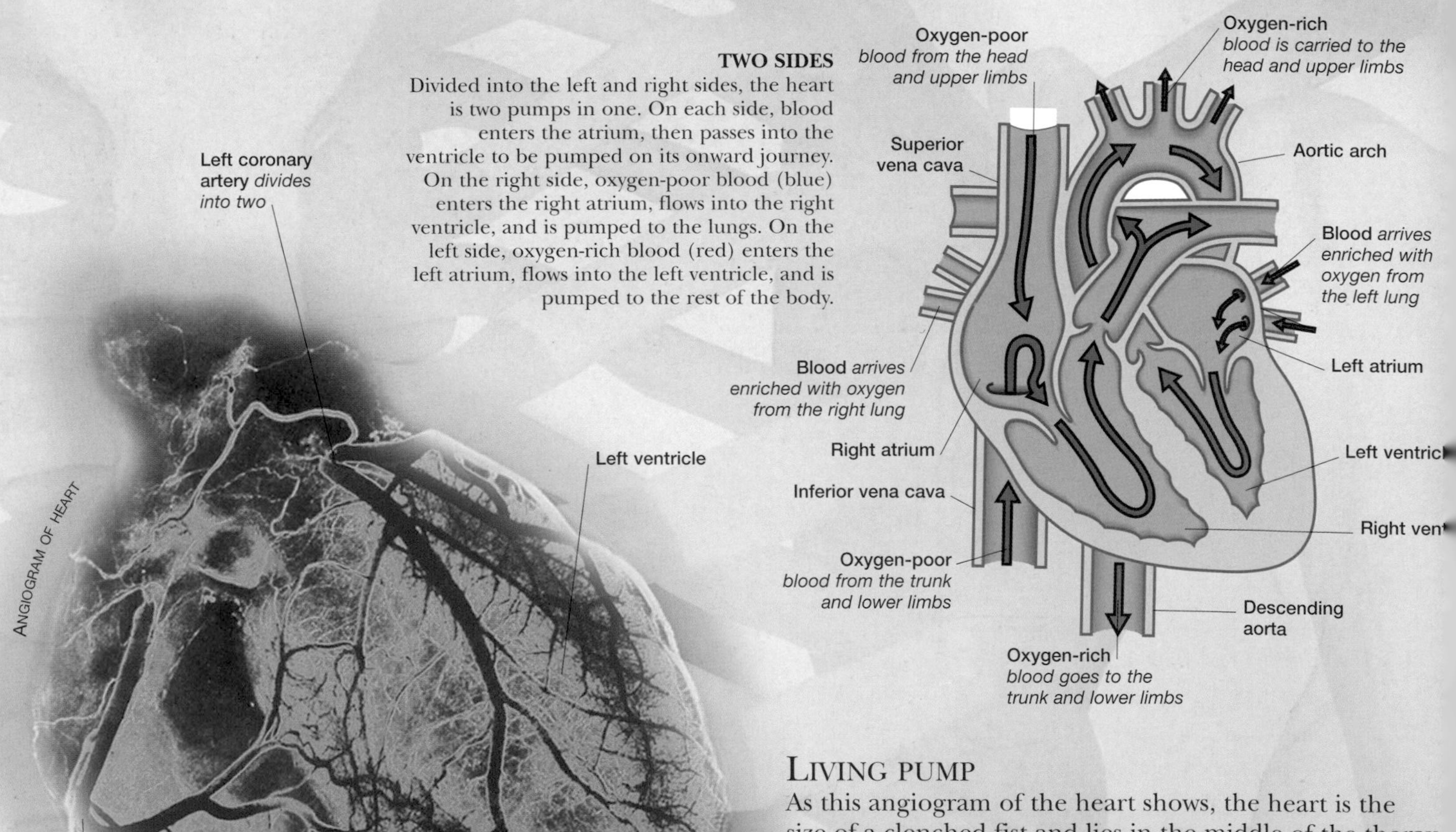

LIVING PUMP

As this angiogram of the heart shows, the heart is the size of a clenched fist and lies in the middle of the thorax (chest) with its apex (tip) pointing downwards and to the left. The angiogram also reveals the coronary blood vessels on the surface of the heart. Because the heart is a pump made out of living tissue, it too requires a supply of blood. However, the blood that gushes through it every second cannot meet the demands of the heart muscle for food and oxygen. To overcome this problem, coronary arteries arise from the aorta as it leaves the heart, and branch to carry oxygen-rich blood to the cardiac muscle in the heart wall (see also pp. 144–5).

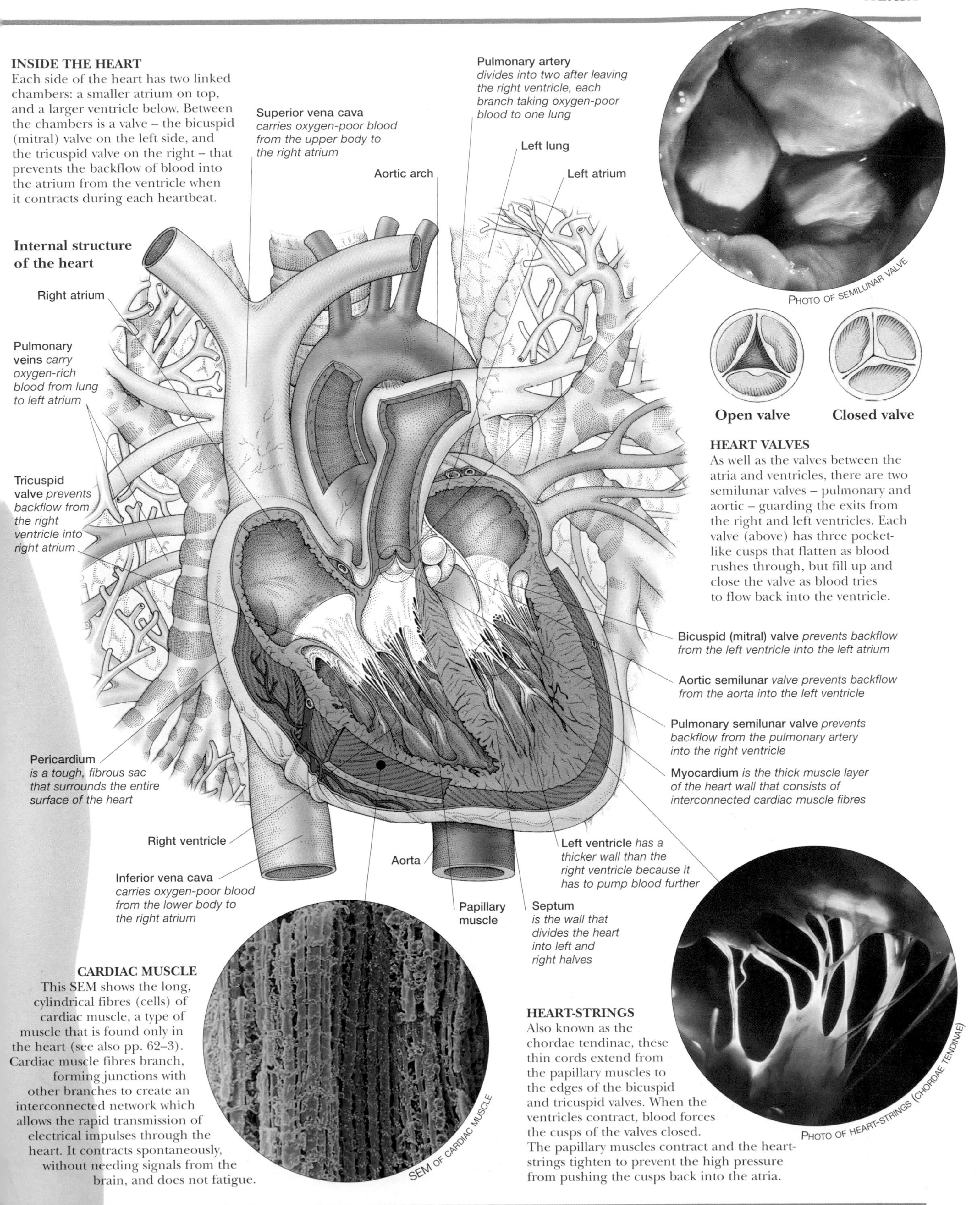

INSIDE THE HEART

Each side of the heart has two linked chambers: a smaller atrium on top, and a larger ventricle below. Between the chambers is a valve – the bicuspid (mitral) valve on the left side, and the tricuspid valve on the right – that prevents the backflow of blood into the atrium from the ventricle when it contracts during each heartbeat.

HEART VALVES

As well as the valves between the atria and ventricles, there are two semilunar valves – pulmonary and aortic – guarding the exits from the right and left ventricles. Each valve (above) has three pocket-like cusps that flatten as blood rushes through, but fill up and close the valve as blood tries to flow back into the ventricle.

CARDIAC MUSCLE

This SEM shows the long, cylindrical fibres (cells) of cardiac muscle, a type of muscle that is found only in the heart (see also pp. 62–3). Cardiac muscle fibres branch, forming junctions with other branches to create an interconnected network which allows the rapid transmission of electrical impulses through the heart. It contracts spontaneously, without needing signals from the brain, and does not fatigue.

HEART-STRINGS

Also known as the chordae tendinae, these thin cords extend from the papillary muscles to the edges of the bicuspid and tricuspid valves. When the ventricles contract, blood forces the cusps of the valves closed. The papillary muscles contract and the heart-strings tighten to prevent the high pressure from pushing the cusps back into the atria.

Heartbeats

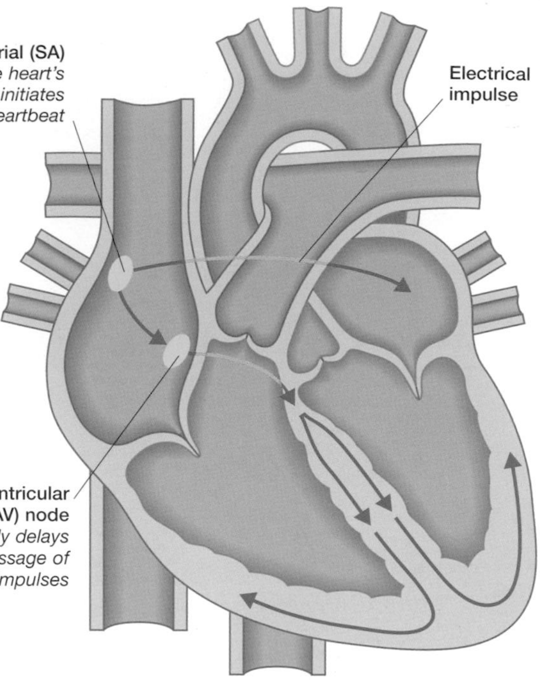

MOST PEOPLE HAVE experienced the feeling of a pounding heart, especially after running fast and then stopping suddenly. Every rhythmic pulsation represents one heartbeat, but not just one event. Each beat is made up of three stages called the heartbeat cycle. The entire cycle is masterminded by a patch of modified cardiac muscle called the sinoatrial (SA) node that acts as a pacemaker, sending out regular electrical impulses to stimulate contraction of the heart's chambers. While heart contraction happens spontaneously, outside control is needed, in the form of the autonomic nervous system (ANS), which alters the rate and strength of heartbeats in order to meet the ever-changing needs of the body.

Sinoatrial (SA) node, *the heart's pacemaker, initiates the heartbeat*

Electrical impulse

Atrioventricular (AV) node *briefly delays passage of impulses*

HEARTBEAT CYCLE

The three diagrams below show the stages of a heartbeat cycle. During diastole (relaxation), the heart relaxes, the atria fill with blood, and the pulmonary and aortic semilunar valves close to prevent backflow. During atrial systole (contraction), the atria contract to push blood into their respective ventricles. During ventricular systole, as the two ventricles contract together to pump blood out of the heart, blood pressure forces the semilunar valves open and the bicuspid and tricuspid valves shut. The three stages correspond to different sections of an ECG recording, an example of which extends across these pages and is explained (right).

ELECTRICAL PATHWAYS

Electrical impulses spread through the atria from the SA node, then are briefly delayed at the atrioventricular (AV) node, giving the atria time to contract before impulses spread through the ventricles and trigger their contraction. This electrical activity, recorded as an electrocardiogram (ECG), below, shows characteristic peaks. The P wave occurs as impulses travel over the atria, the QRS peak as impulses pass through the ventricles, and the T wave as the ventricles relax.

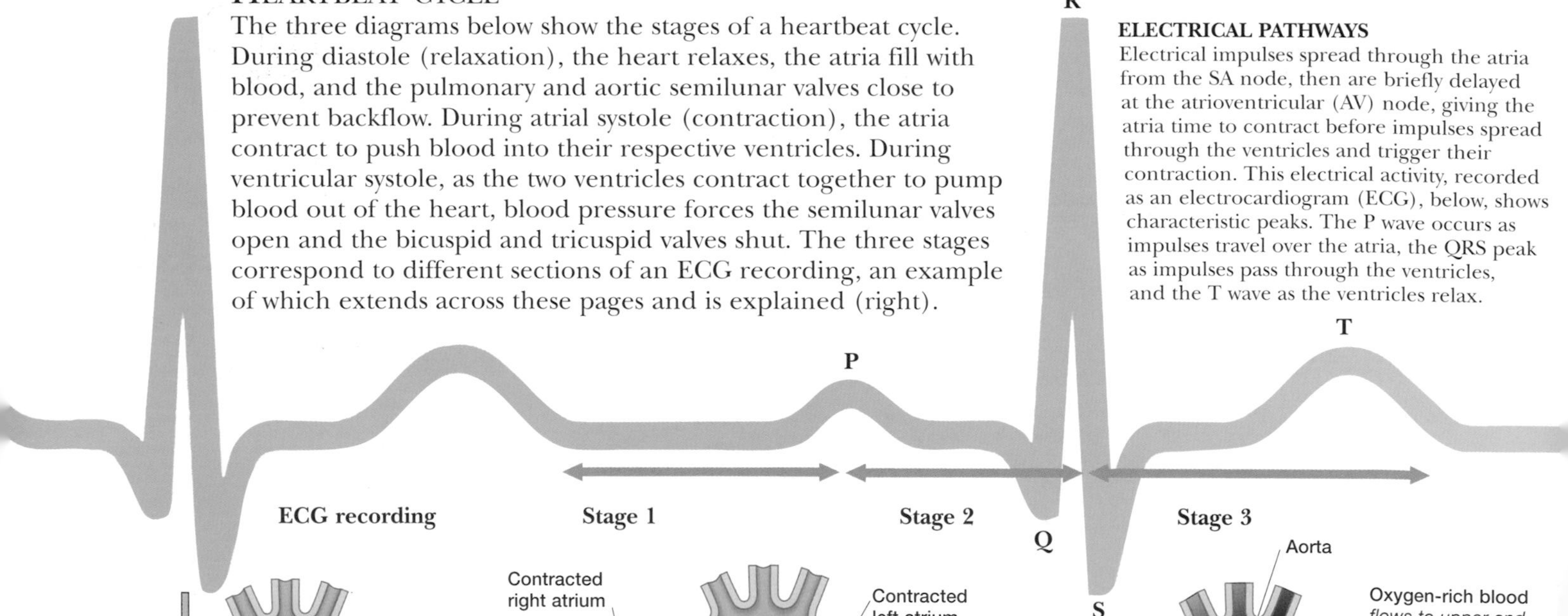

P
R
Q
S
T

ECG recording

Stage 1

Stage 2

Stage 3

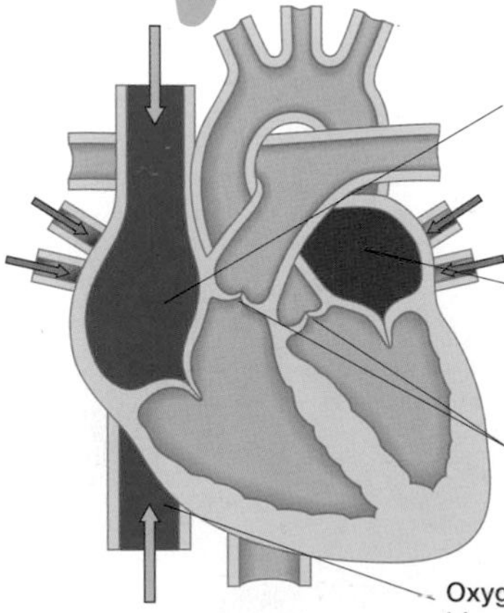

Right atrium *fills with oxygen-poor blood from the body*

Oxygen-rich blood *from lungs fills left atrium*

Semilunar valves *closed*

Oxygen-poor blood *from lower body*

Stage 1. Diastole
The heart muscle is relaxed, allowing oxygen-rich blood from the lungs and oxygen-poor blood from the body to enter the left and right atria respectively.

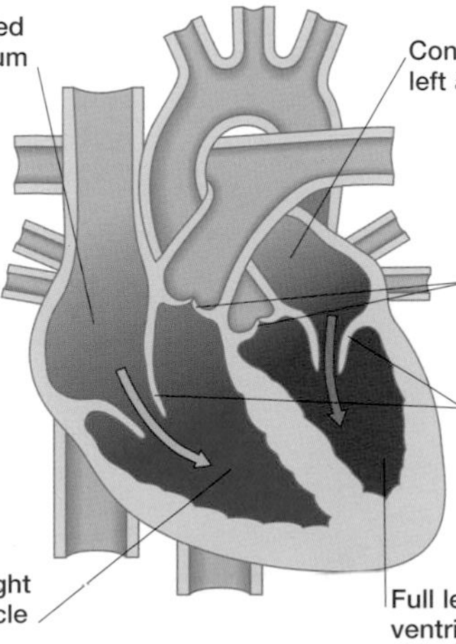

Contracted right atrium

Contracted left atrium

Semilunar valves *closed*

Tricuspid and bicuspid valves *open*

Full right ventricle

Full left ventricle

Stage 2. Atrial systole
The atria contract to squeeze their remaining blood into the ventricles. The tricuspid and bicuspid valves open to allow this, but the semilunar valves remain closed.

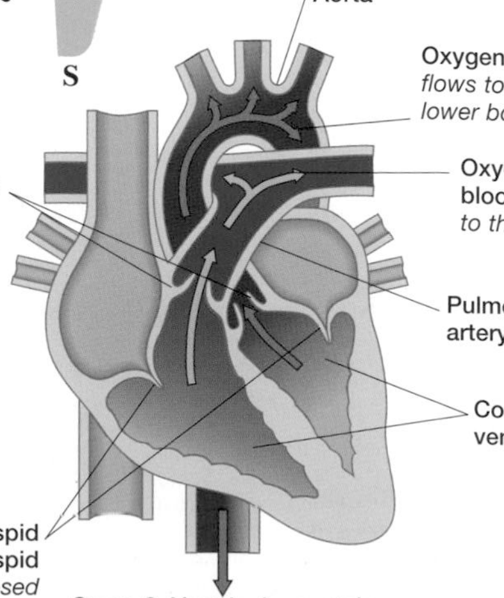

Aorta

Oxygen-rich blood *flows to upper and lower body*

Semilunar valves *open*

Oxygen-poor blood *flows to the lungs*

Pulmonary artery

Contracted ventricles

Tricuspid and bicuspid valves *closed*

Stage 3. Ventricular systole
Contraction of the ventricles sends blood to the lungs and around the body. The semilunar valves are pushed open by the surge of blood, while the tricuspid and bicuspid valves are closed.

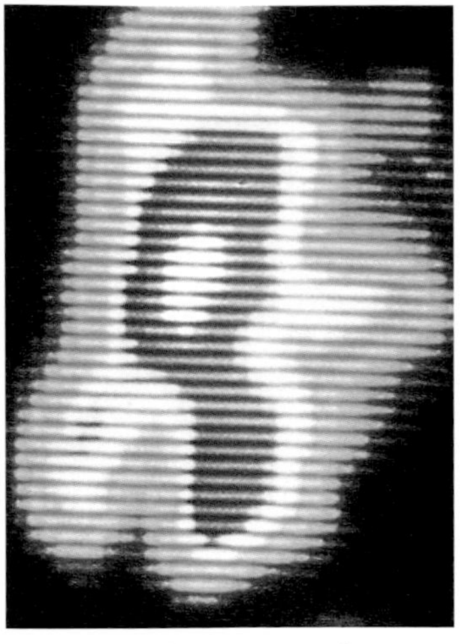

Left ventricle contracted

Left ventricle relaxed

HEART SOUNDS
The gamma camera scans (above) provide one way for doctors to follow the heartbeat cycle in action. Another is to use a stethoscope to listen for the sounds produced when the heart's valves close. Each heartbeat produces two sounds: a longer, louder "lub" sound when bicuspid and tricuspid valves close, and a shorter, sharper "dub" sound when the semilunar valves close. Unusual heart sounds can indicate that a valve is leaking.

LISTENING TO HEARTBEATS

The effective working of the valves can be assessed by listening to the sounds they make as they close. Early physicians listened to the heartbeat by placing an ear to the chest, but a one-ear stethoscope, invented by French doctor René Laënnec (1781–1826) in 1816, made the sounds much clearer and easier to hear. About 40 years later, the principles of this instrument were used to develop the two-ear, or binaural, stethoscope, which is still used today. In addition to assessing the heart, a stethoscope can be used to listen to breathing sounds made by the lungs and to sounds generated by the intestines.

Laënnec-type stethoscope c. 1820

CONTROLLING HEART RATE
Heart rate is controlled by the sympathetic and parasympathetic sections of the ANS (see pp. 98–9), whose nerve fibres terminate in the heart's pacemaking SA node. Under orders from the cardioregulatory centre in the brain stem, which monitors conditions inside the body, sympathetic signals speed up heart rate during exercise or stress, while parasympathetic signals slow heart rate when the body is at rest to about 70 beats per minute.

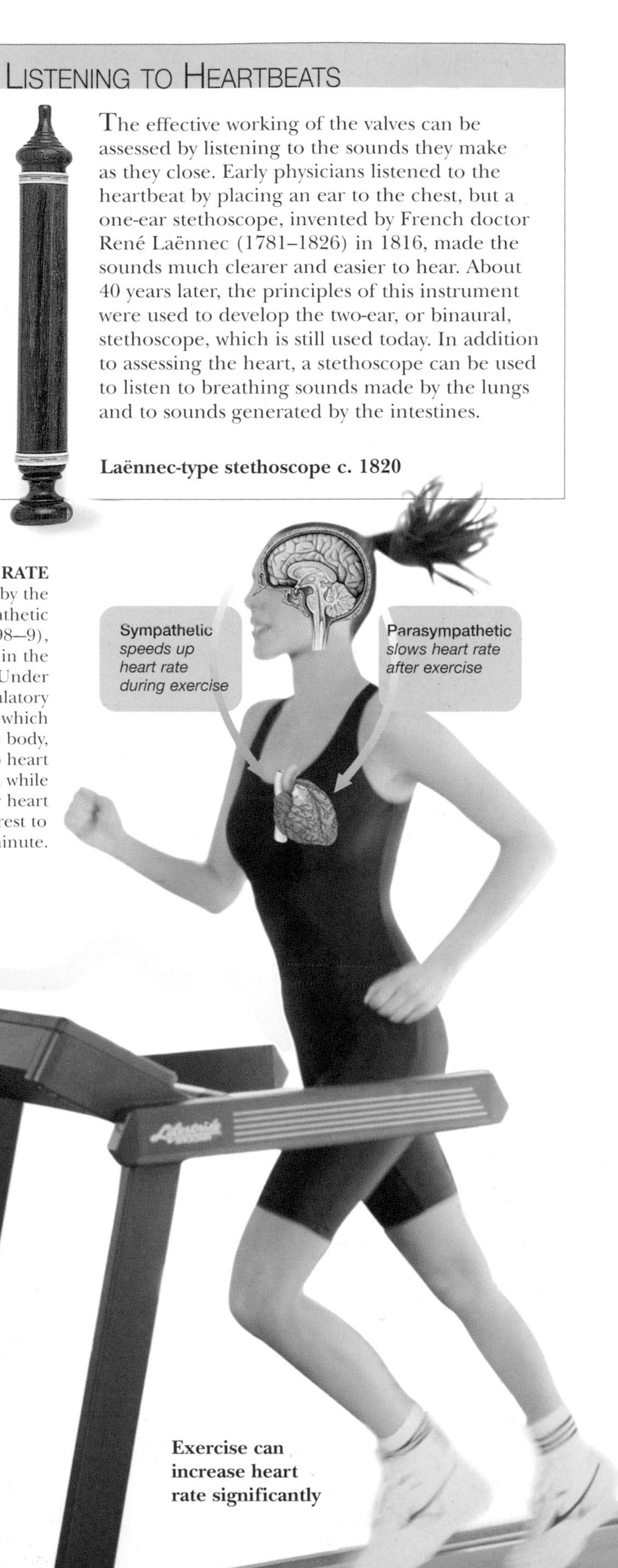

Exercise can increase heart rate significantly

Heart

Fitted heart pacemaker

Lead *connects pacemaker to the heart*

CHEST X-RAY SHOWING HEART PACEMAKER

ARTIFICIAL PACEMAKERS
This X-ray shows an artificial pacemaker implanted under the skin of a person whose SA node has stopped working normally, or whose heart does not conduct electrical impulses properly. Powered by a long-life battery, the pacemaker sends electrical impulses along a wire to stimulate the heart to beat. Some send out impulses at a fixed rate, others switch on if the heart misses a beat, while others vary their rate to match body activity levels.

HEART PROBLEMS

UNSEEN INSIDE THE CHEST, the heart is taken for granted until something goes wrong. A common cause of heart problems is a narrowing or blocking of the coronary arteries, which provide heart muscle with oxygen, a condition known as coronary artery disease. Its main symptom is chest pain, noticed during stress or exercise, when extra demands are put on the heart. The chances of developing this problem are increased by smoking, high blood pressure, a high-fat diet, obesity, and inactivity. But before looking at the consequences and treatment of this disease, consider a drastic measure that can deal with a badly damaged heart.

FIRST HEART TRANSPLANTS
In 1967, Christiaan Barnard led a team of 20 surgeons to perform the first human heart transplant.

HEART TRANSPLANTS

First performed in 1967 in South Africa by Professor Christiaan Barnard (1922–2001), a heart transplant involves the replacement of a diseased heart with a healthy one. Before the damaged heart is removed, the recipient's major blood vessels are connected to a bypass machine that pumps oxygen-rich blood to the brain and other organs. Once in place, the healthy heart is connected to the recipient's vessels and the bypass machine is switched off. Rejection of the new heart by the recipient's immune system is a major risk, and the patient must take preventative medication.

Front view of heart

CORONARY VESSELS
Both main coronary arteries originate from the aorta. The right coronary artery branches to deliver blood mainly to the right side of the heart; the left artery and its branches supply both ventricles and the left atrium, as well as the septum, the muscular wall that divides the heart. The smallest arterial branches connect to the smallest veins by capillaries. Coronary veins return oxygen-poor blood to the coronary sinus, a large vein at the back of the heart that empties into the right atrium.

HEART ATTACKS

Narrowing of coronary arteries can occur as a result of atheroma, a common condition in which fatty plaques are deposited on the lining of arterial walls, causing scarring. If a narrowed artery becomes blocked by a blood clot, the blood supply to an area of the heart is cut off, causing permanent damage to this area. This occurrence is known as a heart attack or myocardial infarction.

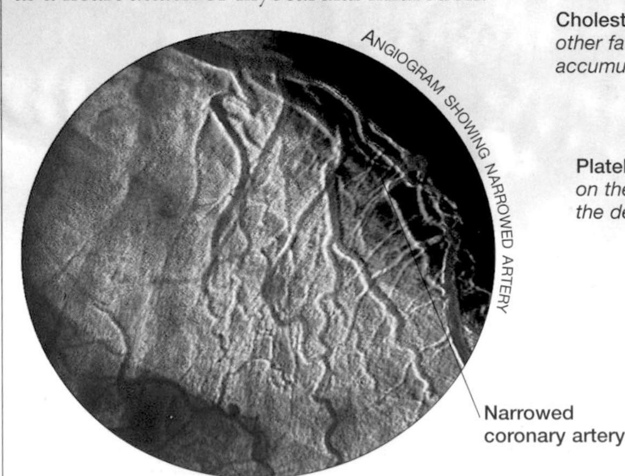

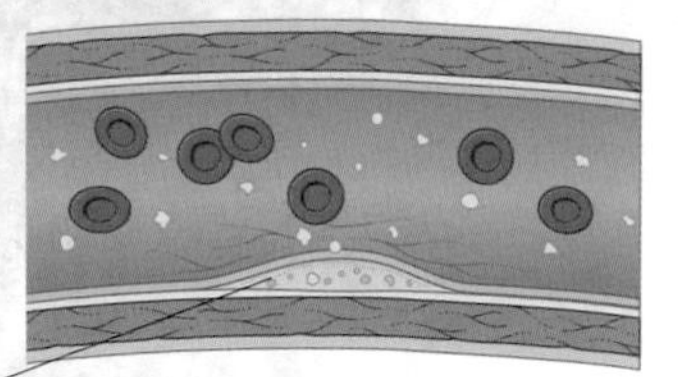

EARLY ATHEROSCLEROSIS
The first signs of atherosclerosis, the narrowing of arteries, are the gradual accumulation of fatty substances in the artery wall.

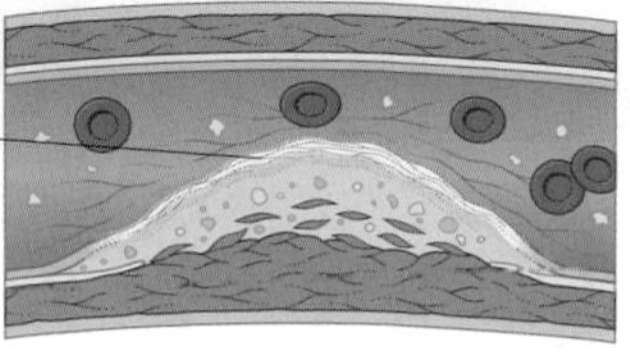

ADVANCED STAGES
The yellow deposits, called atheroma, cause the muscle layer to thicken. They restrict blood flow and narrow the artery.

HEALTHY HEART

The well-being of the heart can be assessed by performing a walking electrocardiogram (ECG), which records the electrical activity within the heart during a period of exercise on a treadmill. The speed of walking is gradually increased – and the treadmill may also be inclined – so that the stress on the heart slowly rises. Some conditions, such as coronary artery disease, may be diagnosed if particular changes are seen on the trace.

PHOTO OF DEFLATED ANGIOPLASTY BALLOON

WIDENING VESSELS

When coronary heart disease has been diagnosed, a doctor may perform balloon angioplasty to prevent a heart attack. A catheter (tube) carrying a deflated "balloon" is inserted, via an artery in the leg, into the affected coronary artery. The balloon is then inflated to enlarge the coronary artery and improve blood flow. Once this is achieved, the balloon is deflated and withdrawn from the body.

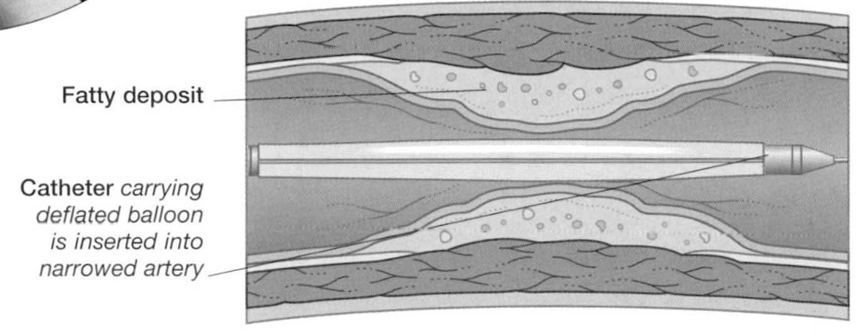
Fatty deposit
Catheter *carrying deflated balloon is inserted into narrowed artery*

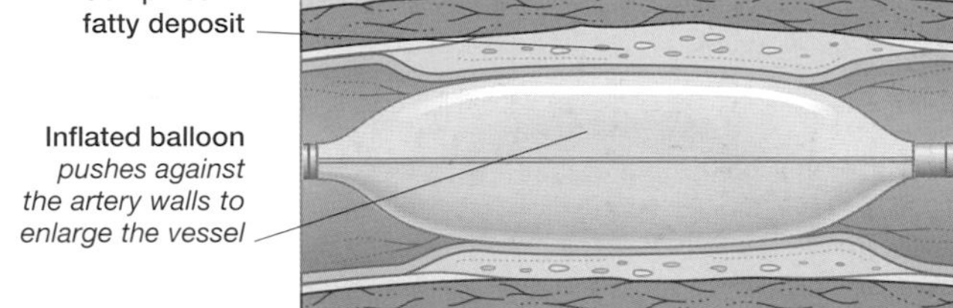
Compressed fatty deposit
Inflated balloon *pushes against the artery walls to enlarge the vessel*

Balloon angioplasty

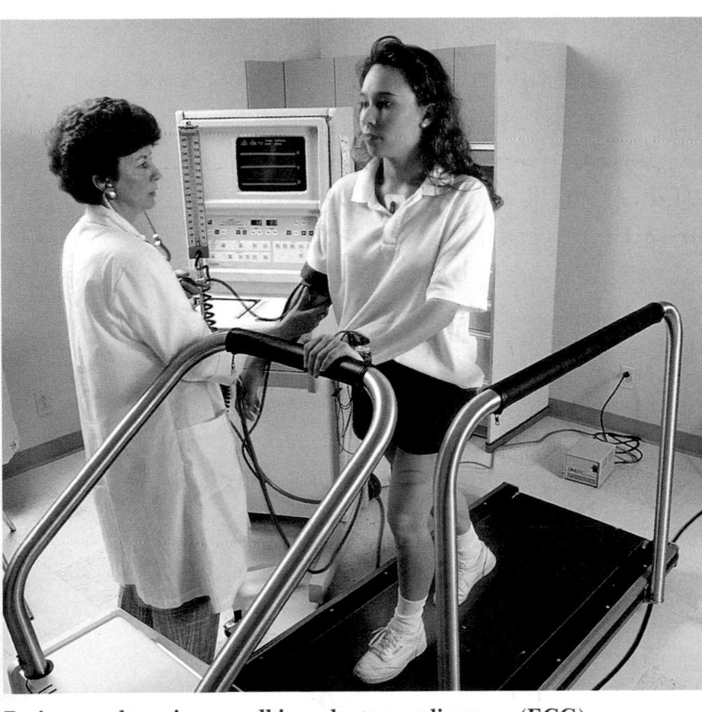

Patient undergoing a walking electrocardiogram (ECG)

EMERGENCY DEFIBRILLATION

Coronary heart disease may result in a heart attack which can cause fibrillation. This occurs when cardiac muscle fibres contract individually and chaotically rather than together, as happens ordinarily. As a consequence, the ventricles cannot pump blood, a condition known as cardiac arrest. To counteract this problem, emergency defibrillation, or cardioversion ("turning the heart"), can be performed by giving the heart a brief electric shock through two metal plates applied to the chest. The shock may restore normal synchronized contractions.

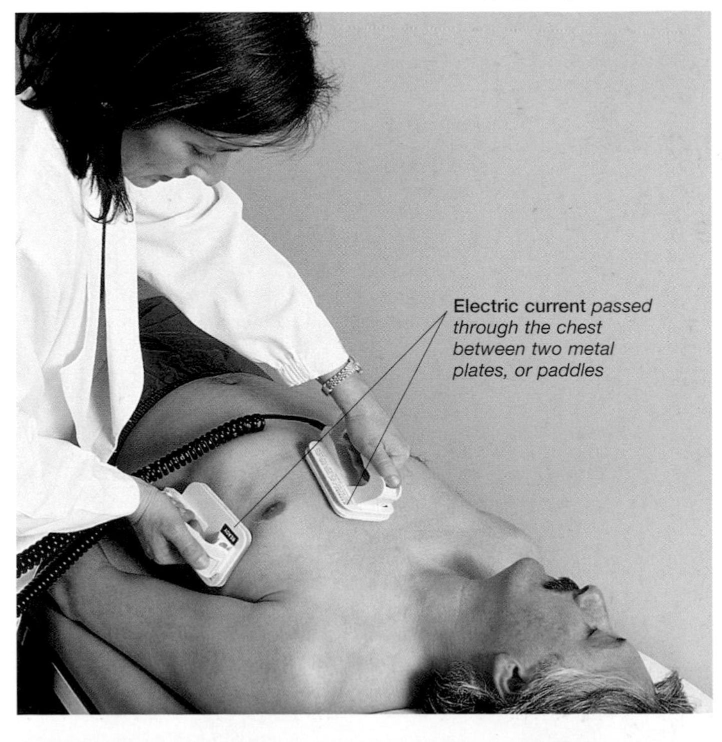
Electric current *passed through the chest between two metal plates, or paddles*

Blood circulation

INSIDE A HUMAN BODY is a system of blood-carrying tubes that, if stretched out, would extend over 150,000 km (93,206 miles), the equivalent of being wrapped around the Earth four times. Some 98 per cent of this incredible distance is made up of the microscopic capillaries that permeate every nook and cranny of the body's tissues to ensure that every cell receives its supplies of food and oxygen and that its waste is removed. Together, the capillaries and other blood vessels – the arteries and veins – form a blood circulation network that follows a figure-of-eight path, first through the lungs to pick up oxygen and then round the body to deliver that oxygen.

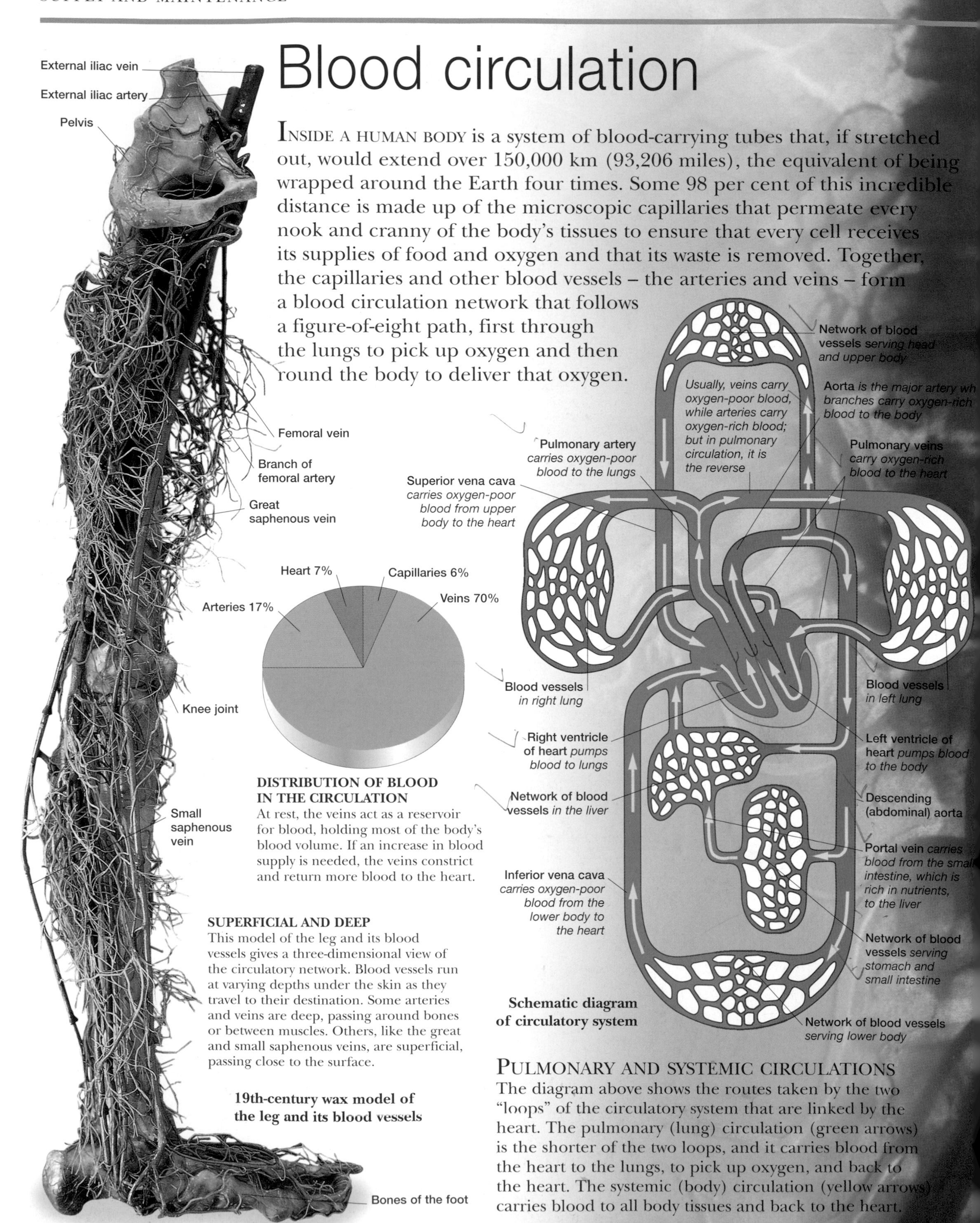
External iliac vein
External iliac artery
Pelvis
Femoral vein
Branch of femoral artery
Great saphenous vein
Knee joint
Small saphenous vein
Bones of the foot
Heart 7%
Capillaries 6%
Arteries 17%
Veins 70%
Network of blood vessels serving head and upper body
Usually, veins carry oxygen-poor blood, while arteries carry oxygen-rich blood; but in pulmonary circulation, it is the reverse
Aorta is the major artery wh
branches carry oxygen-rich blood to the body
Pulmonary artery carries oxygen-poor blood to the lungs
Pulmonary veins carry oxygen-rich blood to the heart
Superior vena cava carries oxygen-poor blood from upper body to the heart
Blood vessels in right lung
Blood vessels in left lung
Right ventricle of heart pumps blood to lungs
Left ventricle of heart pumps blood to the body
Network of blood vessels in the liver
Descending (abdominal) aorta
Portal vein carries blood from the smal
intestine, which is rich in nutrients, to the liver
Inferior vena cava carries oxygen-poor blood from the lower body to the heart
Network of blood vessels serving stomach and small intestine
Network of blood vessels serving lower body

DISTRIBUTION OF BLOOD IN THE CIRCULATION
At rest, the veins act as a reservoir for blood, holding most of the body's blood volume. If an increase in blood supply is needed, the veins constrict and return more blood to the heart.

SUPERFICIAL AND DEEP
This model of the leg and its blood vessels gives a three-dimensional view of the circulatory network. Blood vessels run at varying depths under the skin as they travel to their destination. Some arteries and veins are deep, passing around bones or between muscles. Others, like the great and small saphenous veins, are superficial, passing close to the surface.

19th-century wax model of the leg and its blood vessels

Schematic diagram of circulatory system

PULMONARY AND SYSTEMIC CIRCULATIONS

The diagram above shows the routes taken by the two "loops" of the circulatory system that are linked by the heart. The pulmonary (lung) circulation (green arrows) is the shorter of the two loops, and it carries blood from the heart to the lungs, to pick up oxygen, and back to the heart. The systemic (body) circulation (yellow arrows) carries blood to all body tissues and back to the heart.

Splenic artery

Renal artery

Backbone

Descending (abdominal) aorta

Common iliac artery

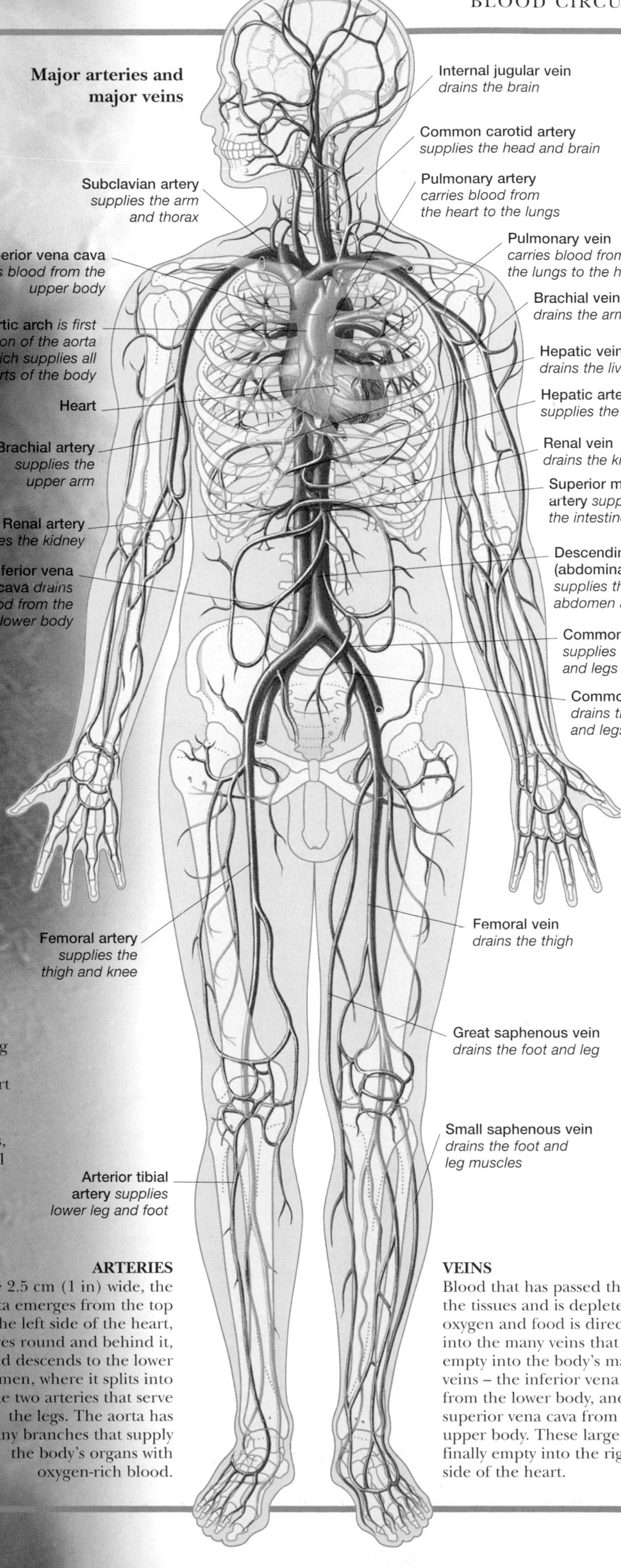

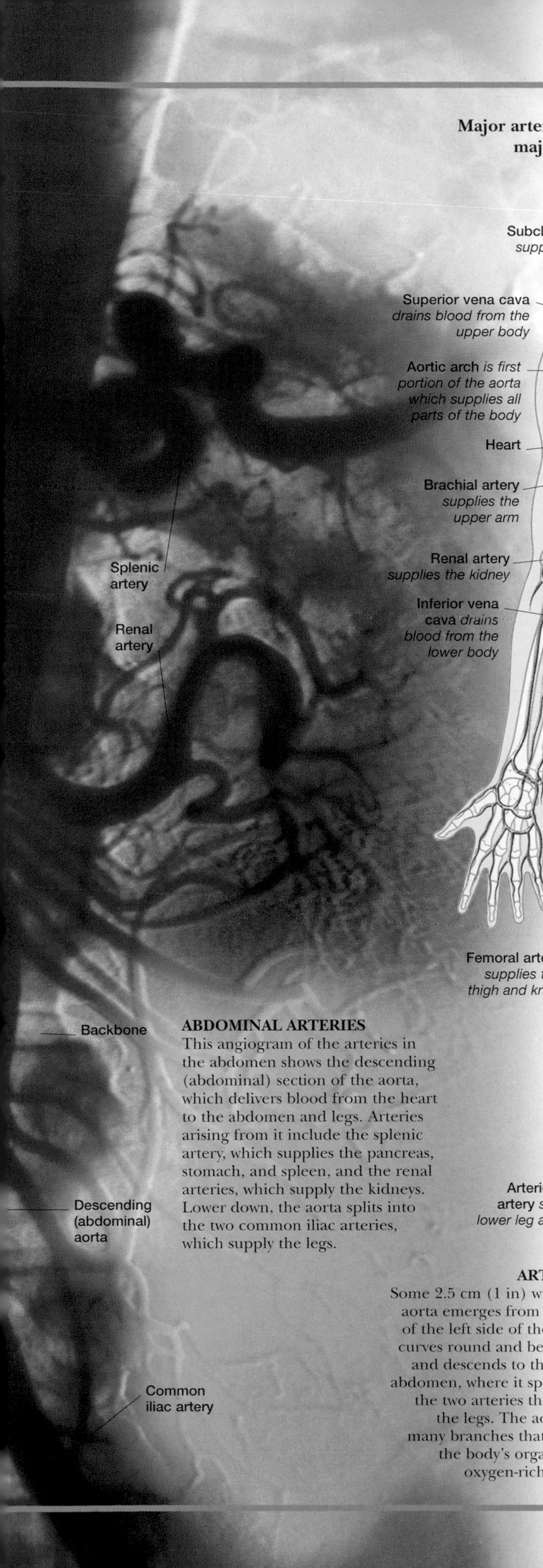

ABDOMINAL ARTERIES
This angiogram of the arteries in the abdomen shows the descending (abdominal) section of the aorta, which delivers blood from the heart to the abdomen and legs. Arteries arising from it include the splenic artery, which supplies the pancreas, stomach, and spleen, and the renal arteries, which supply the kidneys. Lower down, the aorta splits into the two common iliac arteries, which supply the legs.

Major arteries and major veins

Subclavian artery *supplies the arm and thorax*

Superior vena cava *drains blood from the upper body*

Aortic arch *is first portion of the aorta which supplies all parts of the body*

Heart

Brachial artery *supplies the upper arm*

Renal artery *supplies the kidney*

Inferior vena cava *drains blood from the lower body*

Femoral artery *supplies the thigh and knee*

Arterior tibial artery *supplies lower leg and foot*

Internal jugular vein *drains the brain*

Common carotid artery *supplies the head and brain*

Pulmonary artery *carries blood from the heart to the lungs*

Pulmonary vein *carries blood from the lungs to the heart*

Brachial vein *drains the arm*

Hepatic vein *drains the liver*

Hepatic artery *supplies the liver*

Renal vein *drains the kidney*

Superior mesenteric artery *supplies the intestines*

Descending (abdominal) aorta *supplies the abdomen and legs*

Common iliac artery *supplies the pelvis and legs*

Common iliac vein *drains the pelvis and legs*

Femoral vein *drains the thigh*

Great saphenous vein *drains the foot and leg*

Small saphenous vein *drains the foot and leg muscles*

ARTERIES
Some 2.5 cm (1 in) wide, the aorta emerges from the top of the left side of the heart, curves round and behind it, and descends to the lower abdomen, where it splits into the two arteries that serve the legs. The aorta has many branches that supply the body's organs with oxygen-rich blood.

VEINS
Blood that has passed through the tissues and is depleted of oxygen and food is directed into the many veins that empty into the body's main veins – the inferior vena cava from the lower body, and the superior vena cava from the upper body. These large veins finally empty into the right side of the heart.

Blood vessels

BLOOD IS TRANSPORTED around the body in three main types of blood vessel – arteries, veins, and capillaries. Arteries carry oxygen-rich blood from the heart to the tissues. Their walls have a wide middle layer of muscle fibres and elastic tissue that enables them to expand and recoil as blood surges through them at high pressure when the heart contracts. Veins have thinner walls, with valves to prevent backflow as they carry blood under low pressure back to the heart. Capillaries are the smallest, and most numerous, blood vessels. Within walls just one cell thick, they carry blood through the tissues, linking the smallest branches of arteries (arterioles) and veins (venules).

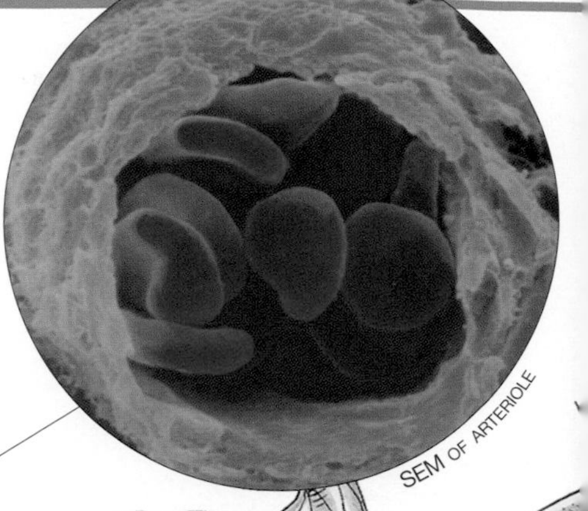

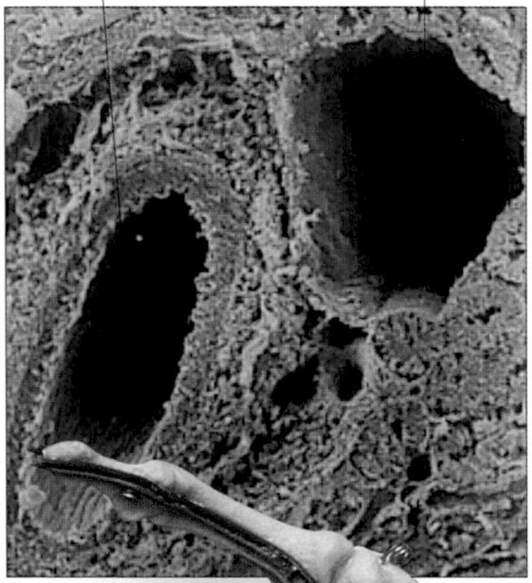

SEM OF ARTERIOLE

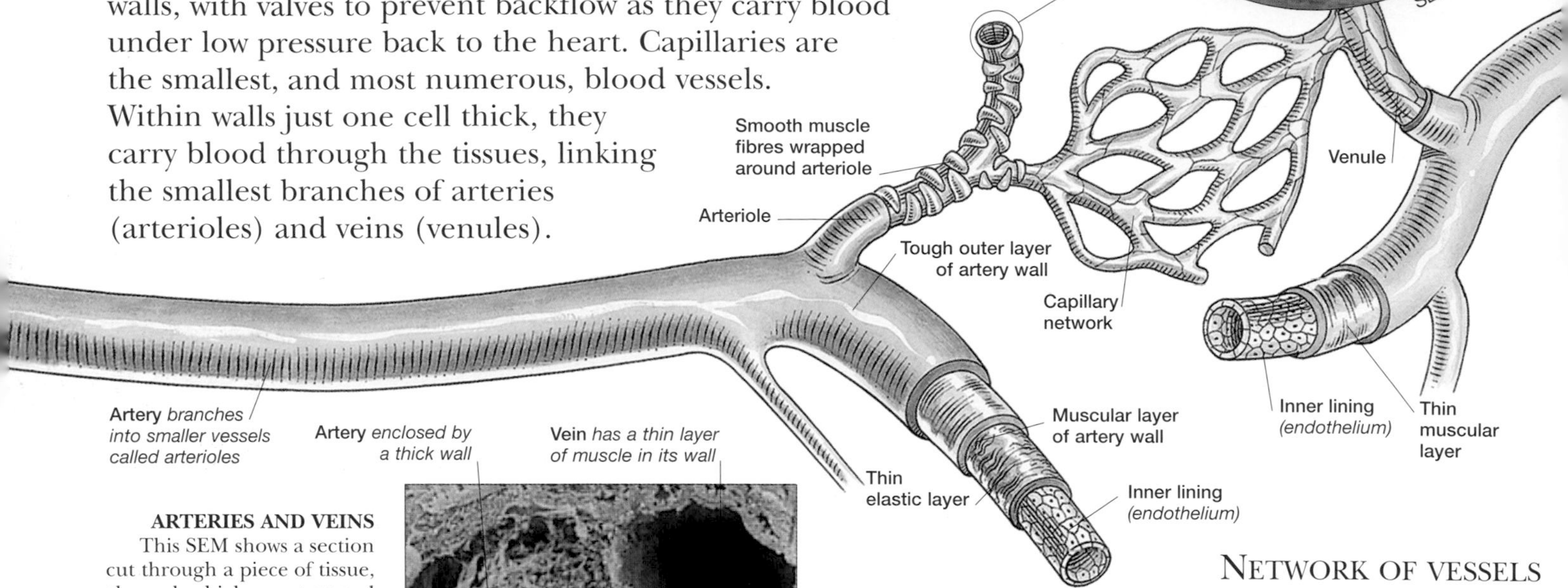

Smooth muscle fibres wrapped around arteriole

Arteriole

Tough outer layer of artery wall

Capillary network

Venule

Artery *branches into smaller vessels called arterioles*

Artery *enclosed by a thick wall*

Vein *has a thin layer of muscle in its wall*

Muscular layer of artery wall

Thin elastic layer

Inner lining *(endothelium)*

Inner lining *(endothelium)*

Thin muscular layer

ARTERIES AND VEINS
This SEM shows a section cut through a piece of tissue, through which an artery and a vein run in parallel. It clearly reveals the difference between arteries and veins. Although they both have about the same diameter, the artery has a narrower lumen (space in the middle through which blood flows) because it has a thicker, more muscular wall, while the vein has a wider lumen but a thinner, less muscular wall.

NETWORK OF VESSELS

The diagram above shows how oxygen-rich blood is carried into the tissues by arteries, which divide to form arterioles less than 0.3 mm (0.001 in) in diameter. The smallest arterioles are wrapped in smooth muscle fibres, which contract or relax to control blood flow. Arterioles divide to form capillaries that pass close to, and serve, tissue cells. The capillaries then unite to form tiny vein branches called venules, which merge to form the larger veins that return oxygen-poor blood to the heart. The model below shows the major blood arteries and veins of the arm, although the capillaries that link them are too small to be visible.

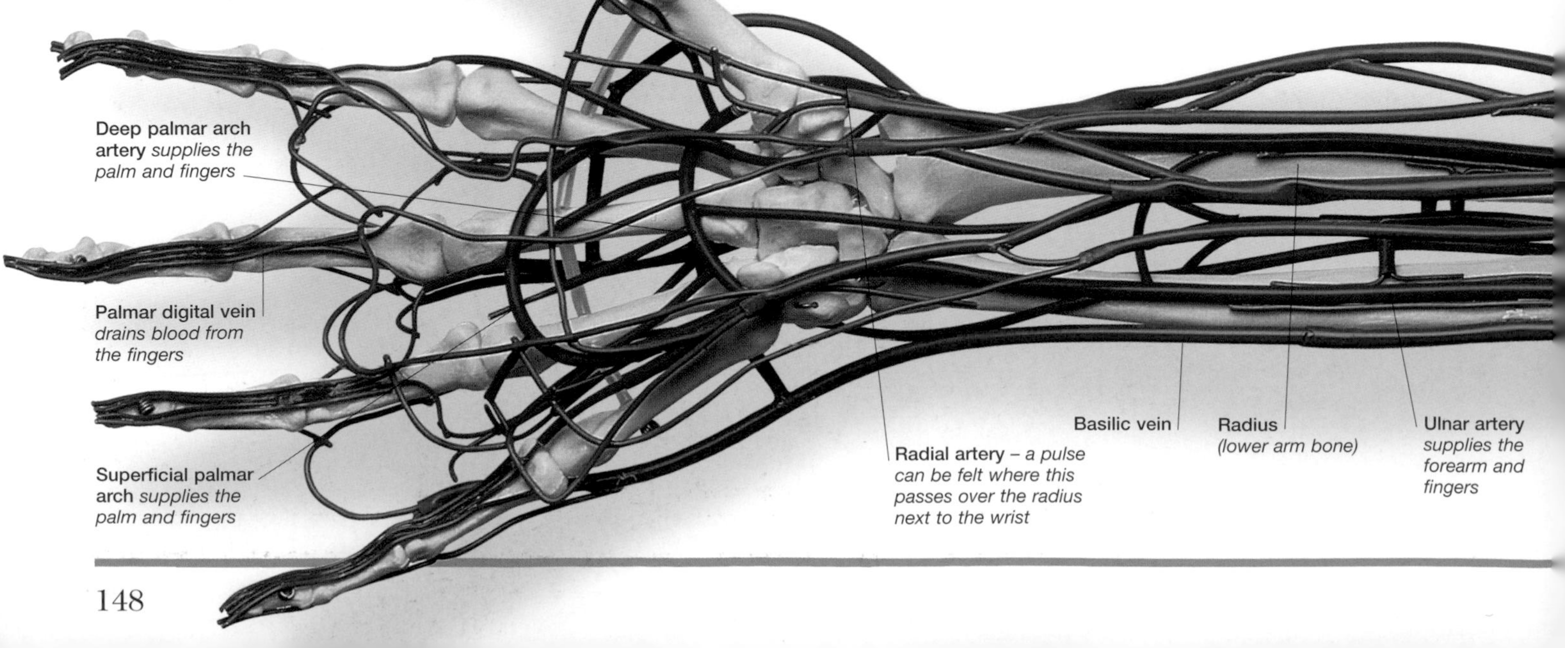

Deep palmar arch artery *supplies the palm and fingers*

Palmar digital vein *drains blood from the fingers*

Superficial palmar arch *supplies the palm and fingers*

Radial artery – *a pulse can be felt where this passes over the radius next to the wrist*

Basilic vein

Radius *(lower arm bone)*

Ulnar artery *supplies the forearm and fingers*

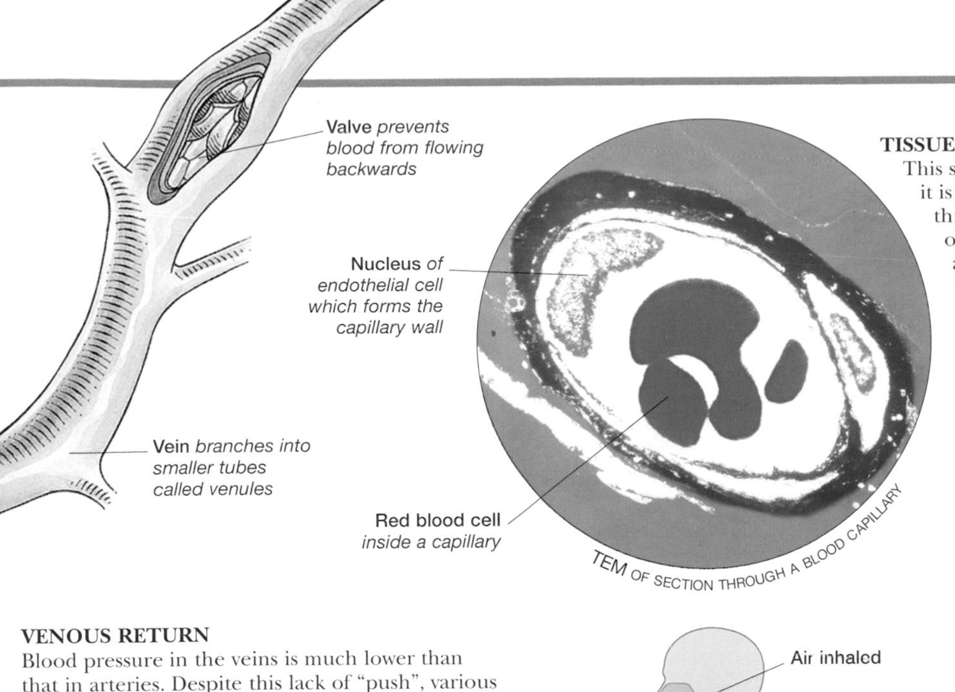

TISSUE DELIVERY
This section through a capillary shows how narrow it is – just wider, in fact, than a red blood cell passing through it. It reveals, too, how thin its wall is, being only one cell thick. The wall is also quite "leaky", allowing fluid carrying food and oxygen to pass out of the blood and into the tissue fluid that surrounds the cells, and return in the opposite direction carrying the cells' wastes.

VENOUS RETURN
Blood pressure in the veins is much lower than that in arteries. Despite this lack of "push", various mechanisms make sure that there is adequate venous return – the return of blood along veins back to the heart. Many deep veins lie within muscles, and when the muscles contract, they squeeze blood back towards the heart. Similarly, low pressure in the thorax (chest) produced during breathing in (see pp. 172–3) also draws blood to the heart.

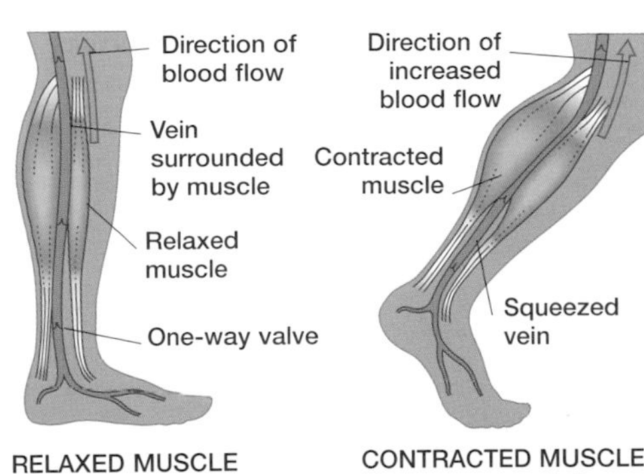

RELAXED MUSCLE
When a person is standing still, the leg muscles surrounding a deep vein are relaxed and blood flow in the vein is sluggish, although one-way valves prevent backflow.

CONTRACTED MUSCLE
During movement, leg muscles surrounding the vein contract and provide a muscular pump that squeezes the vein and pushes blood upwards towards the heart.

Air inhaled
Chest cavity *at low pressure*
Blood *drawn towards heart*
Diaphragm

RESPIRATORY PUMP
During inhalation, pressure inside the thoracic cavity decreases. As pressure is higher in the rest of the body, blood is pushed towards the heart.

Axillary artery *supplies shoulder and arm*

Cephalic vein *drains the hand and arm*

Humerus *(upper arm bone)*

Brachial artery *supplies the upper arm*

Basilic vein *drains the hand and arm*

Elbow vein *is used when doctors need to take a blood sample*

Ulna *(lower arm bone)*

Major blood vessels of the arm

DISCOVERING CAPILLARIES

Italian doctor Marcello Malpighi used a microscope to study the body. A professor at the University of Bologna and personal physician to Pope Innocent XII, Malpighi was the first scientist to describe capillaries in detail and to discover how they connect the arterial and venous blood vessels in the circulatory system. He completed William Harvey's theory of the circulatory system when he observed the passage of red blood cells through capillaries. As a result of his work, his name lives on in several anatomical names, as well as the term "Malpighian layer", a part of the skin.

Marcello Malpighi (1628–94)

Pressure and flow

EACH TIME THE HEART'S ventricles contract, they squeeze out blood, which exerts pressure on artery walls. This pressure provides the impetus that pushes blood around the circulatory system. It changes with each heartbeat, rising as the ventricles contract and falling as they relax. As well as this constant fluctuation, blood pressure increases during exercise as the heart contracts faster and expels a greater volume of blood. Exercise also alters the pattern of blood flow, as blood vessels supplying muscles widen to allow them receive extra oxygen and nutrients. In a healthy person, blood pressure is kept within strict limits by the brain and hormones. If blood pressure remains too high over long periods it can cause health problems.

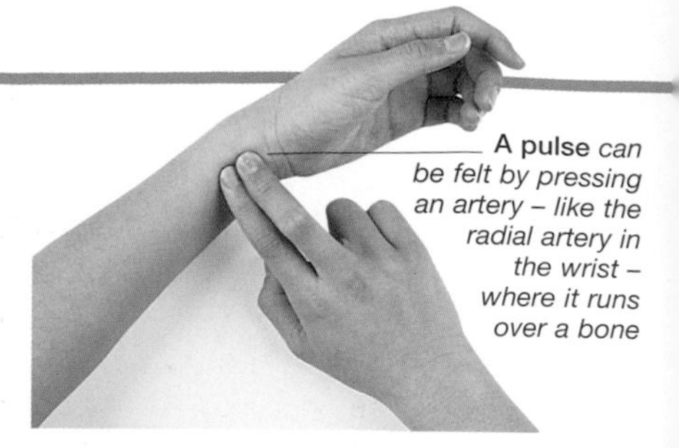

PULSE
After each heartbeat, a pressure wave – or pulse – passes along an artery as its walls bulge and then recoil to withstand the surge of blood. By pressing on an artery, the number of pulses – or heartbeats – per minute can be counted.

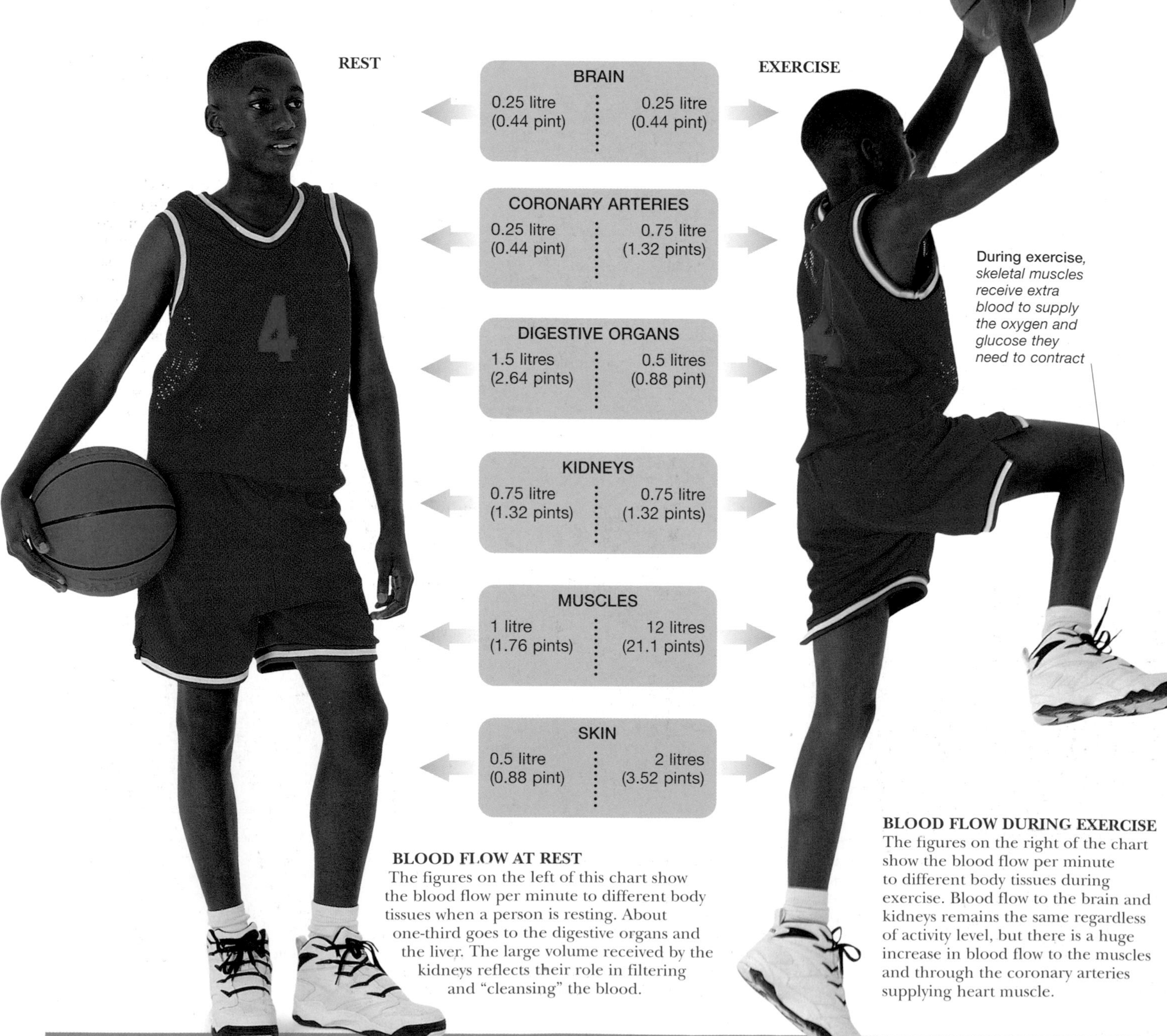

REST
EXERCISE
BRAIN
0.25 litre (0.44 pint)
0.25 litre (0.44 pint)
CORONARY ARTERIES
0.25 litre (0.44 pint)
0.75 litre (1.32 pints)
DIGESTIVE ORGANS
1.5 litres (2.64 pints)
0.5 litres (0.88 pint)
KIDNEYS
0.75 litre (1.32 pints)
0.75 litre (1.32 pints)
MUSCLES
1 litre (1.76 pints)
12 litres (21.1 pints)
SKIN
0.5 litre (0.88 pint)
2 litres (3.52 pints)
During exercise, skeletal muscles receive extra blood to supply the oxygen and glucose they need to contract

BLOOD FLOW AT REST
The figures on the left of this chart show the blood flow per minute to different body tissues when a person is resting. About one-third goes to the digestive organs and the liver. The large volume received by the kidneys reflects their role in filtering and "cleansing" the blood.

BLOOD FLOW DURING EXERCISE
The figures on the right of the chart show the blood flow per minute to different body tissues during exercise. Blood flow to the brain and kidneys remains the same regardless of activity level, but there is a huge increase in blood flow to the muscles and through the coronary arteries supplying heart muscle.

MEASURING BLOOD PRESSURE

Blood pressure is measured using a sphygmomanometer (literally a "pulse pressure measurer"), which consists of an inflatable cuff linked to a pressure gauge, and a stethoscope. In fact, two pressures, not one, are measured (see below) – systolic and then diastolic. The cuff is wrapped around the upper arm and inflated to squeeze the arm until blood flow along the brachial artery (see p. 149) stops. Placing the end of the stethoscope on the skin over the brachial artery, the doctor now deflates the cuff until a pulse can just be heard. The reading on the gauge shows the higher, systolic pressure, which is sufficient to push blood along the narrowed artery. The cuff is deflated further until the pulse sound just disappears, indicating that blood is flowing freely along the artery. The gauge now shows the lower, diastolic pressure.

Inflatable cuff

A stethoscope *is used by a doctor to listen to blood flow*

Gauge *indicates pressure*

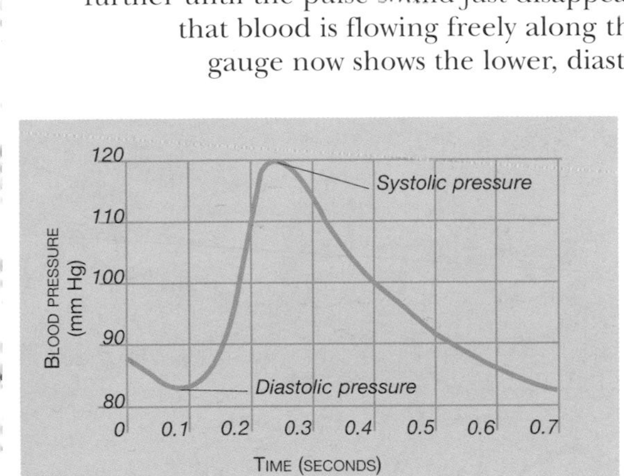

SYSTOLIC AND DIASTOLIC

During each heartbeat cycle (see pp. 142–3), the heart contracts (systole), causing a peak in arterial blood pressure called systolic pressure (see graph, above), then relaxes (diastole), causing a fall to the minimum, diastolic pressure. Pressure, measured in millimetres of mercury (mm Hg), varies according to age, sex, and health, but in a healthy young adult should be about 120/80 (120 mm Hg systolic and 80 mm Hg diastolic).

FROSTBITE

At temperatures below 0°C (32°F), the small arteries supplying the skin and underlying tissues narrow, restricting the supply of blood. If the cold conditions persist, ice is deposited in the tissues, causing damage. The fingers, toes, and nose are particularly susceptible to this condition, known as frostbite. The damage may be permanent, requiring amputation of the affected tissue in very severe cases.

FAINTING

A transient fall in blood pressure causes a momentary reduction in blood flow to the brain and results in fainting (syncope). This may occur when a person stands up suddenly, or following a long period of standing, when blood can pool in the veins of the legs. Lying down and elevating the legs will help restore normal blood pressure.

Areas of blackened skin *indicate tissue death caused by prolonged frostbite*

Lymphoid organs

THE LYMPHOID, or lymphatic organs, which include the lymph nodes, spleen, and tonsils, are the parts of the lymphatic system that fight disease. They have a similar structure, being filled with fibres that support macrophages and lymphocytes. Macrophages ingest and destroy pathogens and cancer cells, while lymphocytes play a key role in the immune system (see pp. 160–1), either by attacking pathogens directly, or by disabling them with antibodies. The most numerous lymphoid organs are the lymph nodes, which are scattered throughout the body and filter pathogens out of the lymph passing through them. The spleen removes pathogens from the blood, while the tonsils intercept those passing towards the throat. The thymus gland plays a vital role during childhood in the development of the immune system.

LYMPH NODES

These small, bean-shaped swellings, each 1–25 mm (0.04–1 in) across, occur along lymph vessels like beads on a string. Lymph nodes filter lymph as it passes through them. Surrounded by a tough capsule, the spaces, or sinuses, inside the lymph node are filled with a network of fibres that support macrophages and lymphocytes. These fibres slow the flow of lymph passing through, while macrophages engulf and destroy bacteria, cancer cells, and debris, and lymphocytes launch their immune defences. During infections, lymph nodes may swell up and become tender, a condition known as "swollen glands".

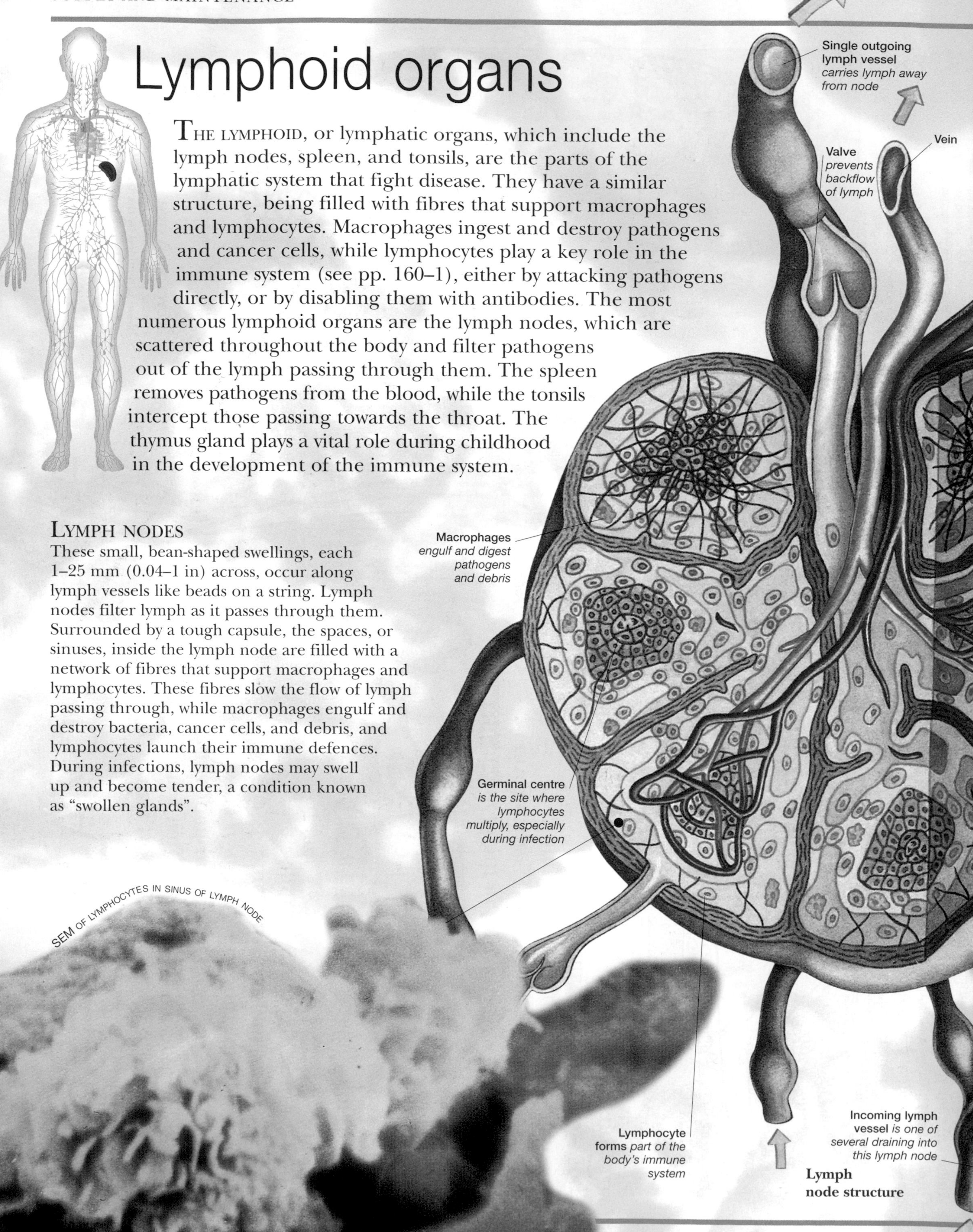
Single outgoing lymph vessel carries lymph away from node
Vein
Valve prevents backflow of lymph
Macrophages engulf and digest pathogens and debris
Germinal centre is the site where lymphocytes multiply, especially during infection
SEM OF LYMPHOCYTES IN SINUS OF LYMPH NODE
Lymphocyte forms part of the body's immune system
Incoming lymph vessel is one of several draining into this lymph node

Lymph node structure

SPLEEN
Largest of the lymphoid organs, the fist-sized spleen lies to the left of the stomach. It receives a rich blood supply through the splenic artery (right) and contains areas of reticular fibres (far right) which support lymphocytes that destroy pathogens. Around these areas are huge numbers of red blood cells, as well as macrophages that process incoming blood by engulfing bacteria, viruses, and worn-out red blood cells.

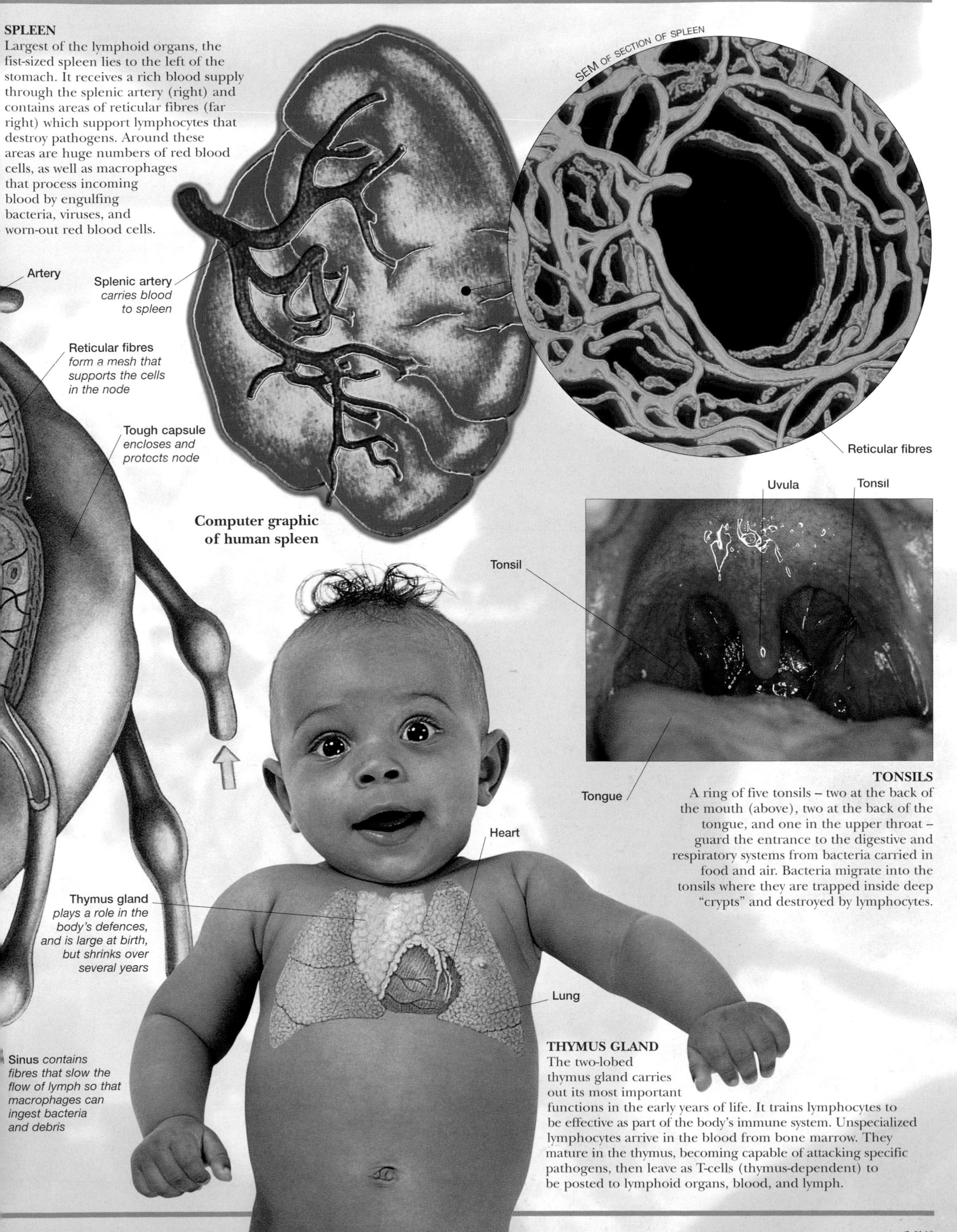
SEM OF SECTION OF SPLEEN
Artery
Splenic artery *carries blood to spleen*
Reticular fibres *form a mesh that supports the cells in the node*
Tough capsule *encloses and protects node*
Reticular fibres
Uvula
Tonsil
Tonsil
Tongue
Heart
Thymus gland *plays a role in the body's defences, and is large at birth, but shrinks over several years*
Lung
Sinus *contains fibres that slow the flow of lymph so that macrophages can ingest bacteria and debris*

Computer graphic of human spleen

TONSILS
A ring of five tonsils – two at the back of the mouth (above), two at the back of the tongue, and one in the upper throat – guard the entrance to the digestive and respiratory systems from bacteria carried in food and air. Bacteria migrate into the tonsils where they are trapped inside deep "crypts" and destroyed by lymphocytes.

THYMUS GLAND
The two-lobed thymus gland carries out its most important functions in the early years of life. It trains lymphocytes to be effective as part of the body's immune system. Unspecialized lymphocytes arrive in the blood from bone marrow. They mature in the thymus, becoming capable of attacking specific pathogens, then leave as T-cells (thymus-dependent) to be posted to lymphoid organs, blood, and lymph.

Diseases

From time to time, the body fails to work properly because one or more of the mechanisms that maintain homeostasis – constant, stable conditions inside the body – goes wrong. Any such breakdown is called a disease. Some diseases are short-lived and easily overcome by the body's natural defence systems, while others are more serious and require outside intervention in the form of drugs. Humans are affected by two types of disease – infectious and non-infectious. Infectious diseases are caused by parasitic organisms called pathogens that break through the body's defences and grow and multiply in its tissues. Most pathogens are micro-organisms. Non-infectious diseases are those not caused by pathogens, and include cancers, nervous system disorders, and autoimmune conditions (see p. 165).

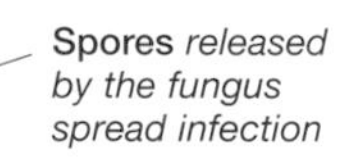

Cross-section of a virus

Viruses

Chemical packages rather than living things, viruses are about one-hundredth the size of bacteria and consist of genetic material, either DNA or RNA, surrounded by a protein coat. Once inside the body, a virus invades a cell, and hijacks its metabolism. It uses this to make copies of itself, which break out and move on to infect other cells. Viral diseases include colds and polio.

Bacteria

The smallest living organisms, bacteria, are found in their trillions living harmlessly in the soil, air, and in water, as well as in or on the human body. But some bacteria – known more commonly as germs – are pathogenic, and cause diseases such as cholera, diphtheria, whooping cough, tuberculosis (TB), and typhoid. Pathogenic bacteria thrive at body temperature, and reproduce by splitting in two about once every 20 minutes. They damage the body by releasing poisons called toxins.

FUNGI

Many fungi are helpful to humans, such as mushrooms or moulds used to produce antibiotics, which feed on dead material. But some fungi are parasitic on humans, including the fungus whose filaments (above) feed on skin and cause athlete's foot (right), and the yeast that causes candidiasis (thrush).

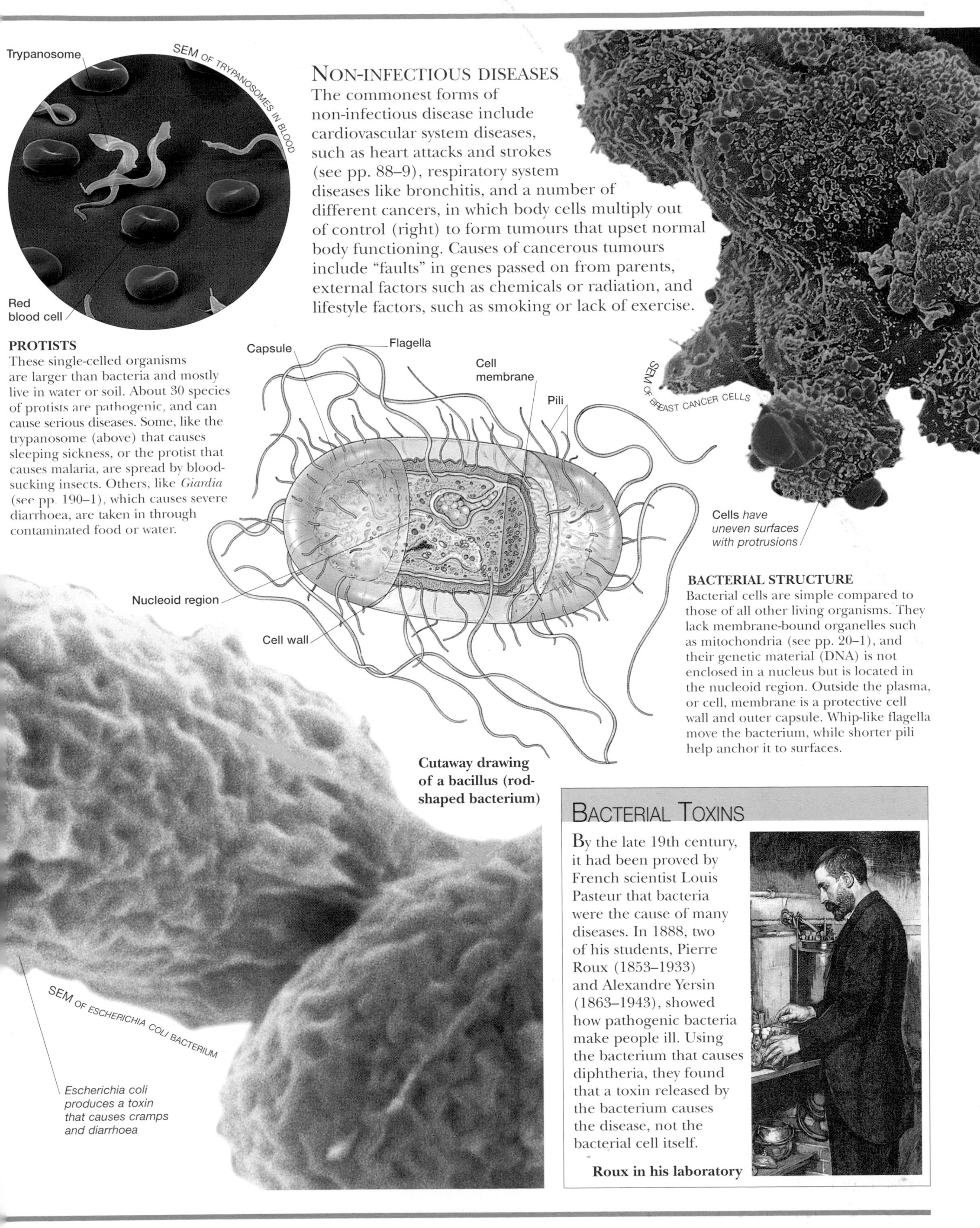

NON-INFECTIOUS DISEASES

The commonest forms of non-infectious disease include cardiovascular system diseases, such as heart attacks and strokes (see pp. 88–9), respiratory system diseases like bronchitis, and a number of different cancers, in which body cells multiply out of control (right) to form tumours that upset normal body functioning. Causes of cancerous tumours include "faults" in genes passed on from parents, external factors such as chemicals or radiation, and lifestyle factors, such as smoking or lack of exercise.

PROTISTS

These single-celled organisms are larger than bacteria and mostly live in water or soil. About 30 species of protists are pathogenic, and can cause serious diseases. Some, like the trypanosome (above) that causes sleeping sickness, or the protist that causes malaria, are spread by blood-sucking insects. Others, like *Giardia* (see pp. 190–1), which causes severe diarrhoea, are taken in through contaminated food or water.

Cutaway drawing of a bacillus (rod-shaped bacterium)

BACTERIAL STRUCTURE

Bacterial cells are simple compared to those of all other living organisms. They lack membrane-bound organelles such as mitochondria (see pp. 20–1), and their genetic material (DNA) is not enclosed in a nucleus but is located in the nucleoid region. Outside the plasma, or cell, membrane is a protective cell wall and outer capsule. Whip-like flagella move the bacterium, while shorter pili help anchor it to surfaces.

BACTERIAL TOXINS

By the late 19th century, it had been proved by French scientist Louis Pasteur that bacteria were the cause of many diseases. In 1888, two of his students, Pierre Roux (1853–1933) and Alexandre Yersin (1863–1943), showed how pathogenic bacteria make people ill. Using the bacterium that causes diphtheria, they found that a toxin released by the bacterium causes the disease, not the bacterial cell itself.

Roux in his laboratory

Non-specific defences

THROUGHOUT LIFE, the body is exposed to an array of infectious pathogens that would, if left unchecked, invade, exploit, and ultimately destroy it. Fortunately, the body has a highly sophisticated defence system, which consists of two main parts: the non-specific defences described here, and the immune system (see pp. 160–1). Non-specific defences are built into the body at birth, and respond in the same way to all invading pathogens. First, skin and other outer defences present a physical barrier. Then, if pathogens breach this, a system of defensive cells and anti-microbial chemicals in blood and tissue fluids springs into action.

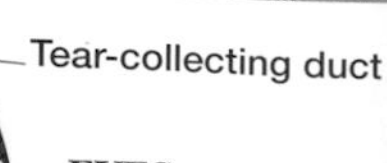

EYES
Tears produced by glands in the eyelids wash away dirt and contain an antiseptic substance.

RESPIRATORY TRACT
Mucus is produced in the lining of the respiratory tract. It traps pathogens and is carried to the throat by cilia.

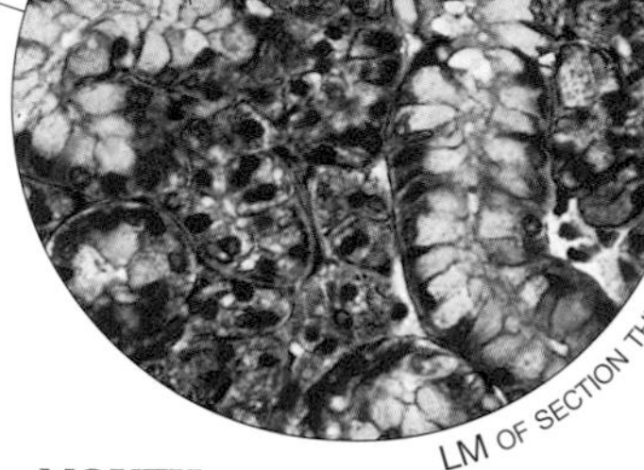

MOUTH
Watery saliva, produced by glands around the mouth, washes the mouth out, and contains bacteria-killing lysozyme.

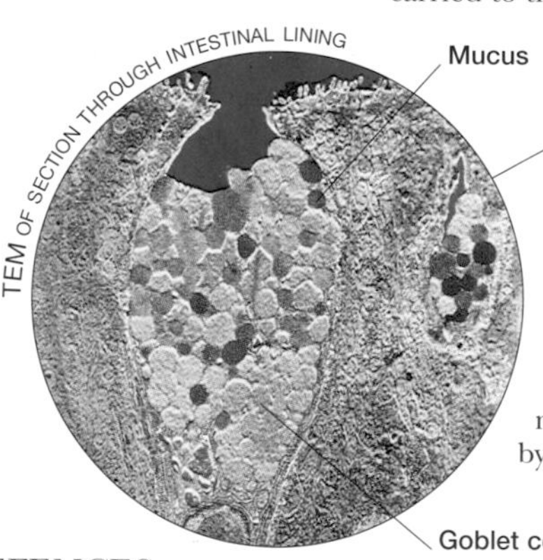

INTESTINES
The lining of the intestines is protected from harmful organisms and chemicals by mucus produced by goblet cells.

STOMACH
Glands in the stomach lining produce hydrochloric acid which kills most invading organisms.

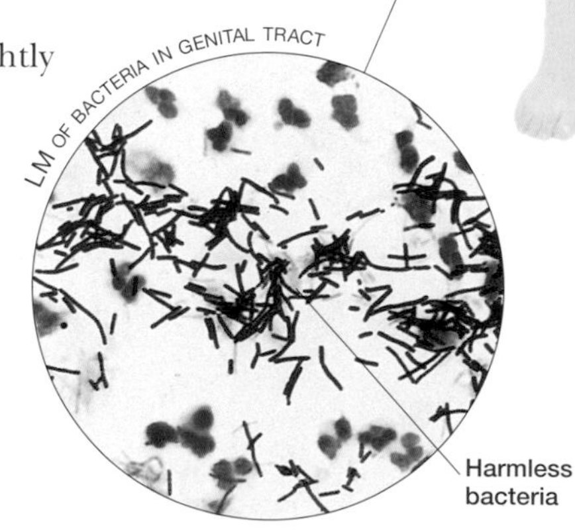

GENITAL AND URINARY TRACTS
The expulsion of urine in the urinary tract and the presence of harmless bacteria in the genital tract prevent harmful organisms from multiplying in these areas.

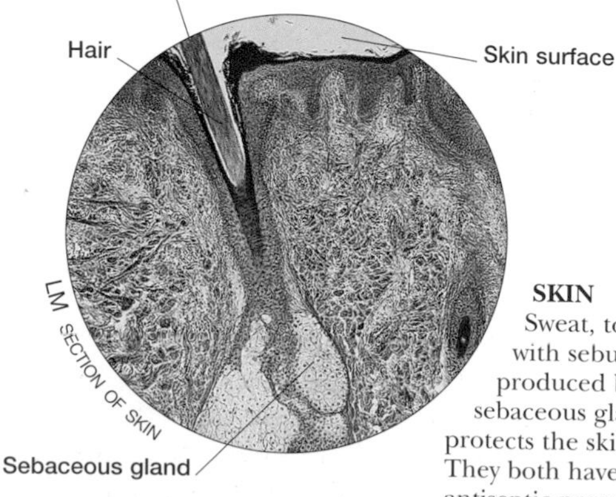

SKIN
Sweat, together with sebum produced by sebaceous glands, protects the skin. They both have mildly antiseptic properties.

OUTER DEFENCES

This diagram shows the main parts of the body's outer defences. The tightly knit cells of the skin, and those of the mucous membrane lining the respiratory, digestive, urinary, and reproductive systems, help stop pathogens entering the tissues. Tears contain a bacteria-killing substance called lysozyme, as do saliva and sweat. Mucus lining the respiratory system traps pathogens, then passes them to the throat for swallowing. Stomach acid destroys most swallowed bacteria. On the skin, and in the digestive and female reproductive system, colonies of harmless bacteria prevent harmful bacteria from establishing themselves.

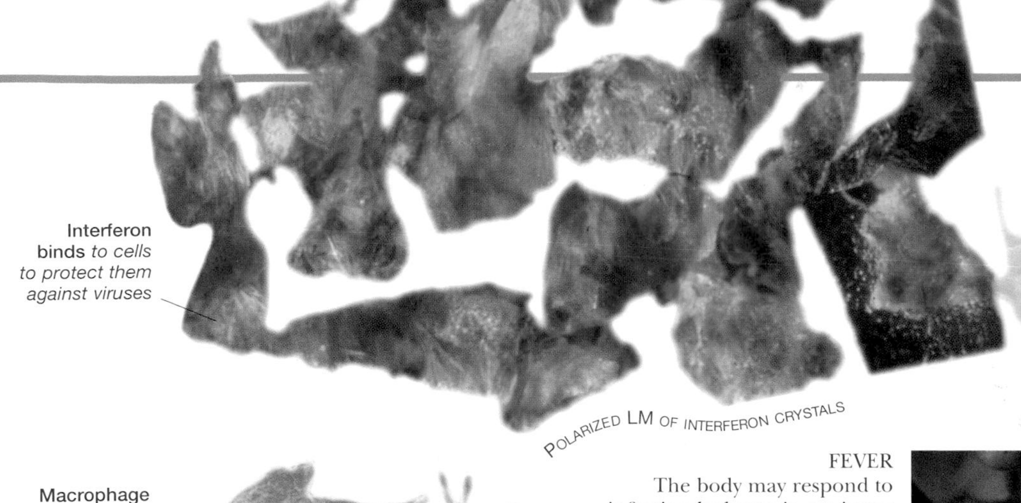

Interferon binds *to cells to protect them against viruses*

POLARIZED LM OF INTERFERON CRYSTALS

ANTIMICROBIAL SUBSTANCES
Two sets of blood proteins – interferon and the complement system – form a key part of the defence force. Interferon is released by body cells already infected with viruses. It stimulates neighbouring cells to protect themselves against viral infection. The complement system of over 20 proteins aids the inflammation process (see below). The proteins attach themselves to bacteria either to make them more "tasty" for phagocytes to eat, or to destroy them by making the cell membranes burst open.

FEVER
The body may respond to infection by bacteria or viruses by raising its temperature above the normal 37°C (98.6°F) in order to stop the invaders multiplying. This strategy, called fever or pyrexia, is often accompanied by sweating, shivering, and a feeling of thirst. It is triggered by white blood cells that release chemicals called pyrogens which reset the body's "thermostat" in the brain's hypothalamus so that body temperature rises above normal.

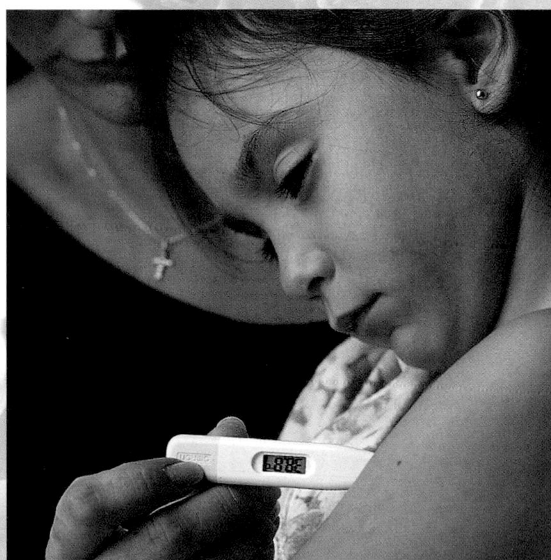

Macrophage *tracks down pathogen*

Extension of macrophage *engulfs protist prior to digestion*

Parasitic protist *that causes the tropical disease leishmaniasis*

SEM OF MACROPHAGE ENGULFING PROTIST

CELL EATERS

Phagocytes are white blood cells that engulf and destroy invading organisms. There are two types: neutrophils and macrophages. Neutrophils circulate in the blood before being transferred to the tissues, where they seek out organisms. Macrophages also start in the bloodstream, where they are known as monocytes, before moving to the tissues, where they become macrophages. Some macrophages stay in one place; others travel around looking for infective organisms. All phagocytes flow around organisms, wrapping them within a membrane that fuses with granules called lysosomes. These granules contain strong chemicals that digest the organisms, producing harmless substances that pass out through the cells' outer membrane.

Chemicals *that attract white blood cells*

Pathogen

Injured skin

Damage

White blood cells *engulf bacteria*

Inflamed tissue

Blood vessel *widens*

Response

INFLAMMATORY RESPONSE
This is the familiar warm, reddish, tender swelling that appears after an injury. At the site of the damage, tissue cells release histamine and other chemicals. They make the blood vessels wider, so that extra blood arrives, and more leaky, so that fluid passes into the tissues, aiding repair and making the area red, swollen, and warm. These chemicals also attract phagocytes that destroy pathogens.

RESPIRATORY system

While the body can do without food or water for a short time, it cannot survive without a continuous supply of oxygen. Its trillions of cells relentlessly consume oxygen in order to release from sugars the energy to power their activities This process – called cell, or internal, respiration – also produces waste carbon dioxide. The body's oxygen supply is provided by the respiratory system, which draws air into the body, transfers its oxygen to the bloodstream, then pushes air out, expelling unwanted carbon dioxide.

SEM OF CILIA INSIDE NASAL CAVITY

NASAL CAVITY
The entrance to the respiratory system is through the nasal cavity, the hollow space behind the nose. Air contains dust and dirt particles that could damage the lungs if they got that far. Fortunately, the nose and nasal cavity provide a filtration service. Hairs guarding the nostrils remove larger particles as air is inhaled (breathed in). Then dust is trapped by sticky mucus, secreted by the nasal cavity lining, which is then moved by cilia to the back of the throat for swallowing. Inhaled air is moistened and warmed as it passes through the nasal cavity.

Lungs and airways

The respiratory system consists of the lungs and the airways, or tubes – nose, pharynx (throat), larynx (voice box), trachea (windpipe), and bronchi – that carry air between the lungs and the outside atmosphere. Inhaled air travels along the nose, pharynx, larynx, and trachea before entering one of two branches – the bronchi – that enter both lungs. Inside the lungs, bronchi divide into smaller and smaller branches that finally end in sac-like alveoli where oxygen and carbon dioxide are exchanged. Exhaled air returns in the opposite direction.

ENDOSCOPIC VIEW OF TRACHEA

SEM OF LINING OF TRACHEA

Air *passes through the nasal cavity, being warmed, moistened, and cleaned as it does so*

Hairs *in nostrils filter out large particles*

Right lung

Cilia *projecting from cells lining the trachea move rhythmically to waft dust-laden mucus up to the throat so it can be swallowed or spat out*

TRACHEA
The trachea (windpipe) is the flexible tube through which air travels from the larynx towards the lungs. At its lower end, it divides into two main bronchi, one for each lung. The trachea is held open by C-shaped rings of cartilage embedded in its walls. The lining of the trachea continues the work of the nasal cavity in removing dust and dirt from air.

RESPIRATORY SYSTEM FUNCTIONS

Ventilation	*Muscle contractions alter the volume of the chest, drawing air along the respiratory tract and into and out of the lungs.*
External respiration	*Within the lungs, oxygen diffuses from the air into the bloodstream, and carbon dioxide diffuses in the opposite direction.*
Internal respiration	*Throughout the body, oxygen diffuses out of the blood into cells, where it is used in the chemical processes that release energy. Carbon dioxide diffuses in the opposite direction.*

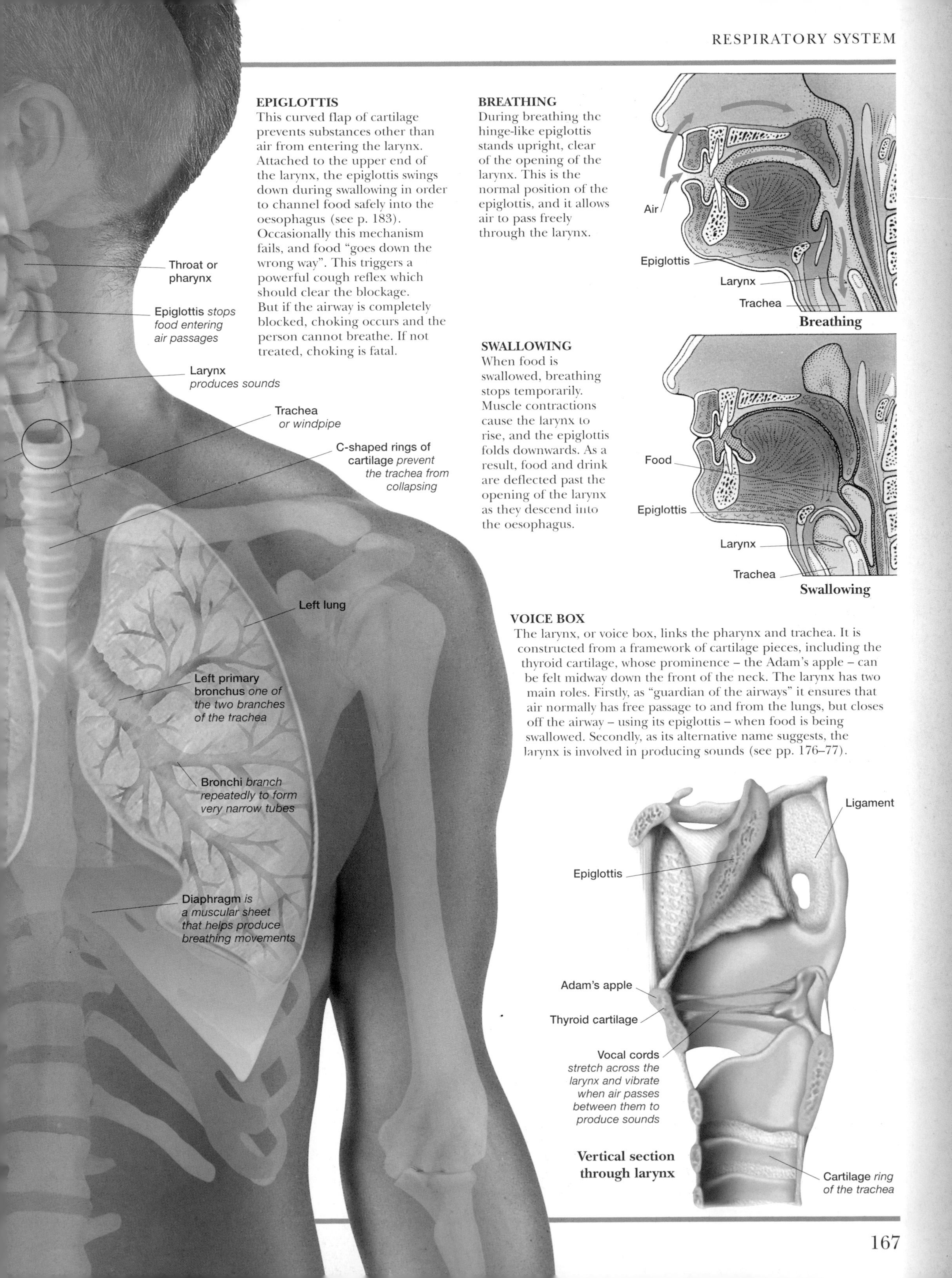

Throat or pharynx
Epiglottis stops food entering air passages
Larynx produces sounds
Trachea or windpipe
C-shaped rings of cartilage prevent the trachea from collapsing
Left lung
Left primary bronchus one of the two branches of the trachea
Bronchi branch repeatedly to form very narrow tubes
Diaphragm is a muscular sheet that helps produce breathing movements
Air
Epiglottis
Larynx
Trachea
Food
Epiglottis
Larynx
Trachea
Ligament
Epiglottis
Adam's apple
Thyroid cartilage
Vocal cords stretch across the larynx and vibrate when air passes between them to produce sounds
Cartilage ring of the trachea

EPIGLOTTIS

This curved flap of cartilage prevents substances other than air from entering the larynx. Attached to the upper end of the larynx, the epiglottis swings down during swallowing in order to channel food safely into the oesophagus (see p. 183). Occasionally this mechanism fails, and food "goes down the wrong way". This triggers a powerful cough reflex which should clear the blockage. But if the airway is completely blocked, choking occurs and the person cannot breathe. If not treated, choking is fatal.

BREATHING

During breathing the hinge-like epiglottis stands upright, clear of the opening of the larynx. This is the normal position of the epiglottis, and it allows air to pass freely through the larynx.

Breathing

SWALLOWING

When food is swallowed, breathing stops temporarily. Muscle contractions cause the larynx to rise, and the epiglottis folds downwards. As a result, food and drink are deflected past the opening of the larynx as they descend into the oesophagus.

Swallowing

VOICE BOX

The larynx, or voice box, links the pharynx and trachea. It is constructed from a framework of cartilage pieces, including the thyroid cartilage, whose prominence – the Adam's apple – can be felt midway down the front of the neck. The larynx has two main roles. Firstly, as "guardian of the airways" it ensures that air normally has free passage to and from the lungs, but closes off the airway – using its epiglottis – when food is being swallowed. Secondly, as its alternative name suggests, the larynx is involved in producing sounds (see pp. 176–77).

Vertical section through larynx

Lungs

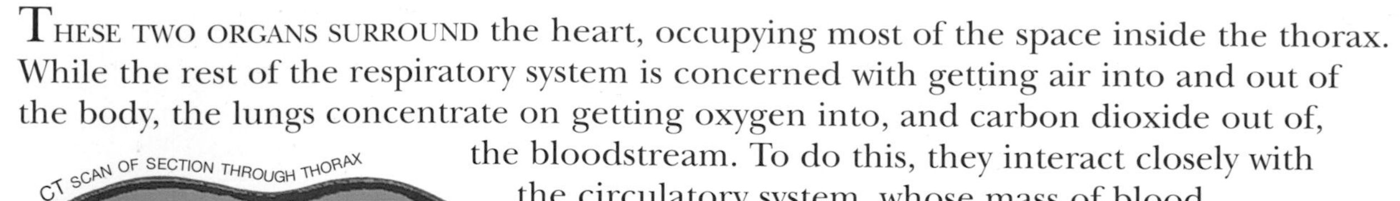

THESE TWO ORGANS SURROUND the heart, occupying most of the space inside the thorax. While the rest of the respiratory system is concerned with getting air into and out of the body, the lungs concentrate on getting oxygen into, and carbon dioxide out of, the bloodstream. To do this, they interact closely with the circulatory system, whose mass of blood vessels gives the lungs their pinky-red colour. The lungs have a spongy feel as a result of their internal structure – a system of air-filled, progressively branching tubes terminating in microscopic "air bags" through which oxygen enters the blood and waste carbon dioxide leaves it.

CT SCAN OF SECTION THROUGH THORAX

Heart

Lung

The right lung is separated into three lobes, the left lung into two

LUNG STRUCTURE

The lungs are light, spongy structures that are approximately conical in shape. The uppermost part of each lung extends above the clavicle (collar bone) into the neck, and their bases rest on the diaphragm. Each lung is divided into separate portions called lobes: the right lung consists of three lobes, whereas the left lung, which is slightly smaller in order to make space for the heart, consists of two. Surrounding the lungs are two membranes called pleura, between which lies a thin layer of pleural fluid which ensures that the lungs expand and shrink smoothly with each breath. The ribcage protects the lungs, and its muscles assist in breathing.

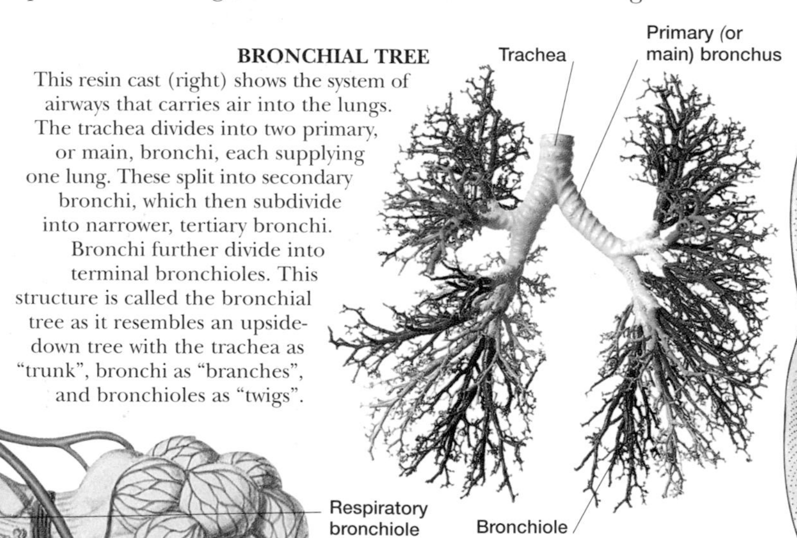

BRONCHIAL TREE
This resin cast (right) shows the system of airways that carries air into the lungs. The trachea divides into two primary, or main, bronchi, each supplying one lung. These split into secondary bronchi, which then subdivide into narrower, tertiary bronchi. Bronchi further divide into terminal bronchioles. This structure is called the bronchial tree as it resembles an upside-down tree with the trachea as "trunk", bronchi as "branches", and bronchioles as "twigs".

ALVEOLAR SACS
As terminal bronchioles penetrate more deeply into the lungs, they divide into microscopic respiratory bronchioles. These lead into blind-ending alveolar sacs resembling bunches of grapes. Each "grape", or alveolus, shares with other alveoli an opening into a duct connecting it to the respiratory bronchiole. Alveoli are the site of gas exchange.

Alveoli *surrounded by blood capillaries*

Tertiary bronchus *branches from the secondary bronchus, which subdivides repeatedly to form terminal bronchioles*

Secondary bronchus *– there are three secondary bronchi in the right lung, each supplying one lobe of the lung*

Right primary bronchus

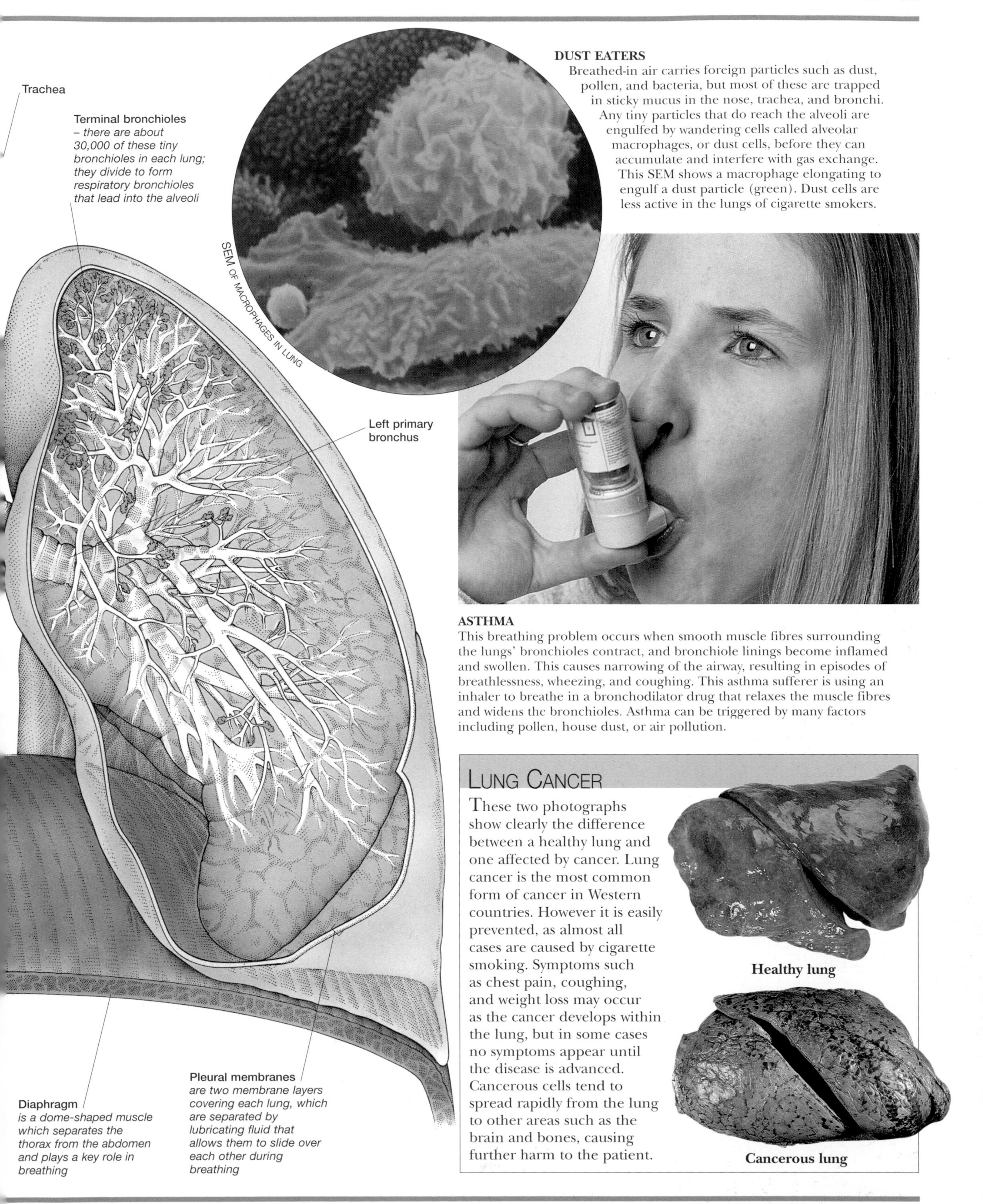
Trachea
Terminal bronchioles – there are about 30,000 of these tiny bronchioles in each lung; they divide to form respiratory bronchioles that lead into the alveoli
SEM OF MACROPHAGES IN LUNG
Left primary bronchus
Diaphragm is a dome-shaped muscle which separates the thorax from the abdomen and plays a key role in breathing
Pleural membranes are two membrane layers covering each lung, which are separated by lubricating fluid that allows them to slide over each other during breathing
Healthy lung
Cancerous lung

DUST EATERS

Breathed-in air carries foreign particles such as dust, pollen, and bacteria, but most of these are trapped in sticky mucus in the nose, trachea, and bronchi. Any tiny particles that do reach the alveoli are engulfed by wandering cells called alveolar macrophages, or dust cells, before they can accumulate and interfere with gas exchange. This SEM shows a macrophage elongating to engulf a dust particle (green). Dust cells are less active in the lungs of cigarette smokers.

ASTHMA

This breathing problem occurs when smooth muscle fibres surrounding the lungs' bronchioles contract, and bronchiole linings become inflamed and swollen. This causes narrowing of the airway, resulting in episodes of breathlessness, wheezing, and coughing. This asthma sufferer is using an inhaler to breathe in a bronchodilator drug that relaxes the muscle fibres and widens the bronchioles. Asthma can be triggered by many factors including pollen, house dust, or air pollution.

Lung Cancer

These two photographs show clearly the difference between a healthy lung and one affected by cancer. Lung cancer is the most common form of cancer in Western countries. However it is easily prevented, as almost all cases are caused by cigarette smoking. Symptoms such as chest pain, coughing, and weight loss may occur as the cancer develops within the lung, but in some cases no symptoms appear until the disease is advanced. Cancerous cells tend to spread rapidly from the lung to other areas such as the brain and bones, causing further harm to the patient.

Gas exchange

EVERY MINUTE, LARGE AMOUNTS OF life-giving oxygen is taken into the bloodstream, while potentially poisonous carbon dioxide is expelled. This happens through a mechanism called gas exchange, which occurs in the lungs' tiny sac-like alveoli. Two features of alveoli make gas exchange fast and efficient. Firstly, the wall of an alveolus is just one cell thick, as is the wall of the blood capillaries that surround it. Where the two meet, they form a respiratory membrane just 0.0005 mm (0.00002 in) wide, across which oxygen can move rapidly into, and carbon dioxide out of, the blood. Secondly, the two lungs contain some 300 million alveoli that collectively provide a surface area for gas exchange of 70 sq m (750 sq ft) – 35 times the surface area of the skin – squeezed into a space inside the chest that is no bigger than a shopping bag.

Before diffusion **After diffusion**

DIFFUSION
The natural tendency of molecules to move randomly from an area of high concentration to one of low until evenly spread out is called diffusion. Some molecules (above, red) can diffuse through cell membranes (green). This is exactly what happens to oxygen during gas exchange in the alveoli, except that breathing brings more oxygen into the alveoli, while blood capillaries carry it away, never allowing it to be evenly distributed.

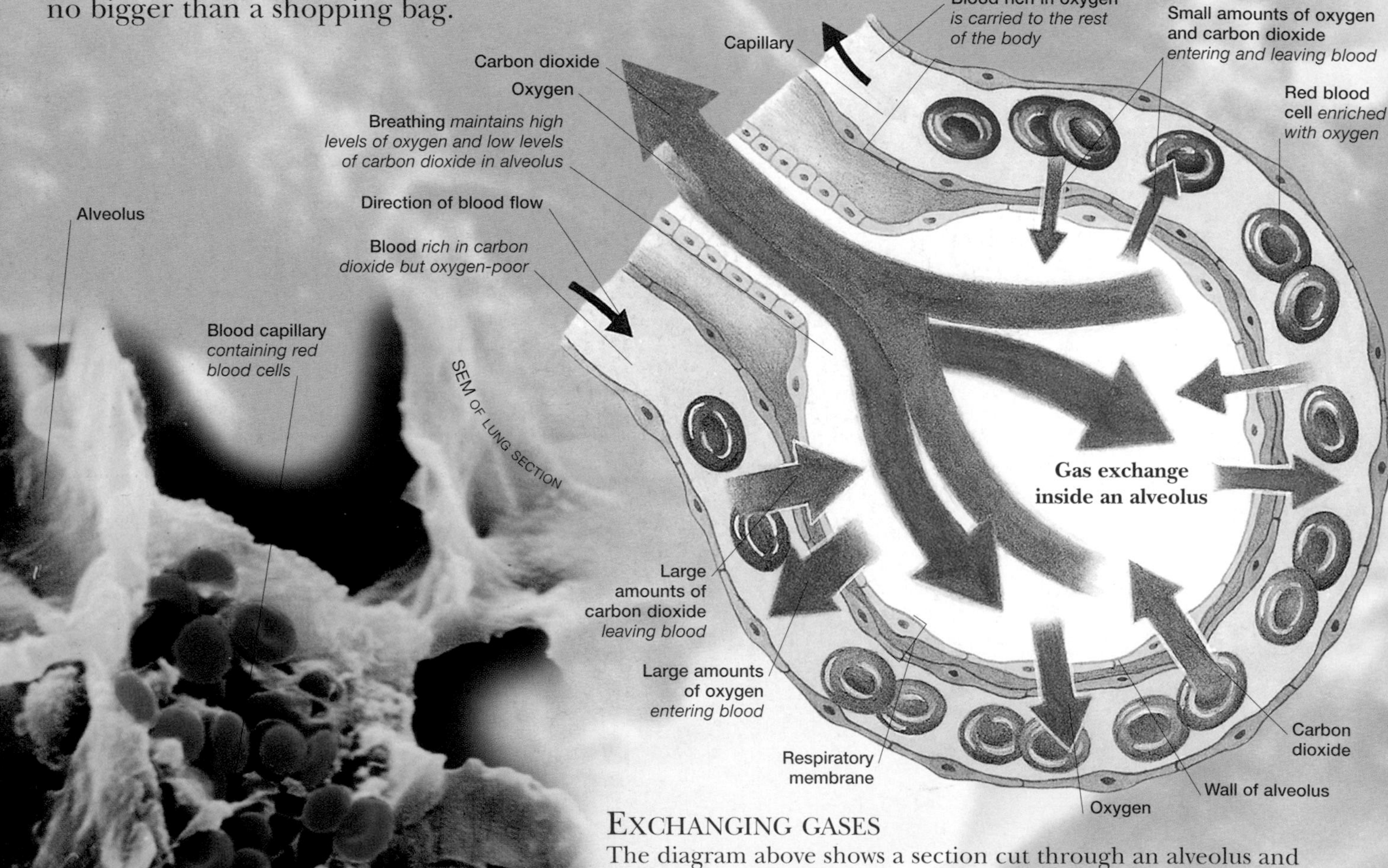

EXCHANGING GASES

The diagram above shows a section cut through an alveolus and the blood capillary that surrounds it. Oxygen in inhaled air diffuses across the thin respiratory membrane, and passes into red blood cells. Inhalation replenishes oxygen supplies in the alveolus, while blood flow removes oxygen-enriched blood. This creates a diffusion "gradient" across the respiratory membrane – high levels of oxygen in the alveolus, low levels in the blood – that ensures a constant flow of oxygen into the bloodstream. The same applies to carbon dioxide, but in the opposite direction. Carbon dioxide diffuses from newly arrived blood into the alveolus, where levels of carbon dioxide are low because it is continually exhaled.

Blood supply

This angiogram of a lung shows a pulmonary artery and its branches. The left and right pulmonary arteries carry the dark red, oxygen-poor blood that is pumped by the heart into the two lungs. Then, they follow the bronchi and bronchioles, branching repeatedly into smaller and smaller vessels until they form the dense networks of capillaries that surround and cling to the alveoli. As they leave the alveoli, these capillaries merge to form progressively larger and larger veins that empty their bright red, oxygen-rich blood into the two pulmonary veins that return blood from each lung to the heart for distribution around the body.

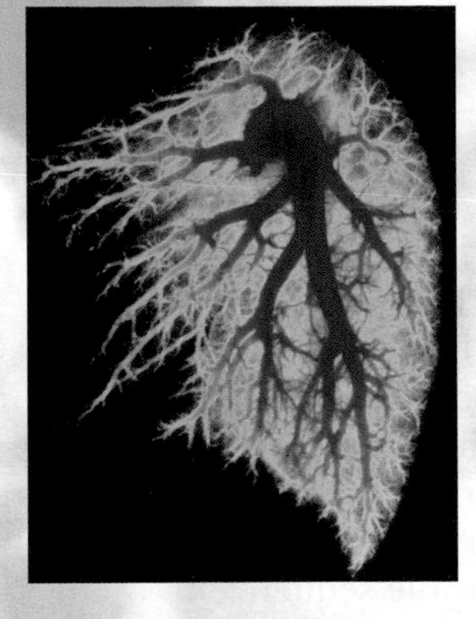

Supplying cells

The final stage of the respiratory process involves the delivery of oxygen to, and the removal of carbon dioxide from, the body's trillions of cells. During gas exchange in the tissues, oxygen diffuses from blood carried by capillaries into the surrounding cells, while carbon dioxide diffuses from tissue cells into the blood.

Respiratory bronchiole

Air space inside alveolus

Wall of alveolus

Capillary

Oxygen diffuses *from the alveolus and binds with haemoglobin in red blood cells*

Carbon dioxide diffuses *from blood plasma into the alveolus*

Gas exchange in the lungs

Oxygen-rich blood *travels from the lungs to the tissues via the heart*

Oxygen-poor blood *returns from the tissues to the lungs via the heart*

Fluid *between cells*

Tissue cell

Blood plasma

Red blood cell

Capillary

Carbon dioxide *diffuses from tissue cells into a capillary and dissolves in plasma, the liquid part of blood*

Oxygen *is unloaded from red blood cells and diffuses from the capillary into tissue cells*

Gas exchange in the tissues

The Bends

As a diver descends underwater, the pressure on their body rises steadily because of the increasing weight of water pushing down on them. Under these conditions, nitrogen in inhaled air dissolves in the blood. If a diver then ascends to the surface too rapidly, the nitrogen comes out of solution in the blood and forms bubbles, like those that appear in a bottle of fizzy drink when the cap is unscrewed. The result is decompression sickness, or the "bends", as nitrogen bubbles cause excruciating pains in joints or muscles, or affect brain function, sometimes with fatal results.

DIGESTIVE system

DURING THEIR lifetime, the average person eats their way through at least 20 tonnes of food. The job of the digestive system is to turn this mountain of nourishment into substances that the body can use, both for energy and for growth and repair. It works like an assembly line in reverse, turning complex nutrients into simpler ones, which the body can then absorb. Food contains three major kinds of nutrient – carbohydrates, fats, and proteins – and the digestive system deals with each kind in a different way. Once these and other nutrients have been extracted, the system gets rid of any undigested waste.

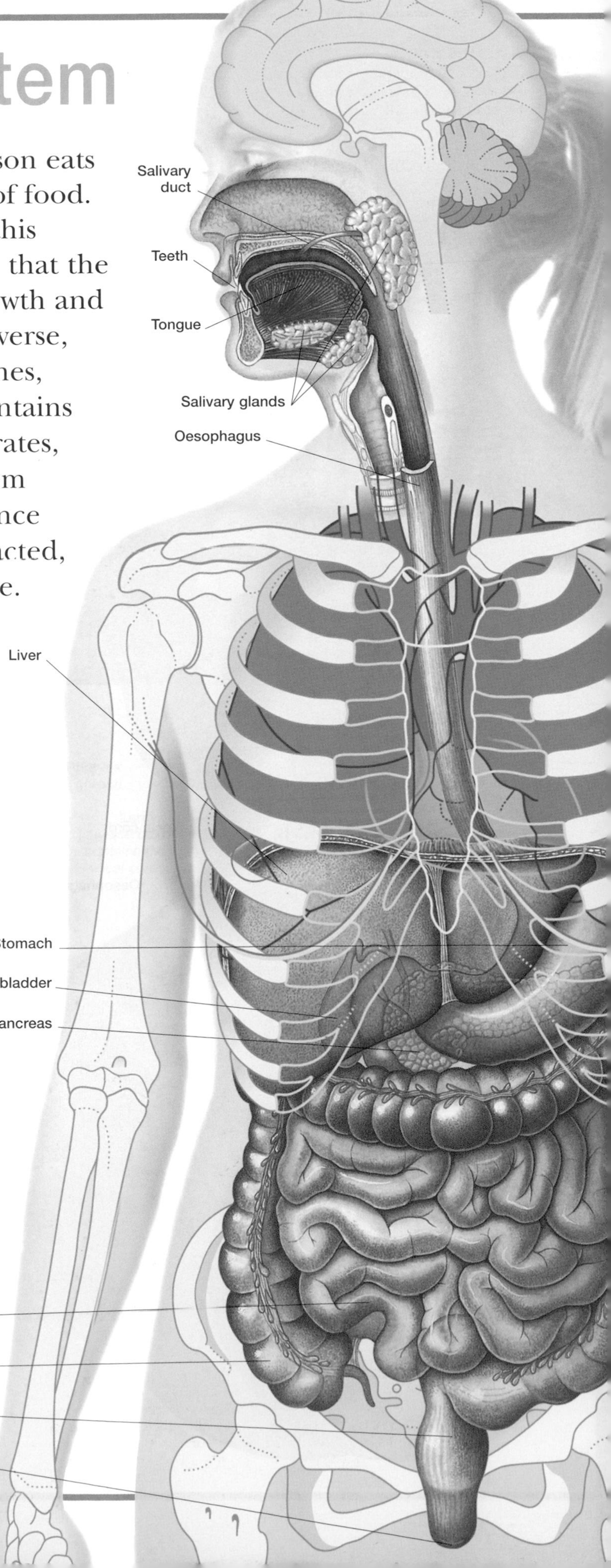

ESSENTIALS AND ACCESSORIES

The core of the digestive system is a long tube called the alimentary canal, or gastrointestinal tract. It runs from the mouth to the anus, and is divided into distinct regions – the oesophagus and stomach, and the small and large intestines – which carry out different tasks. Attached to the tube are a number of accessory organs that help in the process of digestion. They include the teeth, tongue, and salivary glands, as well as the liver, gall bladder, and pancreas. Cells lining the alimentary canal experience a lot of wear and tear, and they often have a working life of just three or four days.

CHEMICAL AND MECHANICAL
This computer-generated image shows a single molecule of pepsin – an enzyme, or chemical catalyst, that breaks proteins down into smaller units called peptides. Pepsin is produced in the stomach, and is one of more than a dozen enzymes that play a part in digestion. The mechanical side of digestion includes chewing and also muscular "churning" by the stomach and intestines. This helps enzymes to get at the substances that they break down.

Salivary duct
Teeth
Tongue
Salivary glands
Oesophagus
Liver
Stomach
Gall bladder
Pancreas
Small intestine
Large intestine
Rectum
Anus

FOOD ON THE MOVE

The alimentary canal operates by muscle power and works at a range of different speeds. Food is only briefly in the mouth and oesophagus, but once it reaches the stomach, it can stay there several hours. From here, food moves into the small intestine, where it is nudged through the lengthy twists and turns at about 1 cm (0.4 in) per minute. Once in the large intestine, it slows down again, particularly if the body is short of water. The times shown below are for a typical meal.

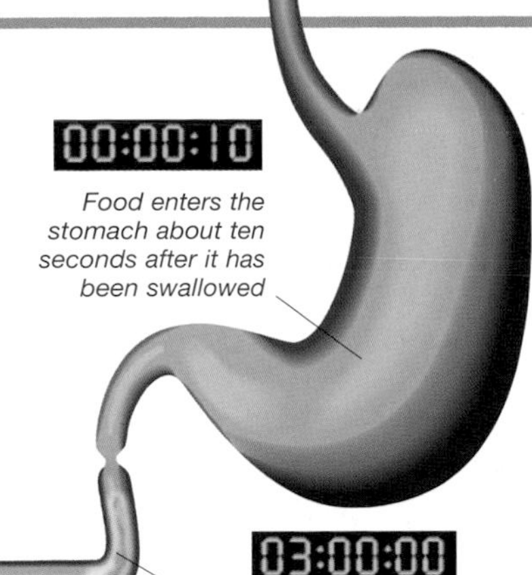

00:00:10

Food enters the stomach about ten seconds after it has been swallowed

03:00:00

If food contains only a small amount of fat, it leaves the stomach within three hours. Fatty or protein-rich food can stay in the stomach for twice as long as this

06:00:00

Semi-digested food, *called chyme, reaches the halfway point of the small intestine about three hours after it left the stomach. By now, many of its nutrients have been absorbed*

Small intestine

08:00:00

About eight hours after being swallowed, watery, indigestible waste completes its journey to the end of the small intestine

Ileocaecal sphincter
(beginning of large intestine)

Large intestine
has a 2 cm (0.8 in) layer of bacteria

20:00:00

By the time it reaches the midpoint of the large intestine, a large proportion of the waste's water has been removed and reabsorbed

During the 12 to 36 hours in the large intestine liquid waste is transformed into semi-solid faeces

32:00:00

Faeces *reach the rectum, the end of the large intestine, between 20 and 44 hours after swallowing*

DIGESTING FOOD

Although food is full of nutrients, most of them are complex molecules that the body cannot absorb. These have to be broken down into smaller and simpler chemicals, which can travel through the lining of the small intestine and into the body itself. These simple substances are formed by enzymes. Enzymes work like chemical scissors, cutting up the large molecules at specific points.

Polysaccharides (starch)

Salivary and pancreatic amylase

Disaccharides (maltose)

Maltase

Monosaccharides (glucose)

Carbohydrate digestion
Complex carbohydrates, such as starch, are broken down by enzymes in saliva and in the small intestine. An enzyme called amylase splits long starch molecules to produce maltose. Maltase then splits maltose molecules to produce glucose. Through this process, long carbohydrate molecules, or polysaccharides, are turned into monosaccharides, or simple sugars.

Protein digestion
The first step in protein digestion takes place in the stomach, where pepsin breaks down protein molecules into smaller units called peptides. In the small intestine, an enzyme called trypsin continues this work, while other enzymes, called peptidases, cut up the peptide molecules to produce individual amino acids.

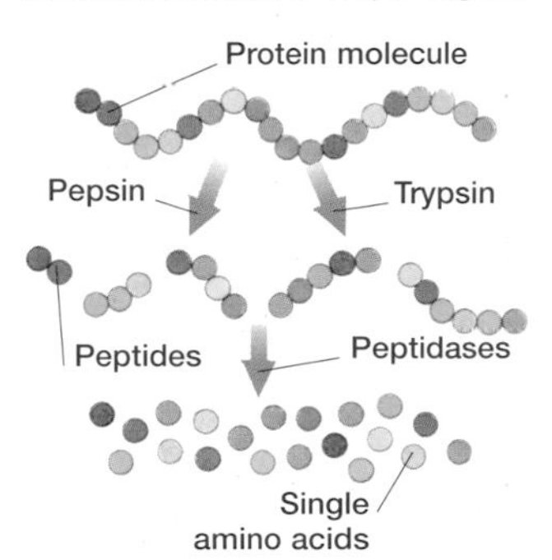

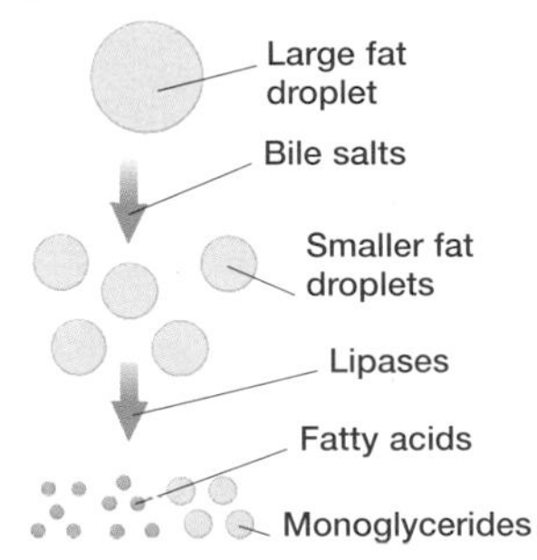

Fat digestion
Fats do not normally dissolve in water, but bile salts turn them into an emulsion of droplets. They are then digested by an enzyme called lipase, which produces fatty acids and monoglycerides. These travel to the sides of the small intestine in microscopic globules, called micelles.

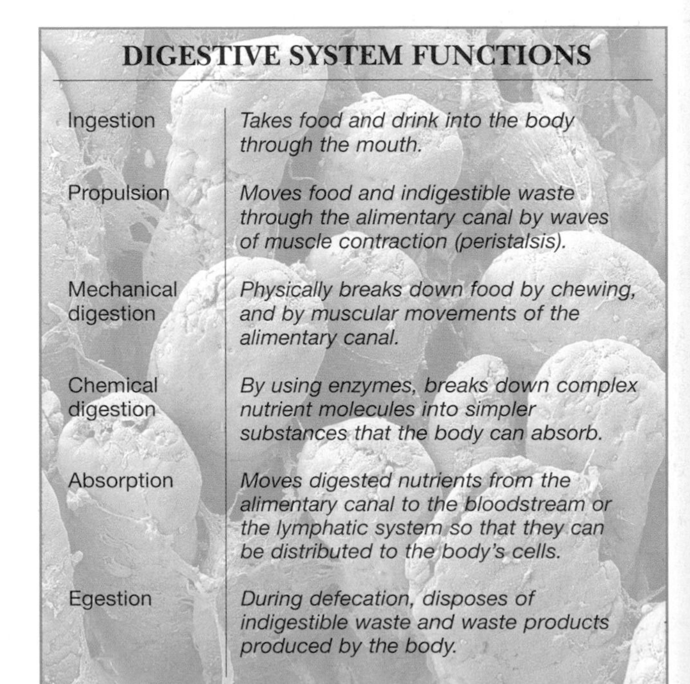

DIGESTIVE SYSTEM FUNCTIONS

Ingestion	*Takes food and drink into the body through the mouth.*
Propulsion	*Moves food and indigestible waste through the alimentary canal by waves of muscle contraction (peristalsis).*
Mechanical digestion	*Physically breaks down food by chewing, and by muscular movements of the alimentary canal.*
Chemical digestion	*By using enzymes, breaks down complex nutrient molecules into simpler substances that the body can absorb.*
Absorption	*Moves digested nutrients from the alimentary canal to the bloodstream or the lymphatic system so that they can be distributed to the body's cells.*
Egestion	*During defecation, disposes of indigestible waste and waste products produced by the body.*

Teeth

LOCATED AT THE ENTRANCE of the digestive system, teeth are the hardest objects in the body. They can withstand tremendous pressure when they bite – thanks not only to their extra-tough crowns, but also to their shock-absorbing roots. They cut, crush, and chew the food that we eat, making it easier both to swallow and digest. Humans have two sets of teeth, and in each, different teeth carry out different work. But teeth all share one important characteristic: once they have appeared, or "erupted", above the gums, their hard outer enamel cannot be repaired or replaced. Enamel can be damaged by acids from food and, if it is breached, the inner part of teeth can decay. But with regular cleaning and a healthy diet, adult teeth can last for life.

TYPES OF TEETH

When someone opens their mouth wide, the differences between individual teeth become easy to see. The incisors, at the front, are the only teeth that have a flat cross-section, with a single cutting edge. They take large chunks out of food, and slice it up. The canines, to each side of them, have a single point, for gripping and tearing. Behind them are the premolars and molars, which are used for chewing food, grinding it down into a paste. These teeth have two or four cusps, or raised edges, and because they are near the back of the jaw, they have an exceptionally powerful bite.

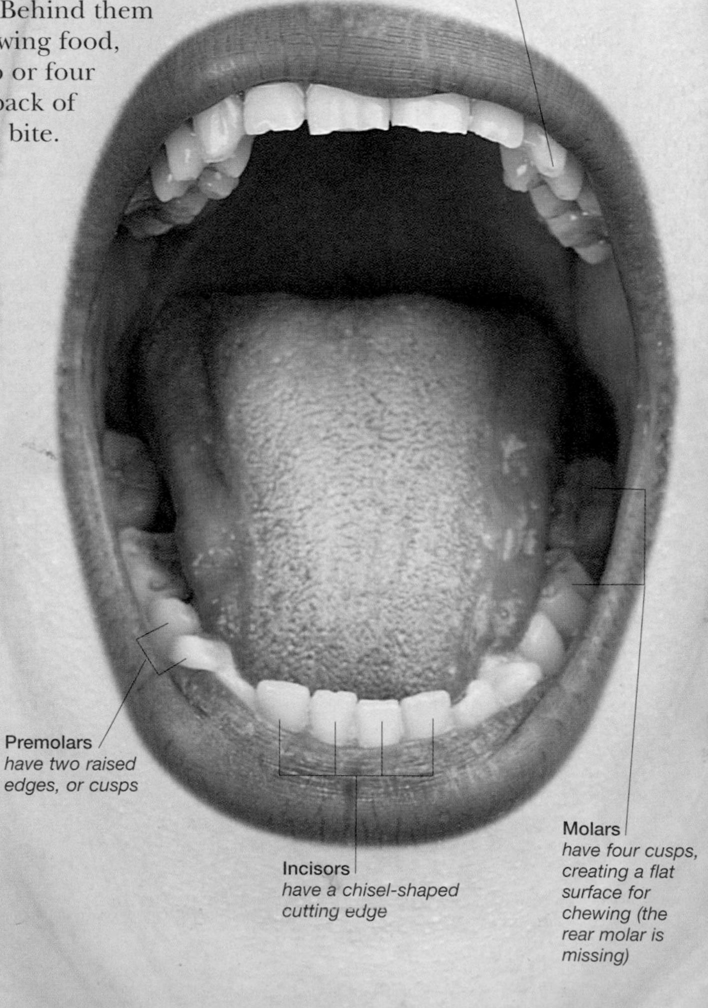

TWO SETS

Milk teeth, also known as deciduous teeth, have small crowns and relatively shallow roots. They begin to appear at about the age of six months, and are complete by about 32 months. From the age of about six years onwards, they are shed and replaced by adult or permanent teeth, which are larger, with longer roots. Most people have 20 milk teeth and 32 adult teeth. Milk teeth appear in a set order, starting with the central incisors and ending with the second molars. Adult teeth start appearing at the front of the jaw, and work backwards as the jaw grows. However, in some people, the third molars, or "wisdom teeth", remain embedded in the jaws.

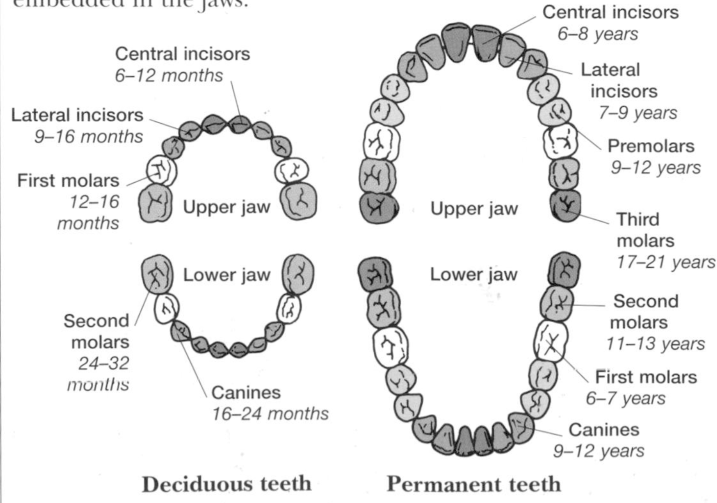

Deciduous teeth **Permanent teeth**

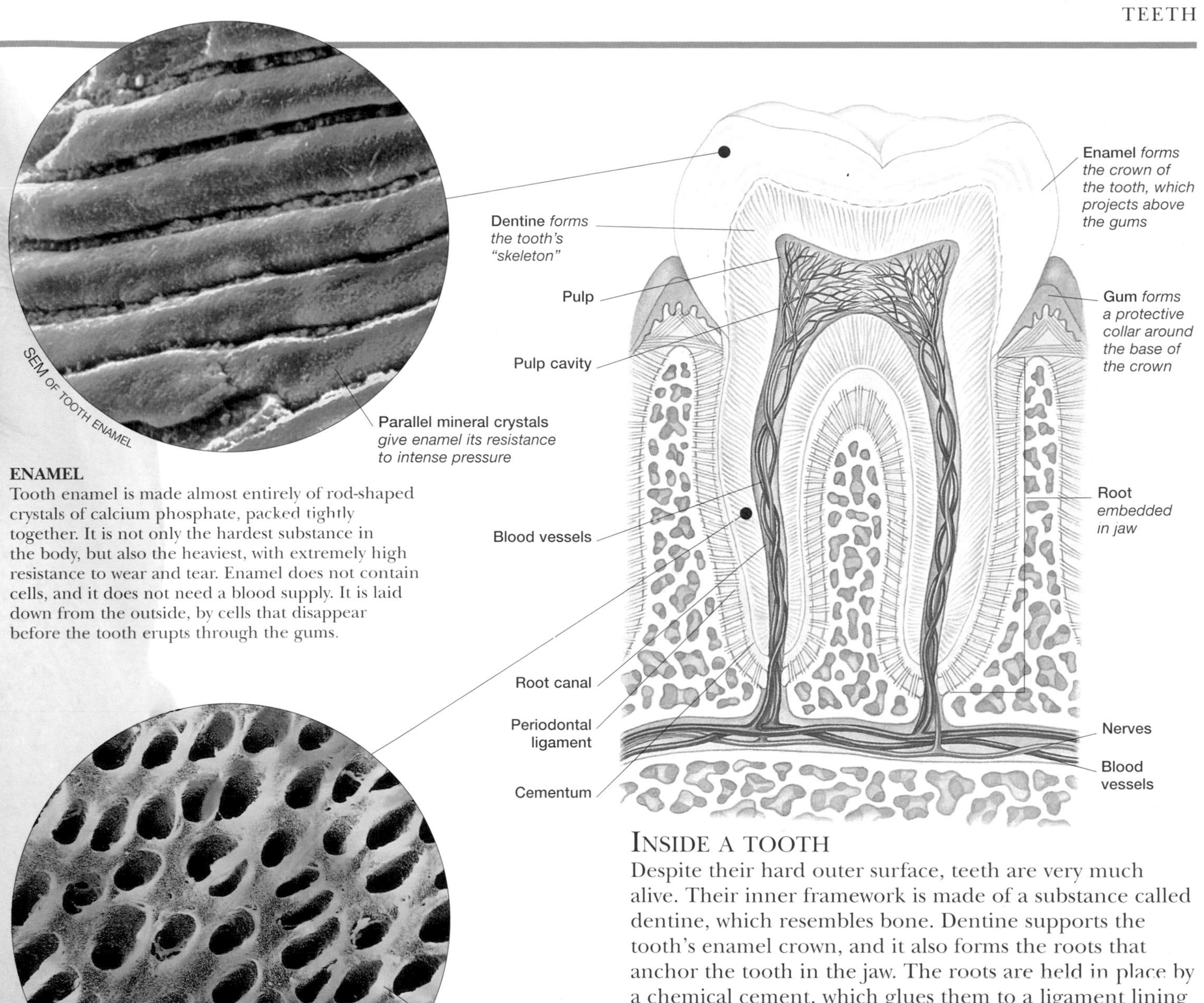
SEM OF TOOTH ENAMEL
Parallel mineral crystals give enamel its resistance to intense pressure
Dentine forms the tooth's "skeleton"
Pulp
Pulp cavity
Blood vessels
Root canal
Periodontal ligament
Cementum
Enamel forms the crown of the tooth, which projects above the gums
Gum forms a protective collar around the base of the crown
Root embedded in jaw
Nerves
Blood vessels
SEM OF DENTINE
Struts in dentine contain a higher proportion of minerals than in bone, making it harder

ENAMEL

Tooth enamel is made almost entirely of rod-shaped crystals of calcium phosphate, packed tightly together. It is not only the hardest substance in the body, but also the heaviest, with extremely high resistance to wear and tear. Enamel does not contain cells, and it does not need a blood supply. It is laid down from the outside, by cells that disappear before the tooth erupts through the gums.

Inside a tooth

Despite their hard outer surface, teeth are very much alive. Their inner framework is made of a substance called dentine, which resembles bone. Dentine supports the tooth's enamel crown, and it also forms the roots that anchor the tooth in the jaw. The roots are held in place by a chemical cement, which glues them to a ligament lining the socket. At the centre of the tooth is a natural cavity, which is full of living tissue, called pulp. This contains blood vessels and nerves, which reach the cavity through hollow root canals. The nerves enable teeth to sense changes in temperature, and, unfortunately, pain.

DENTINE

Compared to enamel, dentine has a more open structure, but it is still heavier and harder than most types of bone. When a tooth bites, it acts like crumple-resistant scaffolding, transmitting the force of the bite between the crown and the jaw. Unlike enamel, dentine is produced by cells in the pulp cavity, and it needs a blood supply to stay alive.

Bacteria *release acids when they break down the sugars in fragments of food*

PROBLEMS WITH PLAQUE

Magnified about 100 times, this photograph shows a layer of plaque on the surface of a tooth. Plaque is a mixture of bacteria and food that builds up on teeth that are not properly cleaned. Bacteria in plaque release acids as they feed, and these can eat through tooth enamel, creating holes that lead to the pulp cavity inside. If this is left untreated, the result is dental caries or tooth decay – an infection that destroys pulp cells, and dentine as well. Plaque can be hard to remove, because many bacteria produce a sticky "glue" that fastens them in place.

Chewing and swallowing

COMPARED WITH SOME ANIMALS, people are slow eaters. The human digestive system is not designed to cope with food in large chunks, so instead, it has to be chewed. Chewing grinds food down into a more manageable form, so that it can be swallowed. Chewing and swallowing happen almost without our noticing, but they both involve complicated movements and rapid reflexes. The tongue manoeuvres food into position between the teeth, while making sure that it does not get bitten itself. The teeth close with exactly the right amount of force, but they immediately stop if they touch something unexpectedly hard. Once the food has been reduced to a pulp, the tongue pushes it to the back of the throat. This triggers swallowing, which sends another mouthful on its way to the stomach.

LM OF A SECTION THROUGH A SALIVARY GLAND

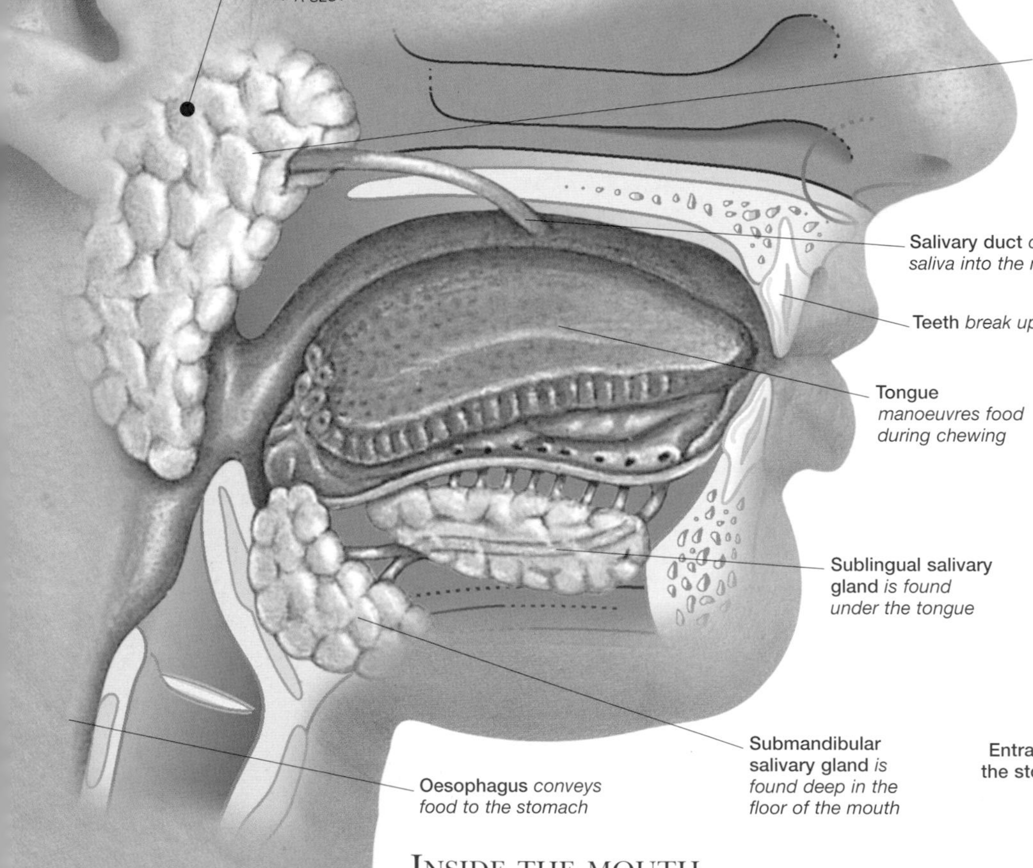

Parotid salivary gland *located in front of the ear*

Salivary duct *carries saliva into the mouth*

Teeth *break up food*

Tongue *manoeuvres food during chewing*

Sublingual salivary gland *is found under the tongue*

Submandibular salivary gland *is found deep in the floor of the mouth*

Oesophagus *conveys food to the stomach*

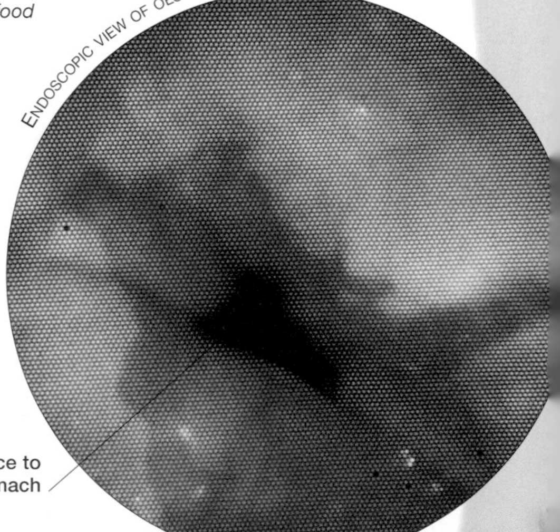

ENDOSCOPIC VIEW OF OESOPHAGUS

Entrance to the stomach

INSIDE THE MOUTH

The mouth is the reception centre of the digestive system, and the place where food gets its initial processing before being passed on. As soon as it arrives, food is given a rapid check by taste buds in the tongue, to make sure that it does not contain anything that might be dangerous. At the same time, the food is bathed in saliva from the three pairs of salivary glands, which moistens it so that it is easier to swallow. Saliva contains the enzyme amylase, and this starts to digest any starch that the food contains. As chewing begins, the lips and cheeks work with the tongue to help guide the food between the teeth.

INSIDE THE OESOPHAGUS

The lining of the oesophagus is coated with mucus, which helps food to slide through on its way to the stomach. Cells lining the oesophagus have tiny folds called microplicae, and these trap mucus, keeping the inner surface slippery. Unlike the trachea, or windpipe, the oesophagus is not reinforced by cartilage because it does not need to be open at all times. When it is not in use, its upper reaches – in the throat – are usually pressed flat.

Ready to swallow

Swallowing is a three-stage process, but only the first part, shown left, is under conscious control. Called the oral stage, it starts when a compact mass, or "bolus", of food has been thoroughly chewed. The tongue then presses against the roof of the mouth, pushing the bolus backwards towards the throat. Meanwhile, the soft palate lifts up, shutting off the opening that leads upwards into the nasal cavity. The bolus is now ready to enter the pharyngeal and oesophageal stages, when reflex actions take over.

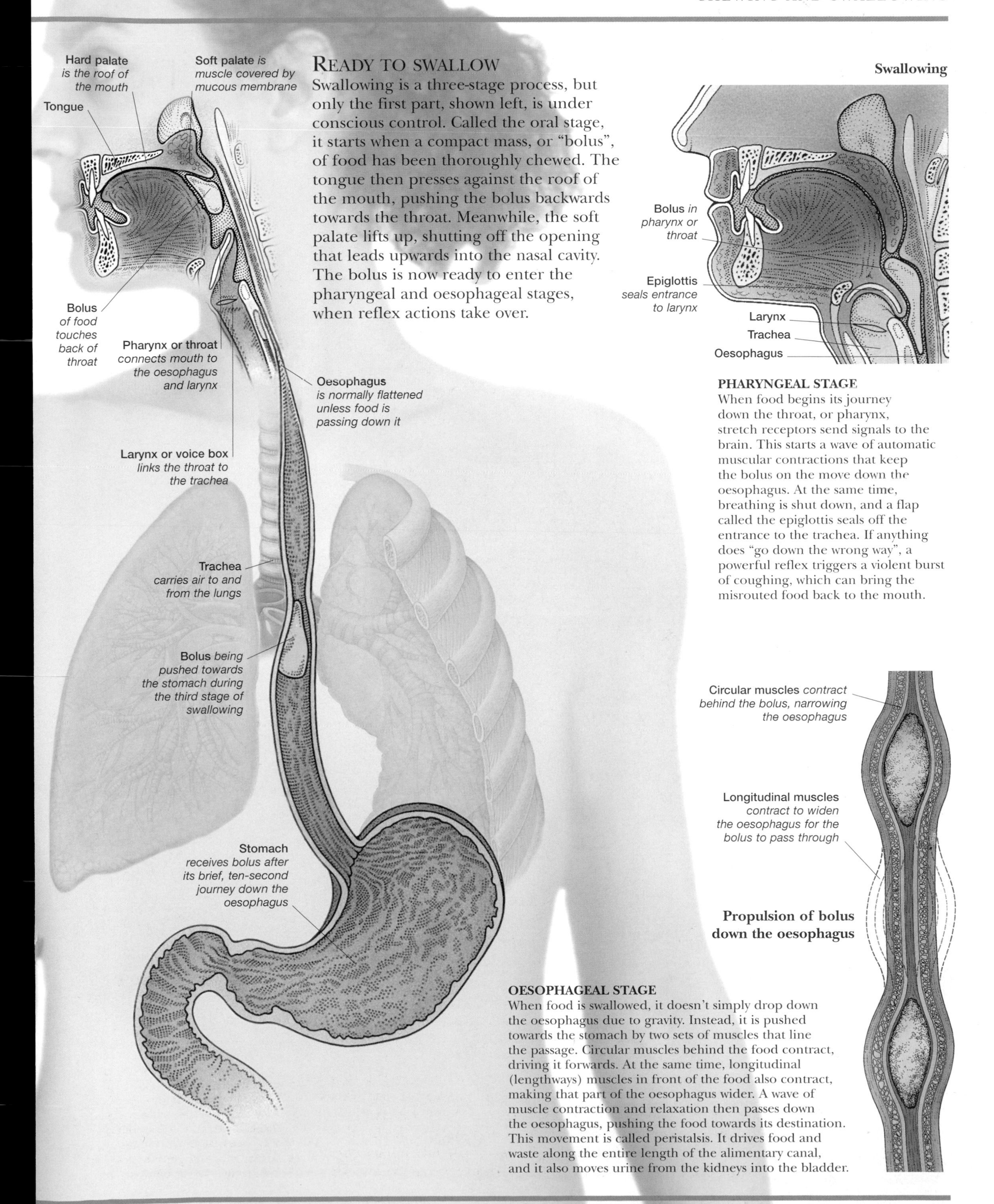

Hard palate *is the roof of the mouth*
Soft palate *is muscle covered by mucous membrane*
Tongue
Bolus *of food touches back of throat*
Pharynx or throat *connects mouth to the oesophagus and larynx*
Oesophagus *is normally flattened unless food is passing down it*
Larynx or voice box *links the throat to the trachea*
Trachea *carries air to and from the lungs*
Bolus *being pushed towards the stomach during the third stage of swallowing*
Stomach *receives bolus after its brief, ten-second journey down the oesophagus*
Swallowing
Bolus *in pharynx or throat*
Epiglottis *seals entrance to larynx*
Larynx
Trachea
Oesophagus
Circular muscles *contract behind the bolus, narrowing the oesophagus*
Longitudinal muscles *contract to widen the oesophagus for the bolus to pass through*
Propulsion of bolus down the oesophagus

PHARYNGEAL STAGE

When food begins its journey down the throat, or pharynx, stretch receptors send signals to the brain. This starts a wave of automatic muscular contractions that keep the bolus on the move down the oesophagus. At the same time, breathing is shut down, and a flap called the epiglottis seals off the entrance to the trachea. If anything does "go down the wrong way", a powerful reflex triggers a violent burst of coughing, which can bring the misrouted food back to the mouth.

OESOPHAGEAL STAGE

When food is swallowed, it doesn't simply drop down the oesophagus due to gravity. Instead, it is pushed towards the stomach by two sets of muscles that line the passage. Circular muscles behind the food contract, driving it forwards. At the same time, longitudinal (lengthways) muscles in front of the food also contract, making that part of the oesophagus wider. A wave of muscle contraction and relaxation then passes down the oesophagus, pushing the food towards its destination. This movement is called peristalsis. It drives food and waste along the entire length of the alimentary canal, and it also moves urine from the kidneys into the bladder.

Stomach

Of all the body's internal organs, the stomach is probably the best known but most misunderstood. This J-shaped bag is tucked beneath the ribs, and although it can stretch to fill with food, it does not absorb any of the nutrients that food contains. Instead, the stomach has two main functions: it gets digestion underway, storing any semi-digested food, and then releases it at a slow and steady rate. In the stomach, highly acidic gastric juice allows enzymes to break down proteins, while powerful waves of muscle contraction churn the food to mix it up. After several hours of this kind of treatment, the runny result – called chyme – is ready to move on.

Inside the stomach

The stomach is the widest and most elastic part of the alimentary canal. When it is empty, it can be smaller than a fist, but its volume can increase by more than 20 times after a meal, because the rugae (deep folds) of its inner surface become smoother as it fills. Unlike the rest of the alimentary canal, the stomach's lining has three layers of smooth muscle, arranged at angles to each other. By contracting in turn, these muscles churn up the food. At the base of the stomach, a ring of muscle called the pyloric sphincter acts like a valve, controlling the release of semi-digested food.

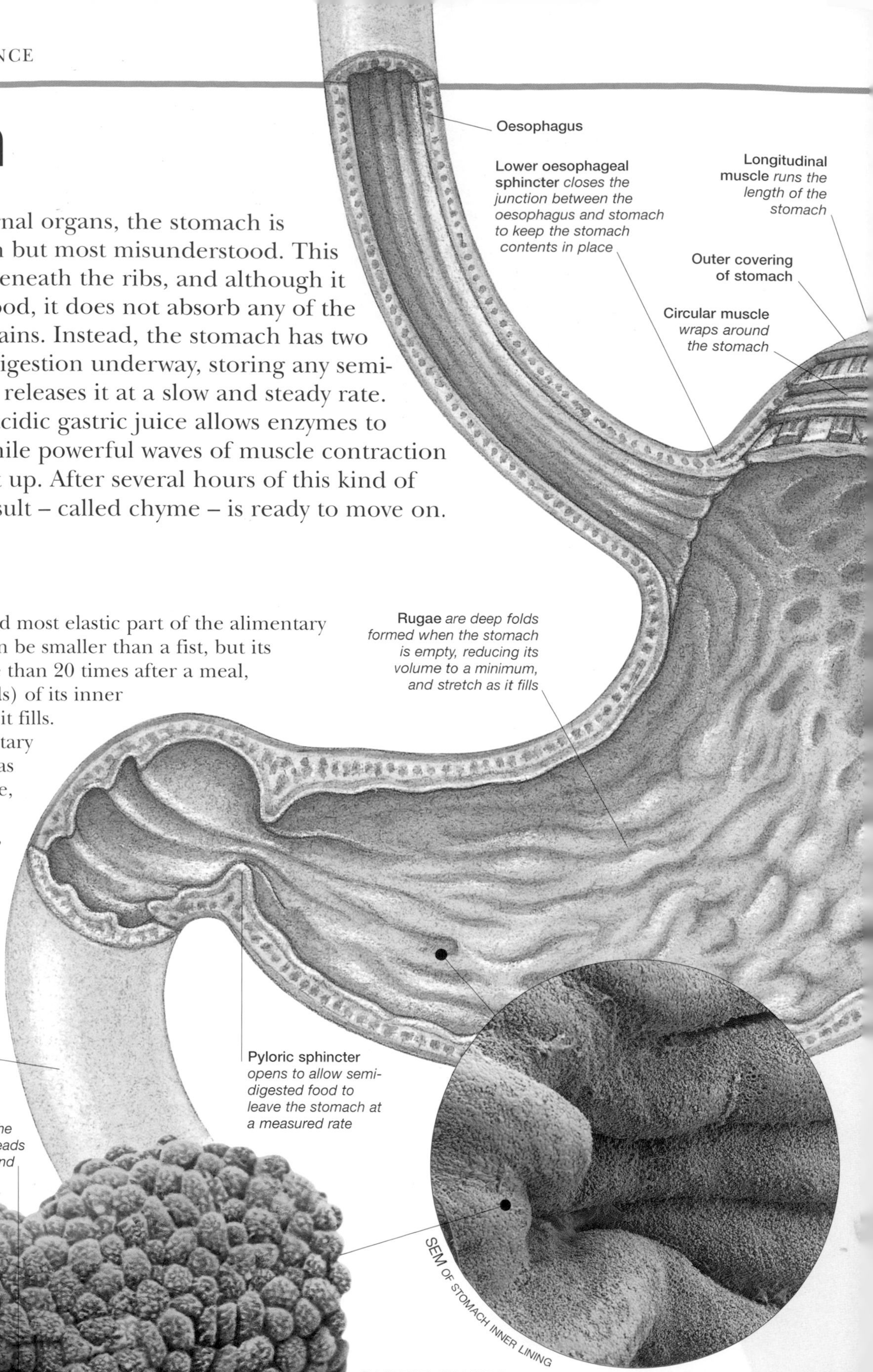

Oesophagus

Lower oesophageal sphincter *closes the junction between the oesophagus and stomach to keep the stomach contents in place*

Longitudinal muscle *runs the length of the stomach*

Outer covering of stomach

Circular muscle *wraps around the stomach*

Rugae *are deep folds formed when the stomach is empty, reducing its volume to a minimum, and stretch as it fills*

Duodenum

Pyloric sphincter *opens to allow semi-digested food to leave the stomach at a measured rate*

Gastric pit *in the stomach wall leads to a gastric gland*

SEM OF STOMACH INNER LINING

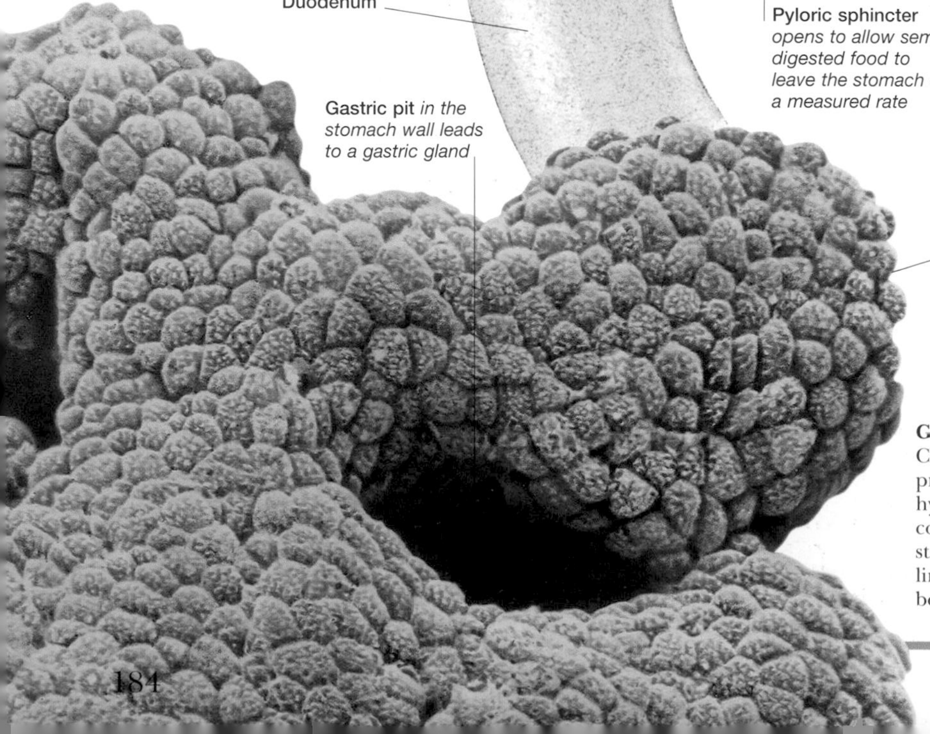

GASTRIC GLANDS

Cells in the stomach's millions of gastric glands produce the components of gastric juice – mucus, hydrochloric acid, and pepsinogen, a substance that is converted into protein-digesting pepsin as it flows into the stomach. The stomach does not digest itself because its lining is covered with protective mucus, and because pepsin becomes active only when it has been "primed" by acid.

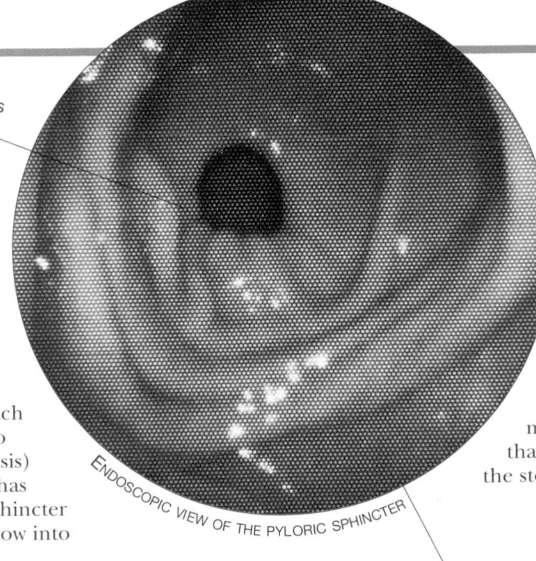
Pyloric sphincter partially relaxes to allow food to pass through
ENDOSCOPIC VIEW OF THE PYLORIC SPHINCTER

FILLING AND EMPTYING
By the time food reaches the stomach, the stomach is ready to receive it because it has been primed by the autonomic nervous system (see pp. 98–99). During its stay here – which can last for up to four hours – nerves and hormones work together to keep the digestive process working smoothly. These two control systems ensure that the stomach secretes enough gastric juice, and also trigger muscular movements (peristalsis) in the stomach wall. When digestion has progressed far enough, the pyloric sphincter relaxes, and the stomach's contents flow into the small intestine.

AUTOMATIC VALVE
Seen through an endoscope, the pyloric sphincter guards the entrance to the duodenum – the first part of the small intestine. When this ring of muscle is tightly contracted, nothing can leave the stomach, but as digestion proceeds, it begins to relax. This relaxation is controlled by a feedback mechanism, which ensures that semi-digested food leaves the stomach at the right rate.

Oblique muscle *runs diagonally*

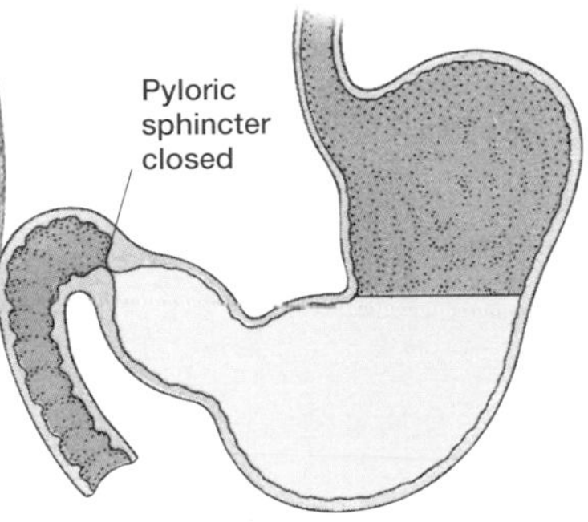
Pyloric sphincter closed

Filling
As it fills, the stomach releases gastric juice, which is mixed with food by waves of muscular contraction, or peristalsis.

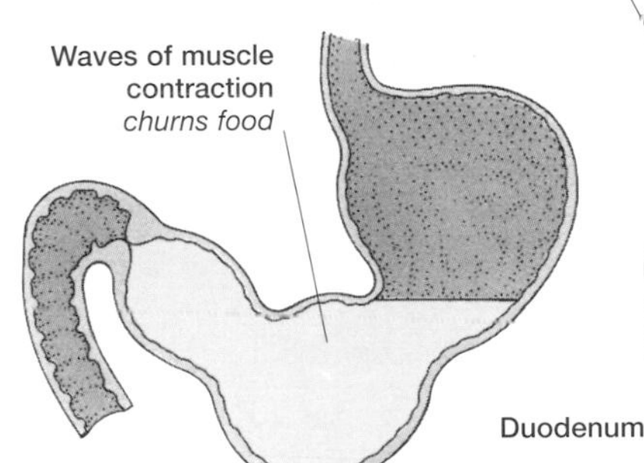
Waves of muscle contraction churns food

Digestion
Vigorous peristalsis churns food as gastric juice digests it into creamy liquid chyme.

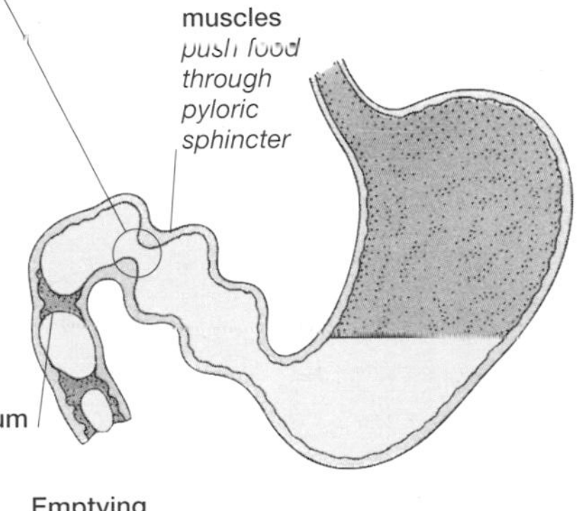
Stomach muscles push food through pyloric sphincter
Duodenum

Emptying
If chyme is liquid enough, the pyloric sphincter relaxes and opens slightly to let small quantities of food pass into the duodenum.

Stomach wall *contains millions of microscopic glands that secrete gastric juice*

ACID ATTACK
When food enters the stomach, it is mixed with a digestive fluid called gastric juice. This juice contains hydrochloric acid, and it is strong enough to dissolve small pieces of bone. These acidic conditions are needed for protein digestion, which is carried out by an enzyme called pepsin.

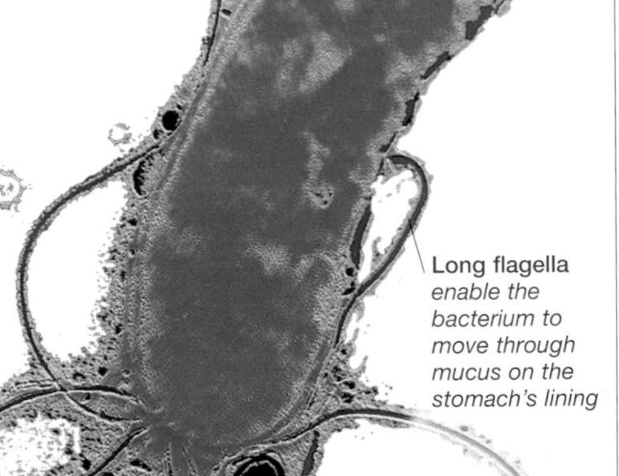
Long flagella enable the bacterium to move through mucus on the stomach's lining

STOMACH BUG
Stomach acid kills most bacteria, but one kind, called *Helicobacter pylori*, manages to thrive in these hazardous conditions. In recent years, this bacterium has been closely studied, and there is mounting evidence that it is linked to two different forms of stomach disease, one of which is stomach cancer. The other disease is gastritis – an inflammation that often leads to ulcers. How the bacterium spreads is not known.

STOMACH STUDY

In 1822, American surgeon William Beaumont treated Alexis St Martin, who had been shot during a hunting trip. Beaumont saved the man's life, but his patient was left with a permanent opening from his stomach to the outside. For the next decade, Beaumont monitored St Martin's stomach, and the fluid that it produced. Although gruesome, the research produced a great deal of useful information, and the patient lived to the ripe age of 82.

Beaumont (1785–1853) examining St Martin's stomach

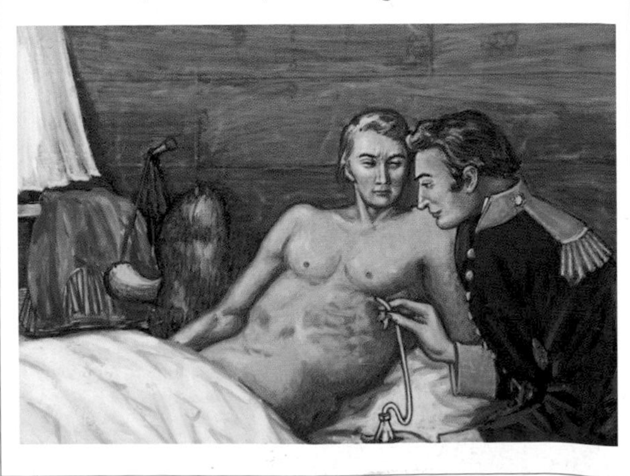

Small intestine

DESPITE ITS MODEST-SOUNDING NAME, the small intestine is the most important part of the entire digestive system. Measuring up to 6 m (19.7 ft) in length, this intricately folded tube is the site where food is fully broken down, and where its nutrients are absorbed. By the time food has been gently squeezed through all its twists and turns, nearly all its useful ingredients have been removed, leaving just watery waste. The small intestine is only about 2.5 cm (1 in) across, but its inner lining has a huge surface area, thanks to microscopic projections called villi. With the help of the pancreas and the liver, the intestine breaks down food into simple substances, and the villi absorb these into the body itself.

ENDOSCOPIC VIEW OF THE SMALL INTESTINE

HOW THE SMALL INTESTINE WORKS

The small intestine is divided into three regions that work in different ways. Starting "upstream", the first part is the duodenum – a 30 cm (12 in) section that receives digestive fluids from the pancreas and the liver. This is where stomach acid is neutralized, and where digestion of food begins in earnest. The second section, called the jejunum, is about 2 m (6.6 ft) long and secretes large amounts of digestive enzymes. The third and longest section, called the ileum, is concerned mainly with absorbing nutrients, rather than breaking food down. All three sections push food along by peristalsis, but they also contract into short segments, ensuring that the food is mixed up.

Duodenum *is the first section of the small intestine, in which chyme is mixed with bile and pancreatic juice*

Jejunum, *the middle section of the small intestine, secretes digestive enzymes*

Circular ridges *increase the surface area of the small intestine*

Villi *are 1 mm (0.04 in) long projections that absorb nutrients*

Ileum, *the final and longest section of the small intestine, has a rich supply of blood and lymph*

VILLI

The small intestine has circular internal ridges, but the greatest boost to its surface area comes from tightly packed villi (singular villus) – finger-like projections that protrude inwards from the intestine's lining. Villi contain capillaries, and also lacteals, which are minute branches of the lymphatic system. Together, these collect the nutrients that the villus absorbs from food, so that they can be carried around the body. Most of the cells that line the villi have even smaller projections, called microvilli. These form "brush borders", which are like chemical worktops. Enzymes are fastened to the brush borders, and they carry out the last stages of digestion before food is absorbed.

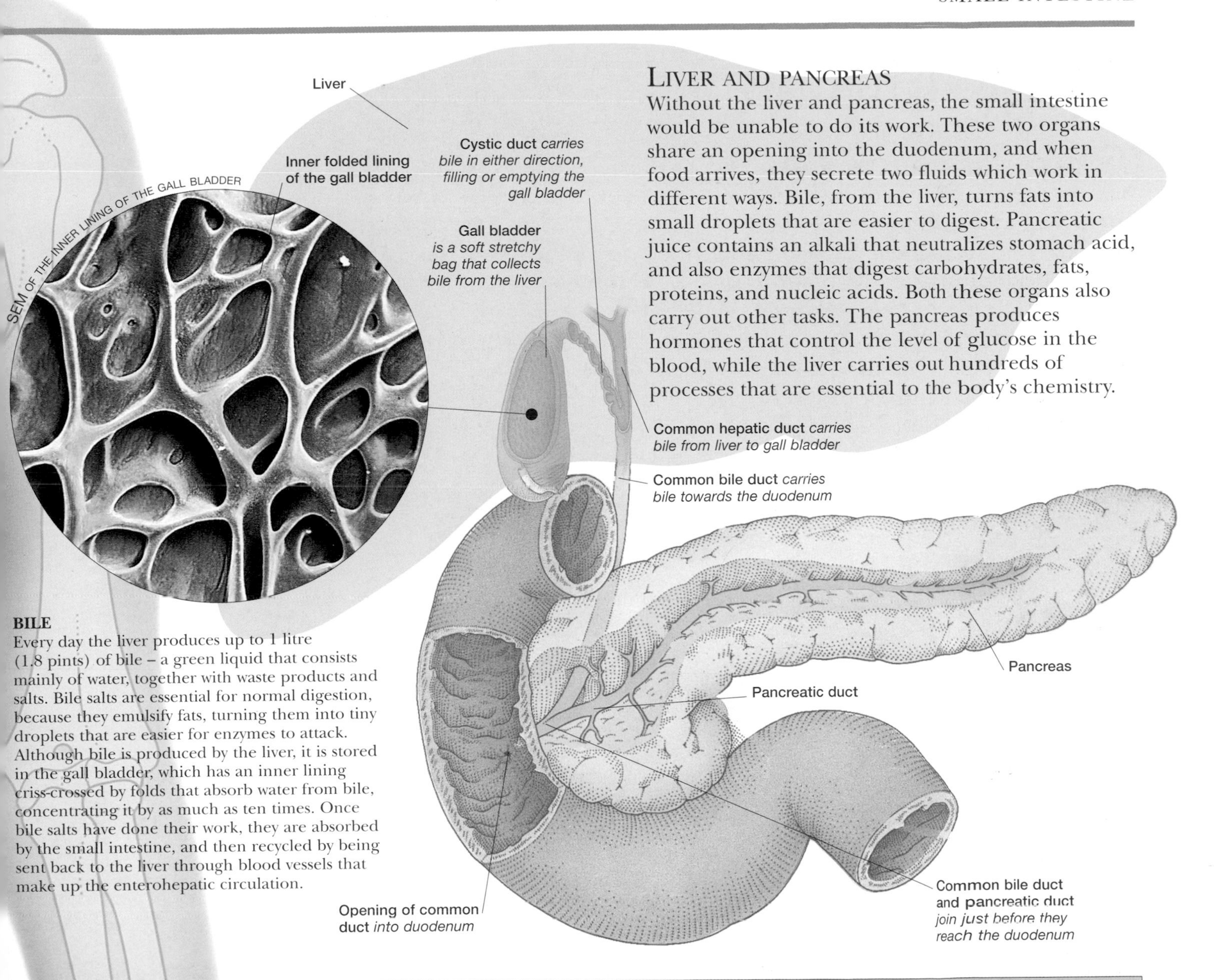
Liver
Inner folded lining of the gall bladder
SEM OF THE INNER LINING OF THE GALL BLADDER
Cystic duct carries bile in either direction, filling or emptying the gall bladder
Gall bladder is a soft stretchy bag that collects bile from the liver
Common hepatic duct carries bile from liver to gall bladder
Common bile duct carries bile towards the duodenum
Pancreas
Pancreatic duct
Opening of common duct into duodenum
Common bile duct and pancreatic duct join just before they reach the duodenum

Liver and pancreas

Without the liver and pancreas, the small intestine would be unable to do its work. These two organs share an opening into the duodenum, and when food arrives, they secrete two fluids which work in different ways. Bile, from the liver, turns fats into small droplets that are easier to digest. Pancreatic juice contains an alkali that neutralizes stomach acid, and also enzymes that digest carbohydrates, fats, proteins, and nucleic acids. Both these organs also carry out other tasks. The pancreas produces hormones that control the level of glucose in the blood, while the liver carries out hundreds of processes that are essential to the body's chemistry.

BILE
Every day the liver produces up to 1 litre (1.8 pints) of bile – a green liquid that consists mainly of water, together with waste products and salts. Bile salts are essential for normal digestion, because they emulsify fats, turning them into tiny droplets that are easier for enzymes to attack. Although bile is produced by the liver, it is stored in the gall bladder, which has an inner lining criss-crossed by folds that absorb water from bile, concentrating it by as much as ten times. Once bile salts have done their work, they are absorbed by the small intestine, and then recycled by being sent back to the liver through blood vessels that make up the enterohepatic circulation.

Enzymes in Action

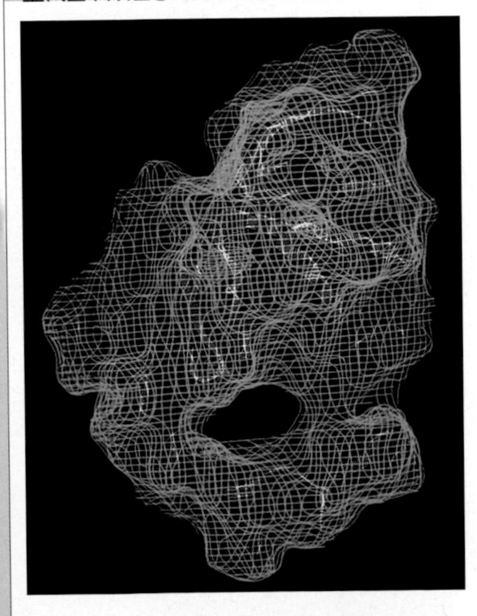

Enzymes are proteins that speed up chemical reactions in the body. If they did not exist, it would take several decades to digest even a single meal. There are two main groups of digestive enzymes: those made by the pancreas and released in pancreatic juice, and those on the "brush borders" of the small intestine. Each enzyme acts on one particular type of nutrient, turning it into smaller and simpler products. Enzyme molecules are not used up when they do their work. This means that they can do the same job thousands or even millions of times in succession.

Computer-generated image of the enzyme amylase

Enzyme	Acts on	Product
Pancreatic		
Amylase	Starch	Maltose
Trypsin	Proteins	Peptides
Chymotrypsin	Proteins	Peptides
Carboxypeptidase	Proteins	
Lipase	Fats and oils	Fatty acids and monoglycerides
Nuclease	Nucleic acids	Pentoses and bases
Brush border		
Peptidases	Peptides	Amino acids
Maltase	Maltose	Glucose
Sucrase	Sucrose	Glucose and fructose
Lactase	Lactose	Glucose and galactose
Nuclease	Nucleic acids	Pentoses and bases

Large intestine

THE LARGE INTESTINE IS the final stretch of the alimentary canal. It is more than twice as wide as the small intestine, but only about one-quarter as long. Instead of twisting and turning, it follows a more straightforward path, with just a handful of sharp bends. The large intestine does not produce any enzymes, and it does not play a direct part in digestion. Instead, its chief function is to reabsorb water to help the body's fluid balance, and to make waste easier to expel. The large intestine also has another role: it absorbs vitamins that are made by bacteria. Huge numbers of microbes thrive in its warm and moist interior, and they break down substances that have escaped digestion, before eventually being expelled themselves.

INSIDE THE COLON
Endoscopes are often used to examine the colon for signs of disease. This view shows the inside of a healthy colon, with a corridor of pockets, called haustra, separated by narrower parts of the intestine wall. The intestine looks triangular in cross-section because it has three taeniae, (bands of muscle) running almost all the way along it.

ENDOSCOPIC VIEW OF THE COLON

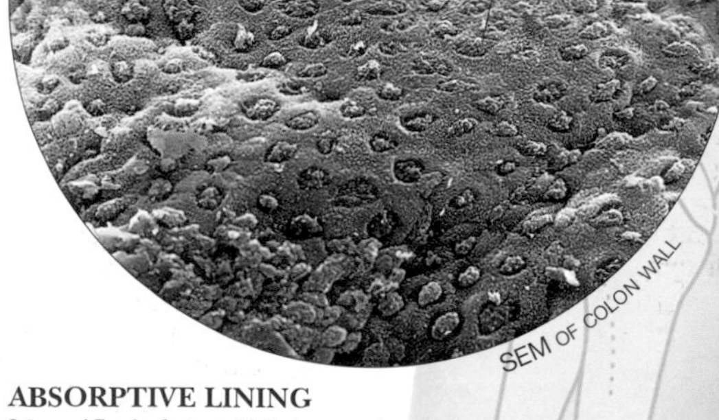
Glands produce mucus to lubricate passage of faeces

SEM OF COLON WALL

ABSORPTIVE LINING
Magnified about 400 times, the lining of the large intestine (above) looks much smoother than other parts of the alimentary canal. Unlike the small intestine, it does not have villi, but it does have small glands, which can clearly be seen in the picture (blue). These contain cells that produce mucus. Water-absorbing cells are spread all over the large intestine's lining, and in the sides of its glands.

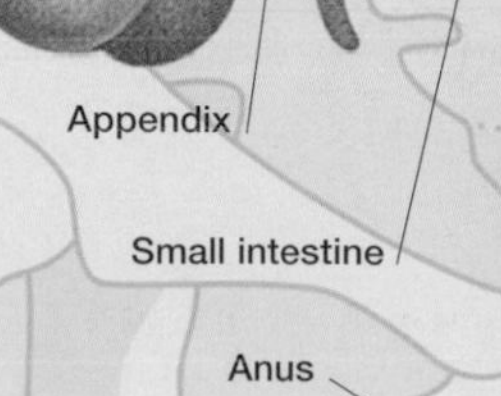
Descending colon *travels down the left side of the abdominal cavity*
Taenia omentalis *is one of three parallel bands of muscle that run along the colon*
Transverse colon *travels across the abdominal cavity, below the liver and spleen*
Ascending colon *travels up the right side of the adominal cavity*
Appendix
Small intestine
Anus
Rectum
Sigmoid colon *leads down to the rectum*

COLON

The large intestine begins at the ileocaecal sphincter, or valve, and it ends at the rectum and anus. The section between these points is called the colon and measures about 1.5 m (5 ft) in length. The colon follows a path shaped like the edge of a shield, travelling up, across, and then down the lower part of the abdominal cavity. Unlike the small intestine, the colon has muscle bands (taeniae), which gather it up into a series of pockets (haustra) that help to compact waste before passing it on. The rectum collects waste once most of its water has been removed, and holds the waste ready for disposal.

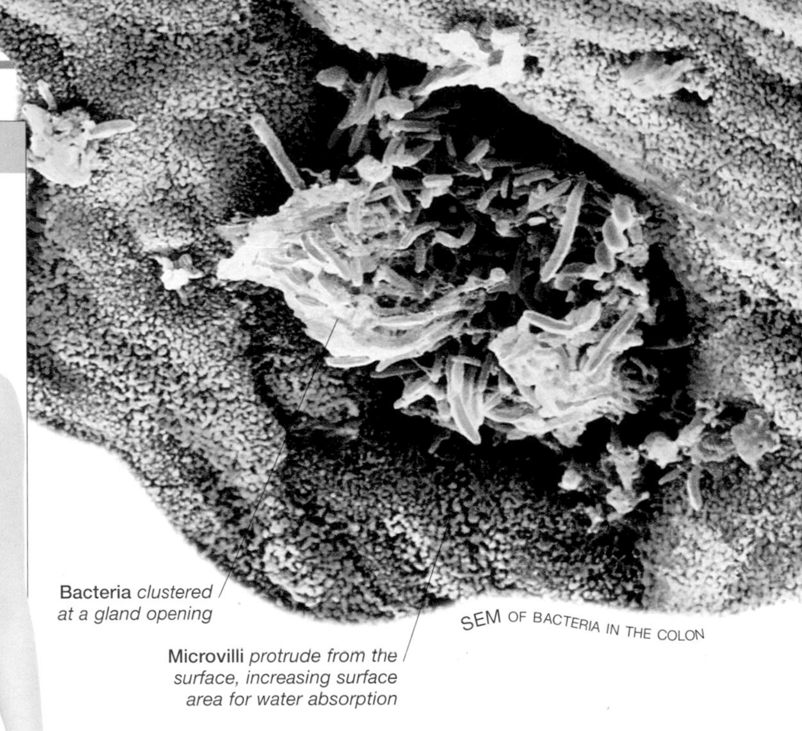
Bacteria clustered at a gland opening
Microvilli protrude from the surface, increasing surface area for water absorption
SEM OF BACTERIA IN THE COLON

Fluid In, Fluid Out

Water entering the alimentary canal	
Saliva	*1 litre (1.8 pints)*
Water in drinks	*2.3 litres (4 pints)*
Bile	*1 litre (1.8 pints)*
Pancreatic juice	*2 litres (3.5 pints)*
Gastric juice	*2 litres (3.5 pints)*
Intestinal juice	*1 litres (1.8 pints)*
TOTAL	9.3 litres (16.4 pints)

Water reabsorbed by the alimentary canal	
Small intestine	*8.3 litres (14.6 pints)*
Large intestine	*0.9 litres (1.6 pints)*
TOTAL	9.2 litres (16.2 pints)
Water lost in faeces	0.1 litre (0.2 pints)

This diagram shows how much water moves in and out of the alimentary canal during a typical day. More than 9 litres (15.8 pints) of water are needed to move food and nutrients, and to create the right conditions for enzymes to work. However, the digestive system reabsorbs and recycles almost all of this water, so that the body does not become dehydrated. Reabsorption is a vital part of the body's water balance: if it breaks down, as happens in dysentery and other diseases that produce severe diarrhoea, people can become seriously ill.

GUT BACTERIA

There are more bacteria in the large intestine than in the rest of the body put together – so many, in fact, that they are estimated to form a layer 2 cm (0.8 in) thick. Their function is to break down organic matter in waste, and to produce gas. Bacteria also make a range of useful chemicals, such as vitamin K, which the body absorbs and uses. Bacteria make up nearly 50 per cent of waste by the time its gets expelled.

DEFAECATION

After spending about five to ten hours in the large intestine, compacted waste, known as faeces, is ready for disposal. Faeces contain a variable amount of water, together with undigested fibre, dead gut cells, and living and dead bacteria. They also contain digested bile pigments, which give them their colour. Faeces are pushed into the rectum by peristalsis, and are expelled through the anus during defecation – an essential final stage in the digestive process. Preparations for defecation are made by automatic reflexes, but it begins when the outer sphincter of the anus relaxes, a movement that is initiated by voluntary control.

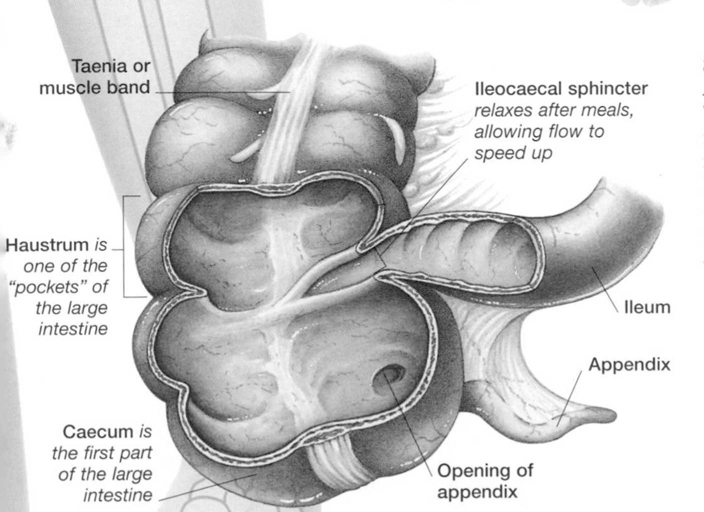
Taenia or muscle band
Ileocaecal sphincter *relaxes after meals, allowing flow to speed up*
Haustrum *is one of the "pockets" of the large intestine*
Ileum
Appendix
Caecum *is the first part of the large intestine*
Opening of appendix

SPHINCTER AND APPENDIX

The ileocaecal sphincter is a muscular ring that controls the flow of digested waste into the large intestine. It joins the side of the intestine, just above a pocket called the caecum. Attached to the caecum is the appendix. Long ago in human evolution, this narrow blind-ended tube played a part in the digestion of plant food, but in modern humans it has no useful function. Inflammation of the appendix, known as appendicitis, is potentially dangerous and is often treated by surgery.

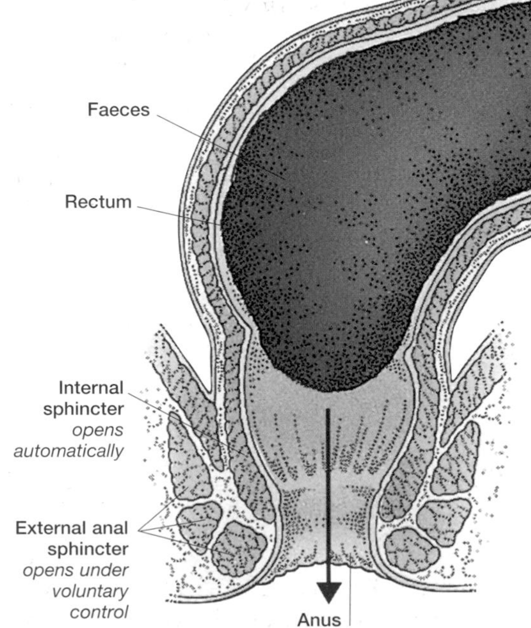
Faeces
Rectum
Internal sphincter *opens automatically*
External anal sphincter *opens under voluntary control*
Anus

Nutrition

NUTRIENTS PROVIDE US with all the substances we need to keep our bodies working normally. Macronutrients, which include carbohydrates, fats, and proteins, make up most of what we eat. They supply energy, as well as building materials that the body uses for growth and maintenance. Micronutrients, which include vitamins and minerals (see pp. 194–95), are usually present in much smaller amounts, but without them, the body's chemistry cannot work. In nature, some animals manage to get all their nutrients from just one kind of food, but humans are not like this. We need to eat a mixture of different foods, to get the nutrients we need in the right amounts. This mixture is known as a balanced diet, and it is one of the most important factors in staying healthy.

Fats, oils, and sugar-rich foods *contain essential lipids, but only small quantities are needed because lipids are present in other foods; sugary foods are tempting, but in a well-balanced diet, most carbohydrate should come from starchy foods.*

WATER AND FIBRE
Two other essential nutrients are water and fibre. Water makes up over 50 per cent of body weight and is constantly being lost. Fibre adds bulk to food, and improves the efficiency of the muscles in the intestinal wall.

NUTRIENTS IN FOOD

For most of human history, nothing was known about the chemistry of food. Today, food labels usually indicate the exact contents of a food. The richest sources of carbohydrates are sugar, cereals, and potatoes, and anything made from them, such as pasta and bread. Protein is found in all kinds of meat, as well as in some plant-based food, such as nuts and beans. Lipids are found in vegetable oils and animal fats, and also in foods that contain butter or milk. Plant-based food is often rich in two other ingredients essential for a healthy diet – water and dietary fibre.

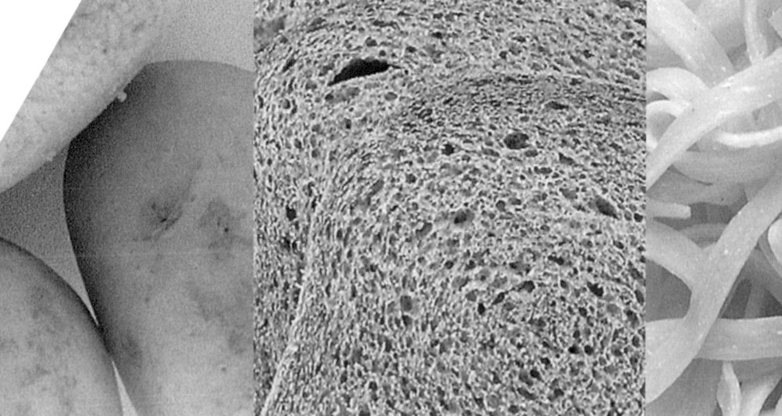

Cereals and potatoes *contain plenty of starch, a complex polysaccharide. During digestion, starch is converted into glucose – the body's main source of energy. Many of these foods also contain vitamins and minerals, such as iron.*

FOOD PYRAMID

We eat many different things, so it can be difficult to decide whether or not a diet is well balanced. This food pyramid helps to solve the problem. It organizes food into categories, according to the main nutrients that they contain, and it also shows how much of each category the body needs. Foods rich in carbohydrates make up the base of the pyramid, because they are needed in large amounts. Fruit and vegetables come next, as sources of vitamins, minerals, and dietary fibre (roughage). Protein-rich food is close to the top, while fats, oils, and sugar-rich foods are at the peak. They contain useful nutrients, but are best eaten in small amounts.

Meat, fish, eggs, and nuts *all supply protein, which is digested to produce amino acids; some animal proteins can supply all the amino acids that the body needs.*

A fast food meal of burger and chips

Dairy products *such as milk, butter, cheese, and yoghurt contain variable amounts of protein and animal fat. They are good sources of calcium, while their fat content ranges from about 70 per cent in butter, to zero in some kinds of yoghurt.*

A healthy stir-fry

FAST FOOD

In today's busy world, fast food (left) can make a convenient and often tasty alternative to something that has to be prepared and cooked. But unlike meals made using a mixture of fresh ingredients (above), fast food does not add up to a balanced diet. One reason for this is that it often contains extremely high levels of sugars and fats, but very low levels of minerals and vitamins, and the dietary fibre needed to keep the digestive system working smoothly.

Vegetables *are a prime source of vitamins and minerals and contain lots of dietary fibre; this indigestible plant matter adds bulk to food and digested waste, helping it to move easily through the alimentary canal.*

Fruit *contains water and dietary fibre, and is often a good source of vitamins. Its sweetness comes from simple carbohydrates or sugars, which are useful as a rapid energy boost.*

FOOD POISONING

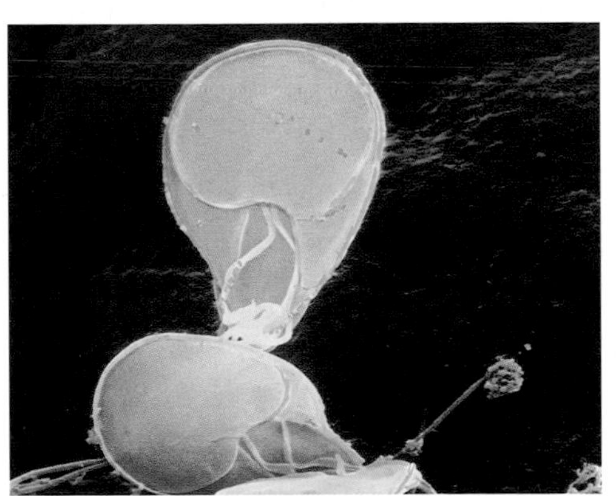

Protist parasite *Giardia*

Even with careful hygiene, food and drinks can be contaminated with micro-organisms, some of which can produce acute food poisoning or long-running illnesses. The protist parasite *Giardia* causes fairly minor symptoms, but some bacteria, such as *Salmonella*, cause severe diarrhoea and vomiting, leaving a person feeling weak and dehydrated. The best way to treat these symptoms is to drink plenty of water, to make up for the fluid that the body has lost.

THE SEARCH FOR VITAMINS

BY THE END OF the 19th century, any uncertainty about the cause of diseases had all but disappeared. Accepted by doctors and scientists alike, Louis Pasteur's germ theory maintained that diseases were caused by bacteria and other micro-organisms. But in the early 1900s, a few innovative scientists proved that some diseases were the result not of germs but instead a lack, or deficiency, of "factors" – later called vitamins – found in tiny amounts in food. The realization that there was a link between diet and disease sparked off a search to identify these "factors" and their deficiency diseases. That poor diet could cause disease had, however, been suspected much earlier.

TREATING SCURVY
On board ship, British naval surgeon James Lind examines sailors he has treated for scurvy to determine which of his remedies has worked best. Lind conducted the first ever serious clinical trial of possible cures for disease, using carefully controlled experiments.

SCURVY AT SEA

In the 18th century, sailors on long sea voyages survived for many months on an unvarying diet of preserved food. Many succumbed to scurvy, a disease that resulted in loose teeth, bleeding gums, bruising, and even death. All remedies failed until Scottish naval surgeon James Lind (1716–94) took an interest in the matter. Lind selected 12 crewmen with scurvy, divided them into pairs, and gave each pair different foods for two weeks. The two sailors fed citrus fruits – oranges and lemons – recovered rapidly. Lind published his results in 1753, but it was not until 1795 that the British Admiralty put them into practice, and it would be another 100 years before any link between diet and disease was made.

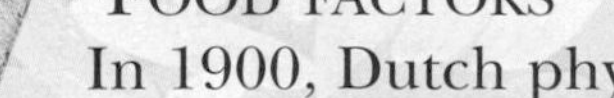

White rice
Brown rice

BEATING BERIBERI
For people whose diet consists mainly of rice, the type of rice consumed could make a big difference to their health. Polished (white) rice has been stripped of the outer husk retained in whole (brown) rice. This husk contains the vitamin B_1, which is necessary to prevent the disease beriberi.

HOPPY'S FACTORS
Known as "Hoppy" by his colleagues at England's Cambridge University, biochemist Frederick Gowland Hopkins was a pioneer of nutrition research. He found that tiny amounts of "accessory food factors" – later called vitamins – were necessary for the body to function normally.

FOOD FACTORS

In 1900, Dutch physician Christiaan Eijkmann (1858–1930) was despatched to Indonesia to investigate the cause of beriberi, a disease that causes numbness and muscle weakness. Eijkmann discovered that if chickens were fed on polished (white) rice with the outer husk removed they developed a disease very similar to beriberi. Remarkably, if fed on whole (brown) rice, they rapidly recovered. Eijkmann concluded that the husk

MALFORMED BONES
A girl with rickets, caused by a lack of vitamin D, pictured with her brother in Budapest, Hungary, around 1920. She shows the characteristic bowing of the legs caused by her upper body weight pushing downwards on weakened leg bones.

contained an "essential food factor". Eijkmann's work was paralleled by that started in 1906 by Frederick Gowland Hopkins (1861–1947). By feeding rats carefully controlled diets, Hopkins showed that to stay healthy they – and presumably humans too – needed tiny amounts of what he termed "accessory food factors".

Naming vitamins

"Food factors" were given a new name – "vitamins" – in 1912 by Polish-American chemist Casimir Funk (1884–1967). Funk's idea that diseases such as beriberi and scurvy are caused by the lack of a particular vitamin set in motion a new era of scientific research. In 1914, Joseph Goldberger (1874–1929) of the US Public Health Service demonstrated that pellagra – a disease that causes dermatitis, diarrhoea, and dementia – was not spread by insects but was the result of poor diet, and could be reversed by a vitamin (niacin) found in protein-rich foods. British physician Edward Mellanby (1884–1955) showed in 1918 that a substance in cod liver oil – later identified by American scientist E.V. McCollum (1879–1967) as vitamin D – could prevent rickets, a deficiency disease characterized by weakened bones.

Vitamin supplements

By the end of the 1930s, scientists had identified vitamins A, C (the scurvy-preventing vitamin in citrus fruits), D, E, and the B vitamins, including B_1, B_2, and B_{12}. This wealth of knowledge about vitamins means that deficiency diseases have been all but eliminated in the developed world.

SUPPLEMENTS
A wide range of vitamin supplements like these are available from pharmacies. How effective vitamin supplements are is the subject of considerable debate. Many nutritionists believe that, for most people, a mixed diet containing fresh produce provides all the vitamins they need.

EXTRA VITAMINS
During the Second World War, food was rationed in Britain, but the government took measures to protect young children from vitamin deficiencies. Here, mothers in wartime London receive free vitamin supplements in the form of orange juice (vitamin C) and cod liver oil (vitamins A and D) for their children.

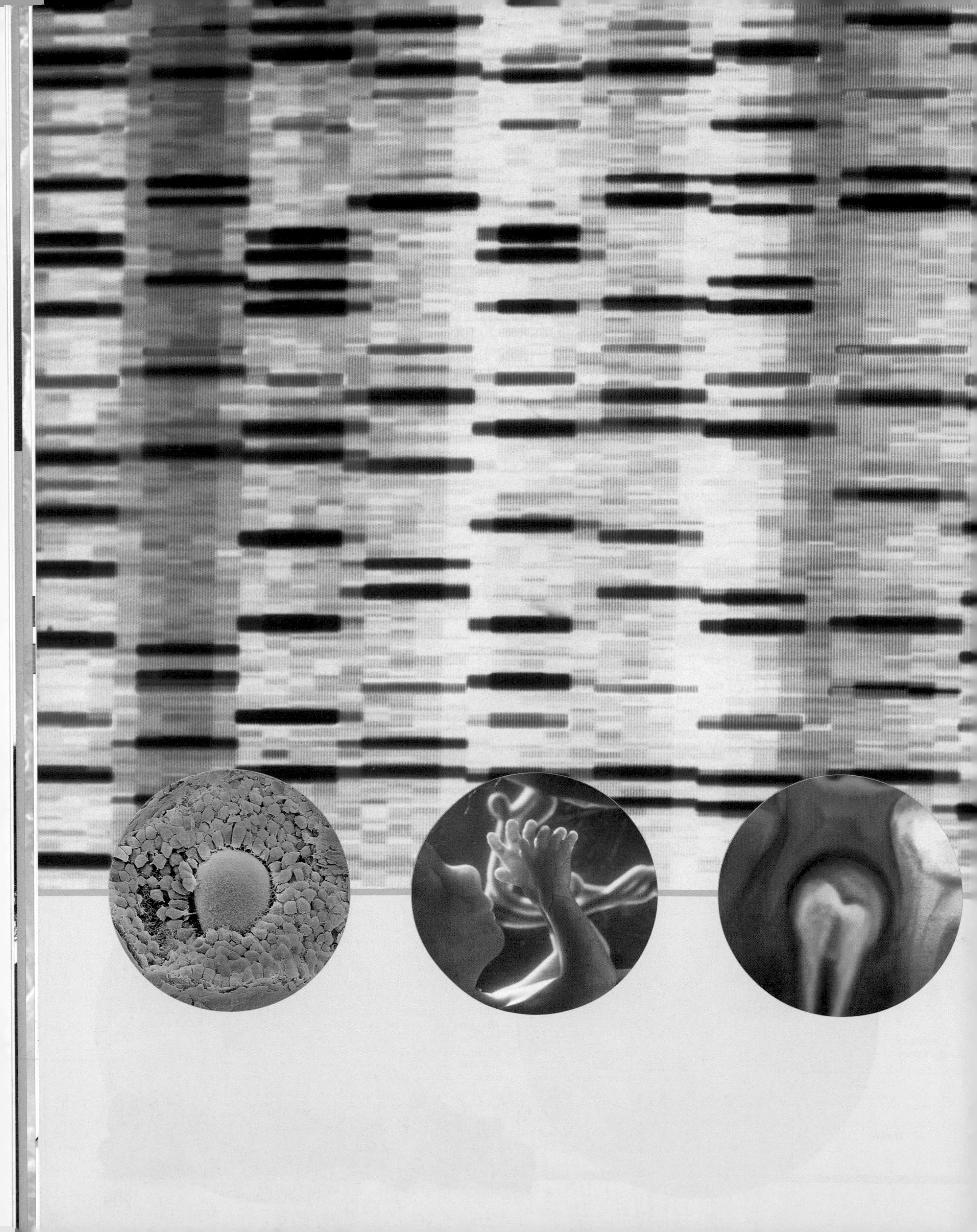

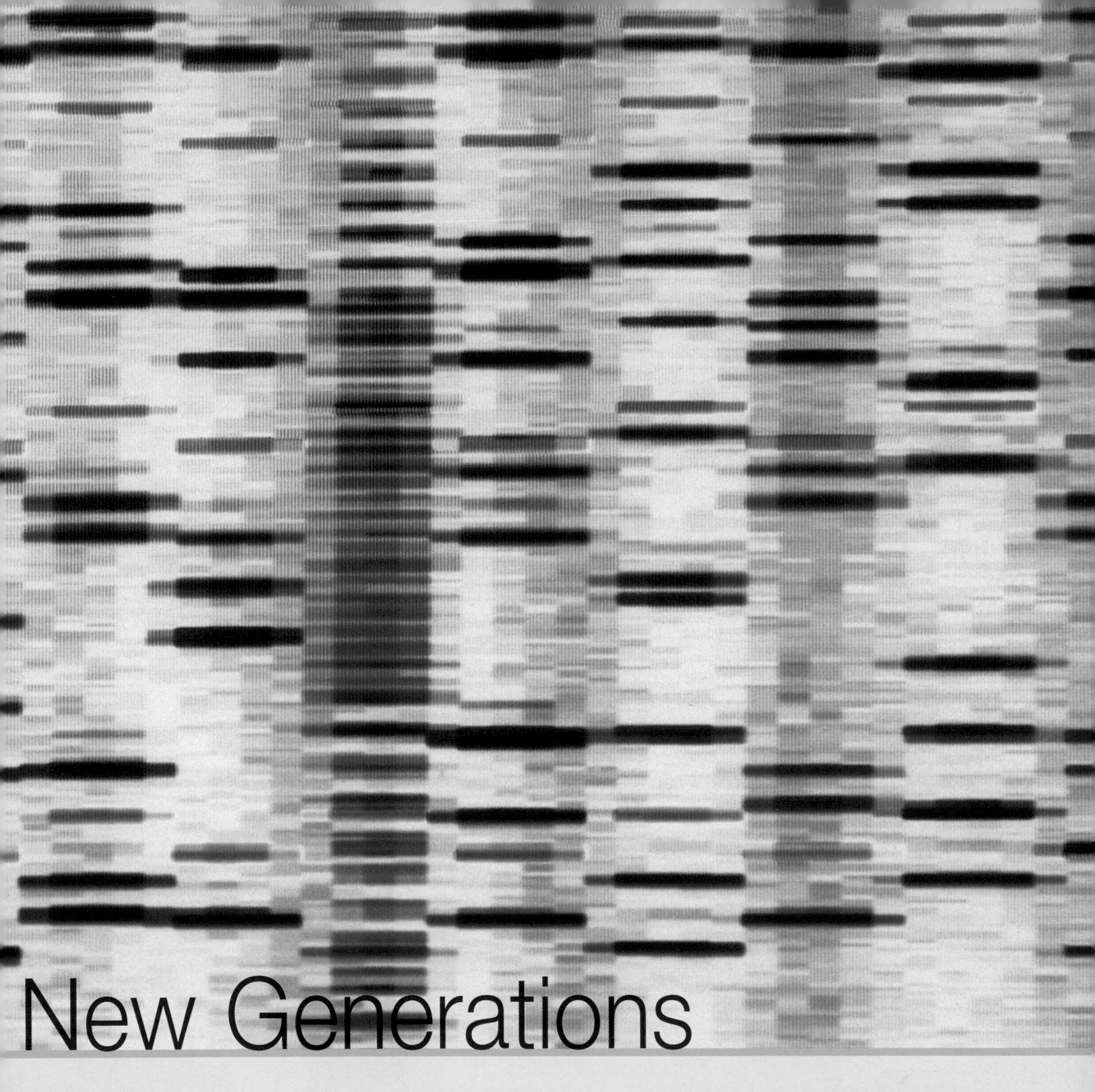

New Generations

EVERY PERSON ON EARTH follows the same life history that starts with their birth and ends with their death. During that life, many men and women reproduce, producing and nurturing a new generation of children who will eventually succeed them. Each child resembles their parents, but is not identical to either of them. These similarities and differences are controlled by thousands of chemical instructions called genes found inside every body cell. The identity and structure of these genes is currently under investigation.

REPRODUCTIVE systems

REPRODUCTION IS A complex process in which genetic material from two parents is brought together to form a new human being. This material is contained within special sex cells – sperm, which are produced by males, and ova (or eggs), produced by females. In order for reproduction to occur, a sperm and an ovum must join to form a single cell containing a full set of genetic material, half each from the father and mother. The reproductive process continues over nine months of pregnancy, when this single cell develops into a fully formed baby, culminating in childbirth.

Female reproductive system

Ovaries *produce ova*

Uterus, *where the fetus develops*

Fallopian tubes *lead from the ovaries to the uterus*

Vagina *receives the penis during sexual intercourse*

Male reproductive system

Penis *delivers sperm inside the female*

Testes *produce sperm*

REPRODUCTIVE SYSTEMS

Males and females each have specialized reproductive organs. In the male these are the testes and the penis, while the female organs are the ovaries, the fallopian tubes, the uterus, and the vagina. The reproductive systems of both sexes produce sex cells, and enable these cells to meet during sexual intercourse so that fertilization can occur. The female system has a further function not found in the male – it provides an environment in which conception and fetal development occur, and from which the baby emerges during birth.

Daughter *has received half her genetic material from each parent*

NEW GENERATIONS

All living organisms have a limited lifespan, and so new generations must be produced to replace those that die. Some species produce new generations rapidly, but humans reproduce relatively infrequently. This is because humans live longer than many other animals, and put a great deal of time and care into nurturing their offspring while they are in the uterus and as they develop following birth. Having a family is not only a biological urge – it also creates strong social bonds of support and love.

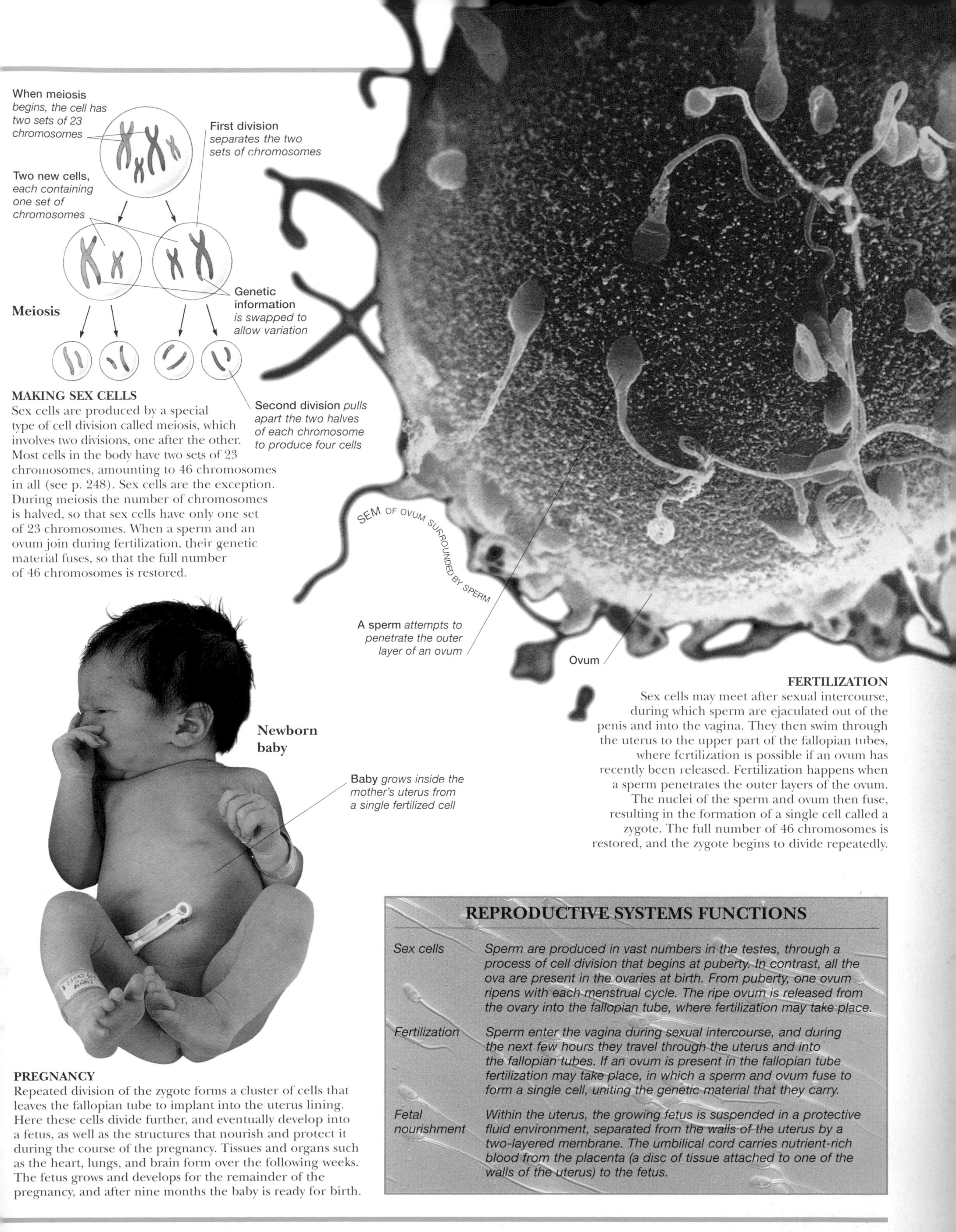
When meiosis begins, the cell has two sets of 23 chromosomes
First division separates the two sets of chromosomes
Two new cells, each containing one set of chromosomes
Genetic information is swapped to allow variation
Meiosis
Second division pulls apart the two halves of each chromosome to produce four cells
SEM OF OVUM SURROUNDED BY SPERM
A sperm attempts to penetrate the outer layer of an ovum
Ovum
Newborn baby
Baby grows inside the mother's uterus from a single fertilized cell

MAKING SEX CELLS

Sex cells are produced by a special type of cell division called meiosis, which involves two divisions, one after the other. Most cells in the body have two sets of 23 chromosomes, amounting to 46 chromosomes in all (see p. 248). Sex cells are the exception. During meiosis the number of chromosomes is halved, so that sex cells have only one set of 23 chromosomes. When a sperm and an ovum join during fertilization, their genetic material fuses, so that the full number of 46 chromosomes is restored.

FERTILIZATION

Sex cells may meet after sexual intercourse, during which sperm are ejaculated out of the penis and into the vagina. They then swim through the uterus to the upper part of the fallopian tubes, where fertilization is possible if an ovum has recently been released. Fertilization happens when a sperm penetrates the outer layers of the ovum. The nuclei of the sperm and ovum then fuse, resulting in the formation of a single cell called a zygote. The full number of 46 chromosomes is restored, and the zygote begins to divide repeatedly.

PREGNANCY

Repeated division of the zygote forms a cluster of cells that leaves the fallopian tube to implant into the uterus lining. Here these cells divide further, and eventually develop into a fetus, as well as the structures that nourish and protect it during the course of the pregnancy. Tissues and organs such as the heart, lungs, and brain form over the following weeks. The fetus grows and develops for the remainder of the pregnancy, and after nine months the baby is ready for birth.

REPRODUCTIVE SYSTEMS FUNCTIONS

Sex cells	*Sperm are produced in vast numbers in the testes, through a process of cell division that begins at puberty. In contrast, all the ova are present in the ovaries at birth. From puberty, one ovum ripens with each menstrual cycle. The ripe ovum is released from the ovary into the fallopian tube, where fertilization may take place.*
Fertilization	*Sperm enter the vagina during sexual intercourse, and during the next few hours they travel through the uterus and into the fallopian tubes. If an ovum is present in the fallopian tube fertilization may take place, in which a sperm and ovum fuse to form a single cell, uniting the genetic material that they carry.*
Fetal nourishment	*Within the uterus, the growing fetus is suspended in a protective fluid environment, separated from the walls of the uterus by a two-layered membrane. The umbilical cord carries nutrient-rich blood from the placenta (a disc of tissue attached to one of the walls of the uterus) to the fetus.*

Male reproductive system

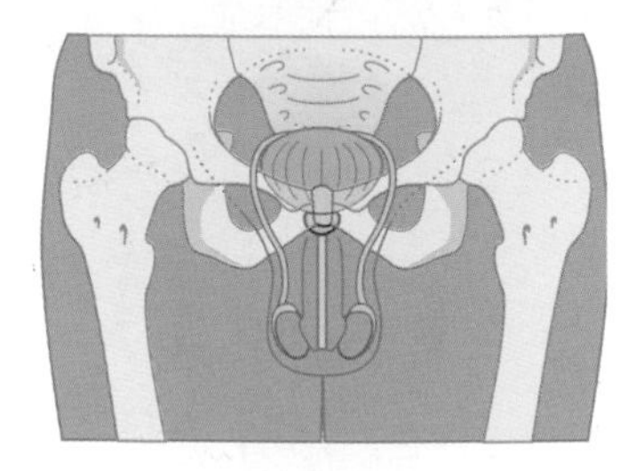

Male reproductive system

THE ORGANS OF A MAN'S reproductive system enable him to make sperm, have sexual intercourse, and fertilize ova produced by a female. The outer, visible parts of the male reproductive system are the penis and a pouch called the scrotum. Inside the scrotum is a pair of oval glands called testes, which are the main reproductive organs of the male. The testes produce millions of minute sex cells, called sperm, and also make male sex hormones. During sexual intercourse, sperm are transferred into the female's vagina by the penis. The sperm are carried from the testes through a series of internal tubes. They are released from the body in a fluid called semen, which is formed from liquids secreted by the seminal vesicles and the prostate gland.

REPRODUCTIVE ORGANS

The testes – the sperm-makers – hang outside the body, inside the scrotum and below the penis. A complicated tube system delivers sperm from the testes to the penis. Small tubes lead from each testis into the long, tightly coiled epididymis, located behind the testis. Each epididymis leads into another tube, called the vas deferens, which joins with a small tube from one of the seminal vesicles to form the ejaculatory duct. The two ejaculatory ducts then join the urethra as it passes through the prostate gland. The urethra is the channel that passes from the bladder to the tip of the penis, through which sperm are released.

PENIS

The penis contains three columns of spongy tissue. During sexual arousal, the spongy columns fill with blood to make the penis larger, firm, and erect, ready for sexual intercourse. Sperm is ejaculated through the urethra, a tube that passes through the middle of the penis to its tip. Urine from the bladder also passes out of the body through the urethra. The tip of the penis is covered by the foreskin, which may be removed during an operation called circumcision.

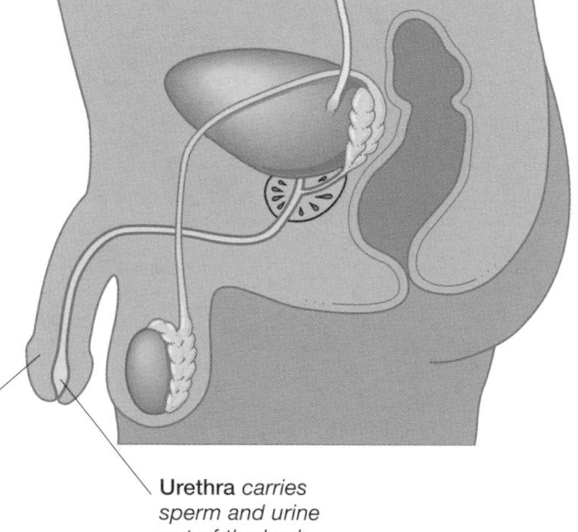
Penis *becomes erect when blood fills the spongy tissue inside*
Urethra *carries sperm and urine out of the body*

Artery *brings blood to the spongy tissue*

Outer skin

Spongy tissue *of corpus cavernosum swells as it fills with blood, making the penis erect*

Urethra *travels through the centre of the spongy tissue of the corpus spongiosum*

Cross-section of penis showing internal structure

Vas deferens *is a tube that carries sperm from a testis to an ejaculatory duct*

Penis *delivers sperm into the female's vagina during sexual intercourse*

Urethra *is a tube that carries sperm and urine to the outside of the body*

Corpora cavernosa *are two parallel cylinders of tissue that fill with blood to make the penis erect*

Corpus spongiosum *is spongy tissue that surrounds the urethra and fills with blood to make the penis erect*

Foreskin *covers and protects the head of the penis*

Glans penis *is the head of the penis*

Testes *produce sperm and sex hormones*

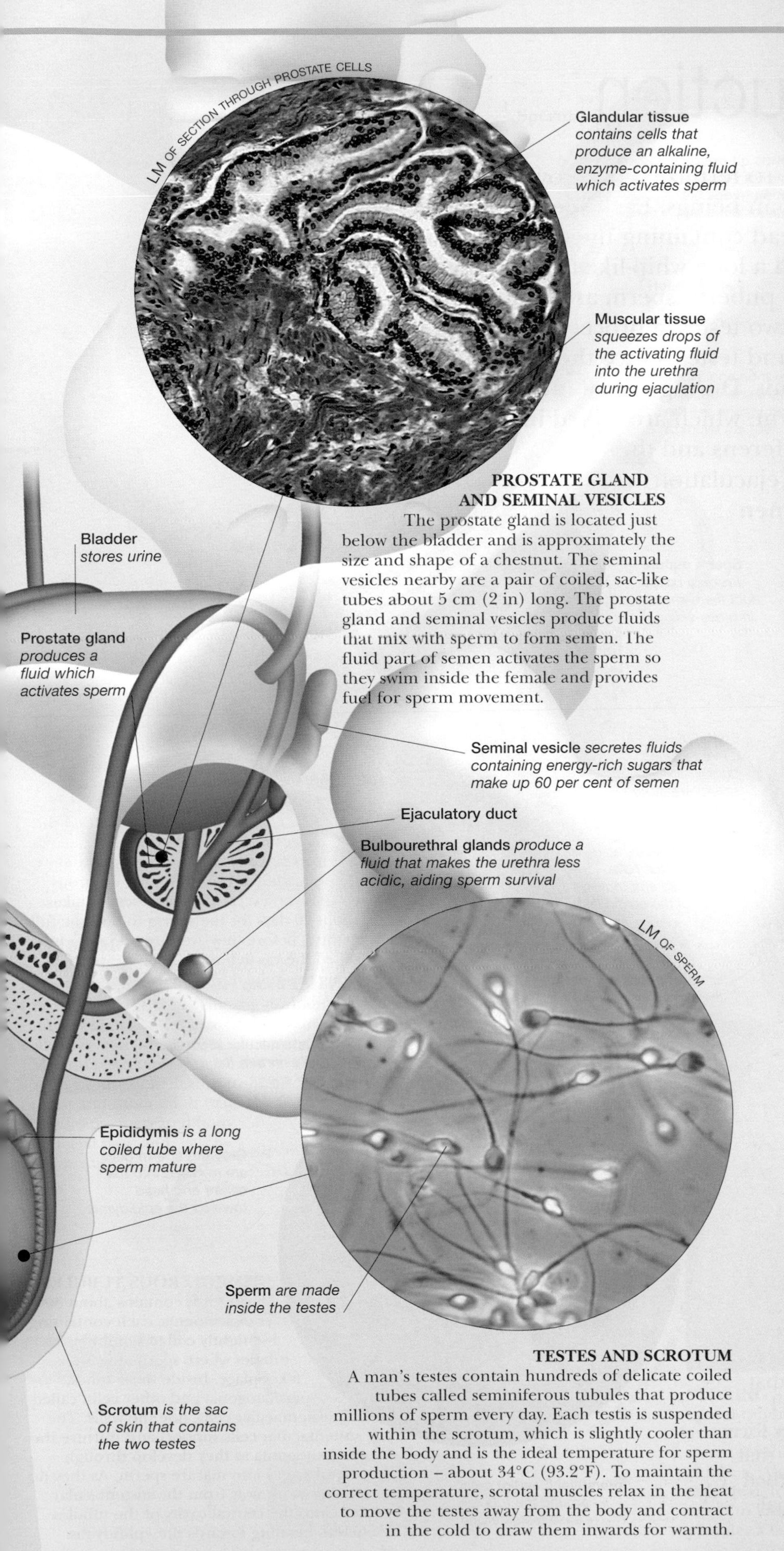
LM OF SECTION THROUGH PROSTATE CELLS
Glandular tissue contains cells that produce an alkaline, enzyme-containing fluid which activates sperm
Muscular tissue squeezes drops of the activating fluid into the urethra during ejaculation
Bladder stores urine
Prostate gland produces a fluid which activates sperm
Seminal vesicle secretes fluids containing energy-rich sugars that make up 60 per cent of semen
Ejaculatory duct
Bulbourethral glands produce a fluid that makes the urethra less acidic, aiding sperm survival
LM OF SPERM
Epididymis is a long coiled tube where sperm mature
Sperm are made inside the testes
Scrotum is the sac of skin that contains the two testes

PROSTATE GLAND AND SEMINAL VESICLES

The prostate gland is located just below the bladder and is approximately the size and shape of a chestnut. The seminal vesicles nearby are a pair of coiled, sac-like tubes about 5 cm (2 in) long. The prostate gland and seminal vesicles produce fluids that mix with sperm to form semen. The fluid part of semen activates the sperm so they swim inside the female and provides fuel for sperm movement.

TESTES AND SCROTUM

A man's testes contain hundreds of delicate coiled tubes called seminiferous tubules that produce millions of sperm every day. Each testis is suspended within the scrotum, which is slightly cooler than inside the body and is the ideal temperature for sperm production – about 34°C (93.2°F). To maintain the correct temperature, scrotal muscles relax in the heat to move the testes away from the body and contract in the cold to draw them inwards for warmth.

BEING MALE

This man and his son are both readily identifiable as being male. Their maleness was determined during the first weeks of development inside the uterus. At this stage, the genetic material of the tiny embryo may or may not stimulate the growth of testes. If testes do grow, they release male sex hormones that stimulate the development of a complete male reproductive system. Years later, at the onset of puberty, the same hormones cause the testes to start producing sperm and make a boy grow into a man.

TINY BODIES

Sperm were discovered in 1677 by a Dutch microscopist, Antoni van Leeuwenhoek (1632–1723). During the next century, scientists called animalculists believed that a complete pre-formed miniature human – sometimes called a homunculus – was enclosed within a sperm. Another group, called ovists, believed that pre-formed offspring were contained inside the ova, or eggs, of females. It was later recognized that babies are not pre-formed but develop gradually after a sperm has fertilized an ovum.

An illustration of a homunculus, based on the drawing by N. Hartsoeker (1656–1725) made in 1694.

Female reproductive system

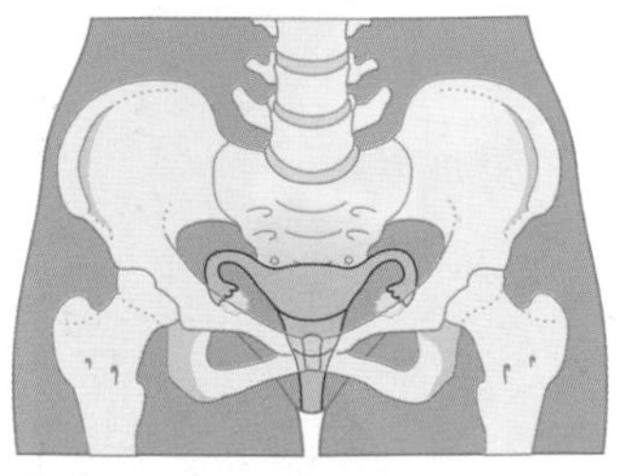

Female reproductive system

THE ORGANS OF A FEMALE'S REPRODUCTIVE system produce female sex cells called ova (eggs), and enable her to have sexual intercourse, become pregnant, and give birth. The organs are all located inside the pelvis except for the outer visible part, called the vulva. The most important organs are a pair of almond-shaped glands called ovaries. These produce ova and make sex hormones. Each month, a single ovum is released from an ovary. If the ovum is fertilized by a sperm from a male, the female may become pregnant. During pregnancy, the fertilized ovum develops into a baby inside the uterus (womb). The baby is born through the vagina, which is a tube of stretchy muscle. The vagina also receives the male's penis during sexual intercourse. After birth, the baby can be fed milk from the mother's breasts.

Pelvic girdle *protects the reproductive organs and other parts of the lower abdomen*

REPRODUCTIVE ORGAN

The main female reproductiv organs are the ovaries, which contain thousands of immatur female sex cells, called ova. The other organs include both internal and external parts. Externally, the vulva contains fol of skin (the labia, or labium as singular) which surround and protect the entrance to the vagina, and the clitoris. Internally, the vagina leads into the uterus, which is hollow and has muscular walls. The uterus connects to a pair of narrow tubes, called the fallopian tubes, one on each side. The outer end of each tube has a funnel-shaped opening which lies very close to an ovary, ready to "catch" an ovum when it is released.

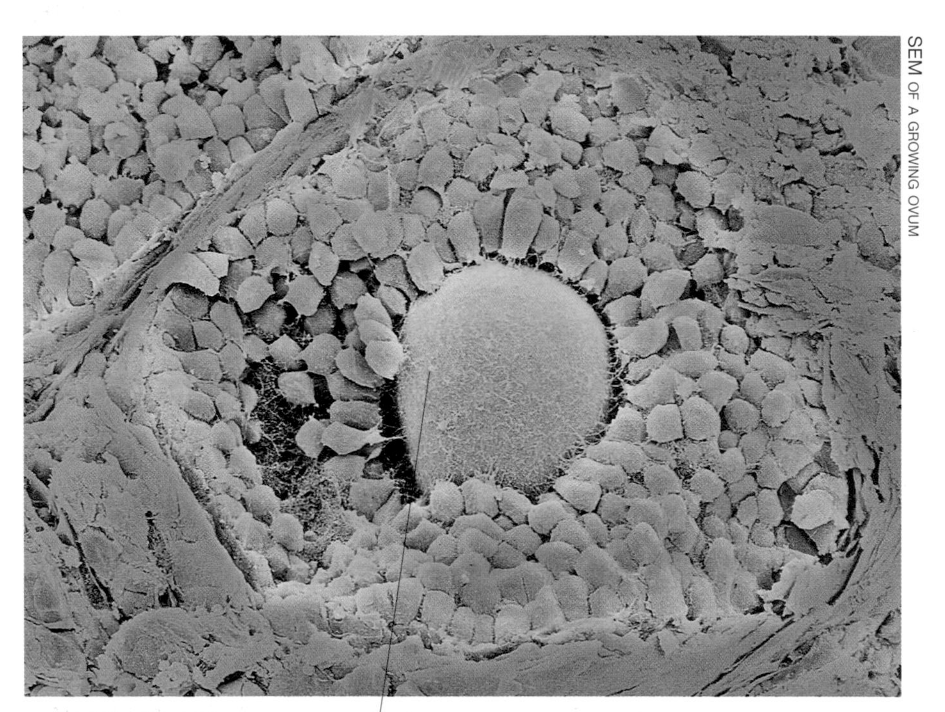

SEM OF A GROWING OVUM

Ovum *matures inside a follicle*

OVARIES AND OVA

At birth, each ovary, which is about 3 cm (1.2 in) long, contains thousands of immature ova. Each ovum (centre, above) is enclosed within a "bag" called a follicle (pink), that also contains nurturing tissue called granulosa cells (blue). After puberty, several ovarian follicles start to mature each month and become fluid-filled bubbles on the ovary's surface. Only one follicle matures fully and its ovum bursts out during ovulation. Each ovary usually alternates to release one ovum a month until the woman is about 50.

DISCOVERING OVA?

The Dutch biologist Regnier de Graaf (1641–73) made a detailed study of the human reproductive system and published a famous book describing the female reproductive organs. De Graaf discovered what he thought was the human ovum in the ovary. In fact, what he had found was a maturing follicle, not the actual smaller ovum inside. Maturing follicles are still sometimes called Graafian follicles in his memory.

Regnier de Graaf

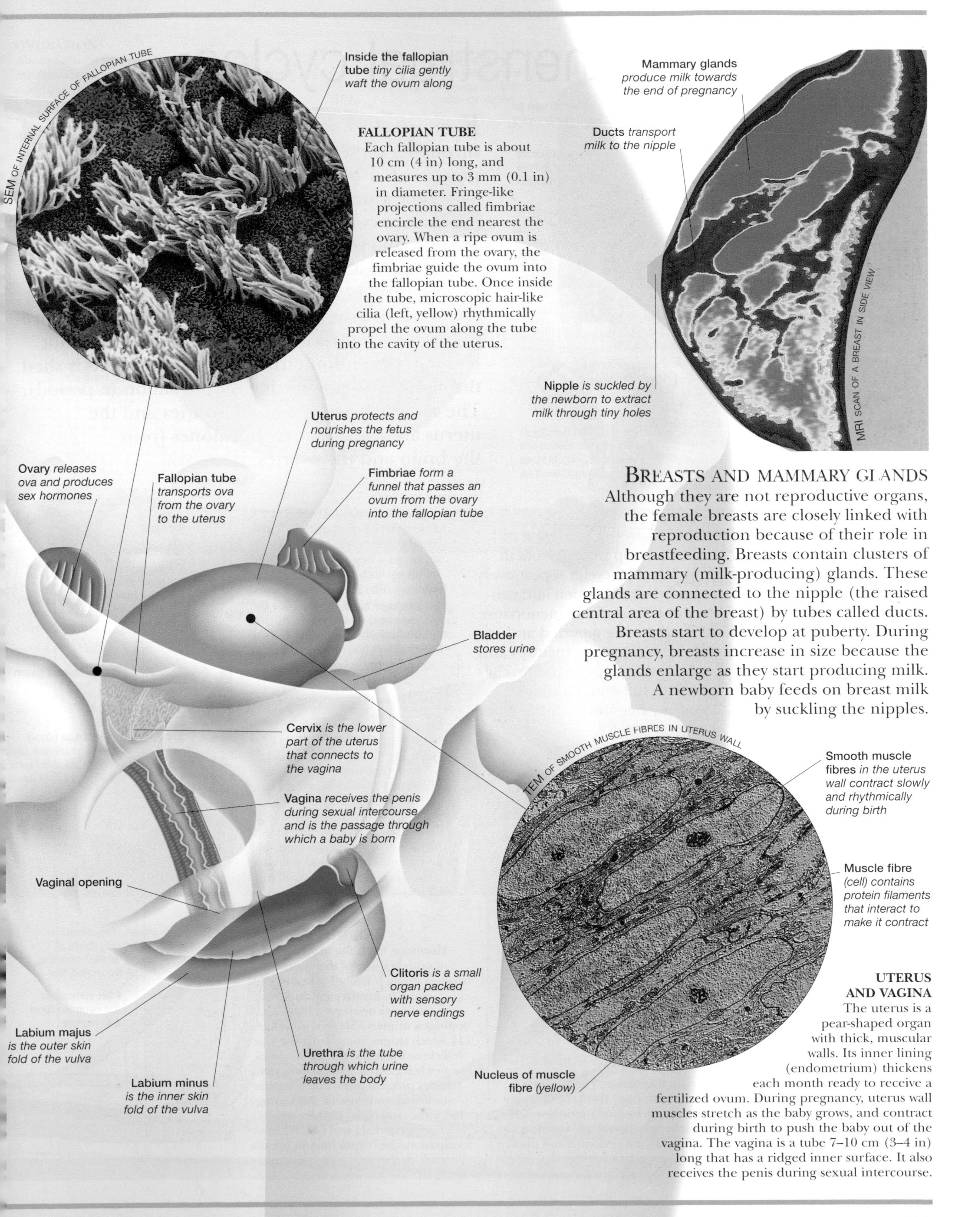
SEM OF INTERNAL SURFACE OF FALLOPIAN TUBE
Inside the fallopian tube tiny cilia gently waft the ovum along
Mammary glands produce milk towards the end of pregnancy
Ducts transport milk to the nipple
Nipple is suckled by the newborn to extract milk through tiny holes
MRI SCAN OF A BREAST IN SIDE VIEW
Uterus protects and nourishes the fetus during pregnancy
Ovary releases ova and produces sex hormones
Fallopian tube transports ova from the ovary to the uterus
Fimbriae form a funnel that passes an ovum from the ovary into the fallopian tube
Bladder stores urine
Cervix is the lower part of the uterus that connects to the vagina
Vagina receives the penis during sexual intercourse and is the passage through which a baby is born
Vaginal opening
Clitoris is a small organ packed with sensory nerve endings
Labium majus is the outer skin fold of the vulva
Labium minus is the inner skin fold of the vulva
Urethra is the tube through which urine leaves the body
TEM OF SMOOTH MUSCLE FIBRES IN UTERUS WALL
Smooth muscle fibres in the uterus wall contract slowly and rhythmically during birth
Muscle fibre (cell) contains protein filaments that interact to make it contract
Nucleus of muscle fibre (yellow)

FALLOPIAN TUBE
Each fallopian tube is about 10 cm (4 in) long, and measures up to 3 mm (0.1 in) in diameter. Fringe-like projections called fimbriae encircle the end nearest the ovary. When a ripe ovum is released from the ovary, the fimbriae guide the ovum into the fallopian tube. Once inside the tube, microscopic hair-like cilia (left, yellow) rhythmically propel the ovum along the tube into the cavity of the uterus.

BREASTS AND MAMMARY GLANDS

Although they are not reproductive organs, the female breasts are closely linked with reproduction because of their role in breastfeeding. Breasts contain clusters of mammary (milk-producing) glands. These glands are connected to the nipple (the raised central area of the breast) by tubes called ducts. Breasts start to develop at puberty. During pregnancy, breasts increase in size because the glands enlarge as they start producing milk. A newborn baby feeds on breast milk by suckling the nipples.

UTERUS AND VAGINA
The uterus is a pear-shaped organ with thick, muscular walls. Its inner lining (endometrium) thickens each month ready to receive a fertilized ovum. During pregnancy, uterus wall muscles stretch as the baby grows, and contract during birth to push the baby out of the vagina. The vagina is a tube 7–10 cm (3–4 in) long that has a ridged inner surface. It also receives the penis during sexual intercourse.

FERTILITY

MODERN MEDICINE HAS DEVELOPED ways to control human fertility – both to prevent it and aid it. The fact that women can produce large numbers of children over 30 or 40 fertile years, and the need to give each child a long period of parental care, has led to the search for methods of contraception to control the number of births and the length of time between them. Conversely, some people have difficulty conceiving children due to problems in their reproductive systems. Modern fertility treatments have made it possible for many childless couples to have a family, by joining ova and sperm in a laboratory.

DR MARIE STOPES
Dr Marie Stopes campaigned for the right for women to use birth control, publishing two bestselling books on the subject. In 1921, she opened the first British birth control clinic.

BIRTH CONTROL PILL
There are two main types of birth control pill, the combined pill and the progestogen-only pill. The hormones in them usually prevent a woman from ovulating. This means that no ova are released, so they cannot be fertilized by sperm.

BARRIER METHODS

The earliest attempts at contraception centred on the use of physical barriers that stopped sperm from reaching the ovum, for example condoms that cover the penis. Their widespread availability today, both as a contraceptive and a barrier to sexually transmitted diseases, was first pioneered by Margaret Sanger (1879–1966) in America and Marie Stopes (1880–1958) in Britain. They both overcame strong social and religious opposition to make contraception freely available, so that women could choose when they would become pregnant.

IUD AND BIRTH CONTROL PILLS

More sophisticated methods of contraception, based on treatments which interfere with the processes of ovum production and fertilization, became available in the second half of the 20th century. Small interuterine devices (IUDs), implanted inside the uterus, work by preventing fertilized ova from implanting. More recent interuterine systems (IUS) contain the hormone progestogen, which stops sperm entering the uterus.

The use of hormones to regulate ovulation was pioneered by several scientists, whose studies explained how hormones controlled human fertility. They developed artificial hormones that stop ova maturing, and these formed the basis of birth control pills, taken orally. Their widespread introduction in the 1960s meant that women could enjoy sexual intercourse without the fear of unwanted pregnancy. Although

Fallopian tube is blocked by a clip during sterilization

FEMALE STERILIZATION
This method of contraception is permanent and involves surgery. Two small incisions are made into the abdomen. The fallopian tubes are then sealed with clips or cut and tied, so that sperm cannot travel through the tubes to fertilize ova. Males can be sterilized by cutting the vas deferens.

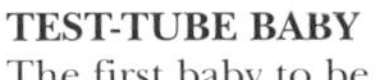

this led to a relaxation in attitudes towards extramarital sex and promiscuity, it also led to an upsurge in the spread of sexually transmitted diseases.

INFERTILITY TREATMENTS

Until late last century, the inability to conceive, either through failure of ovulation, blockage of the female reproductive tract, or defective sperm production in men, meant that many couples could not have their own children. Then an understanding of how hormones control ovulation allowed doctors to use fertility drugs to trigger ovulation.

TEST-TUBE BABY
The first baby to be born as a result of successful IVF was Louise Brown (pictured right, with her father). She was born on 12 August, 1978, in the UK. IVF was pioneered by Dr Patrick Steptoe (1913–88).

Ova could then be collected, fertilized outside the body, and re-implanted into the womb. This technique, called in vitro fertilization (IVF), led to the birth of "test-tube" babies. Male infertility caused by sperm with low mobility can now be overcome by intracytoplasmic sperm injection (ICSI), where microscopic needles inject sperm into an ovum in a laboratory. The fertilized ovum is then re-implanted into the uterus.

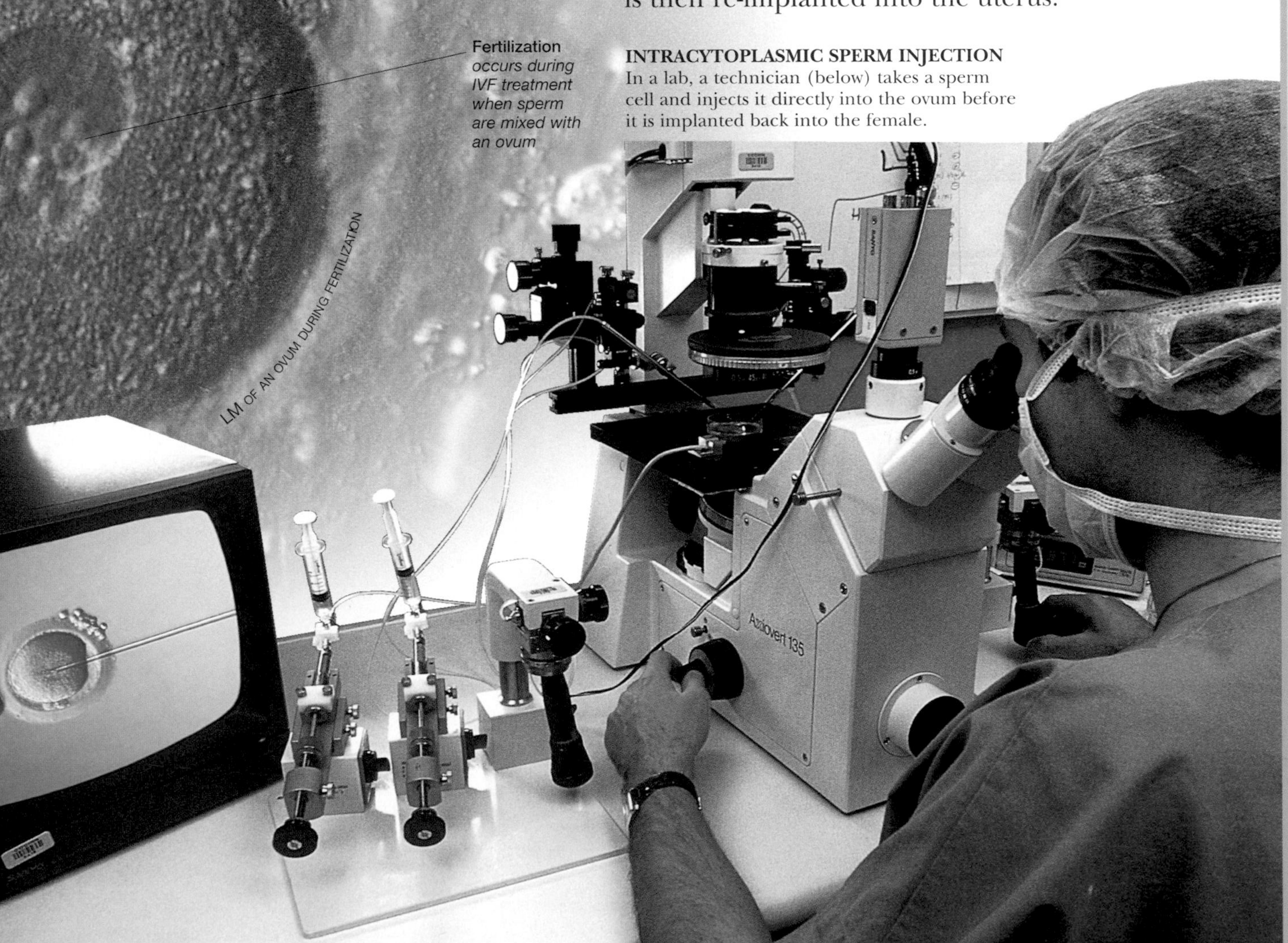

Fertilization *occurs during IVF treatment when sperm are mixed with an ovum*

LM OF AN OVUM DURING FERTILIZATION

INTRACYTOPLASMIC SPERM INJECTION
In a lab, a technician (below) takes a sperm cell and injects it directly into the ovum before it is implanted back into the female.

Early pregnancy

DURING THE FIRST TWELVE WEEKS OF pregnancy (the first trimester), the fertilized ovum develops from a single cell into a complex structure in which the internal organs, face, and limbs have all completed their basic development. At first, the developing baby is called an embryo, and bears little resemblance to a human – at four weeks its face has not yet formed, and it has a tail-like extension in place of legs. However, its heart has already started beating, and other internal organs have begun to form. Once eight weeks have passed, the baby is referred to as a fetus, and by the end of the first trimester it has limbs, a face that is recognizably human, and its internal organs are well developed. Meanwhile, the placenta has become well established in the wall of the uterus, and the mother has begun to experience many of the effects of pregnancy.

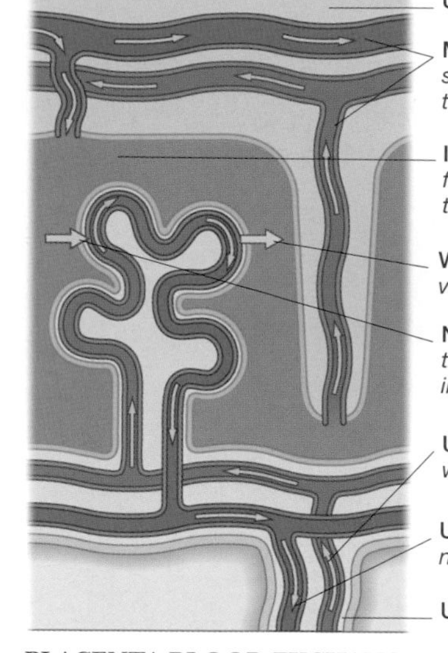
Uterus wall
Maternal blood vessels *supply fresh blood and take away wastes*
Intervillous space *filled with pools of the mother's blood*
Wastes *pass through thin villi walls into the placenta*
Nutrients *pass through thin villi walls into blood vessels*
Umbilical artery *carries wastes away from the fetu*
Umbilical vein *carries nutrients to the fetus*
Umbilical cord

PLACENTA BLOOD EXCHANGE
Inside the placenta, blood from the fetus is distributed into a complex network of small vessels (villi) that protrude into chambers called intervillous spaces through which maternal blood flows freely. These vessels and the tissue that surrounds them are thin, so nutrients and waste products have only a short distance to cross between fetal and maternal blood. Unfortunately, potentially harmful substances, such as drugs and viruses, are also able to cross with little difficulty.

PREGNANCY IN THE FIRST TRIMESTER

Complex changes take place within a woman's body as it works hard to adjust to pregnancy and provide for the developing embryo and placenta. The metabolic rate (see p. 22) increases by 10–25 per cent, and the pulse and breathing rate rise as more oxygen and other nutrients are sent to the fetus, and more carbon dioxide is exhaled. The fluid content of the body increases, as does the mother's weight. Even though the pregnancy is not yet far advanced, the mother's body begins to prepare for the birth – the walls of the uterus become more muscular, and the breasts begin to swell and form new milk-producing ducts.

Breasts *become tender and swell, and the area around the nipples darkens*

Internal organs *such as the lungs, heart, and kidneys work harder*

Abdomen *begins to swell slightly as the fetus grows*

Uterus *protects the growing fetus, while the placenta and umbilical cord nourish it*

Hormones *cause menstruation to stop. An increase in hormones can also cause nausea (morning sickness)*

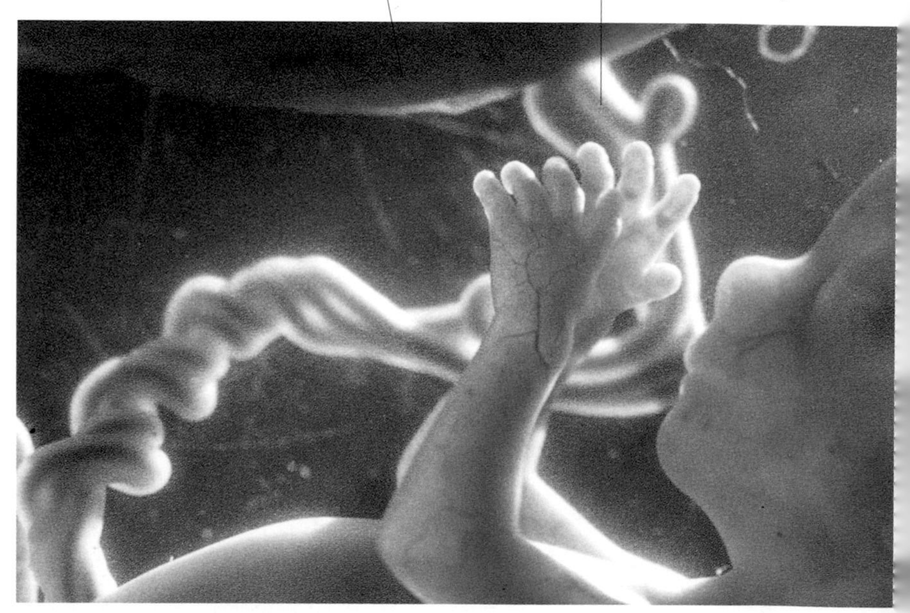
Placenta *provides an interface between mother and fetus. About 600 ml (1.1 pints) of maternal blood passes through the placenta per minute*
Umbilical cord *takes blood to and from the fetus. Its flexibility allows the baby to move around freely*

PLACENTA AND UMBILICAL CORD

The growing fetus has no direct contact with the maternal circulatory system. Instead, it is nourished by the placenta, a disc of tissue located on the internal surface of the uterus. The fetus and placenta are connected by the umbilical cord. In the placenta, blood from the fetus flows through an elaborate network of vessels that are bathed in the mother's blood. This allows oxygen and other nutrients to pass from the mother's blood into that of the fetus, while waste products travel in the opposite direction. Nutrient-rich, refreshed blood then returns to the fetus along the umbilical cord, which is about 60 cm (24 in) long.

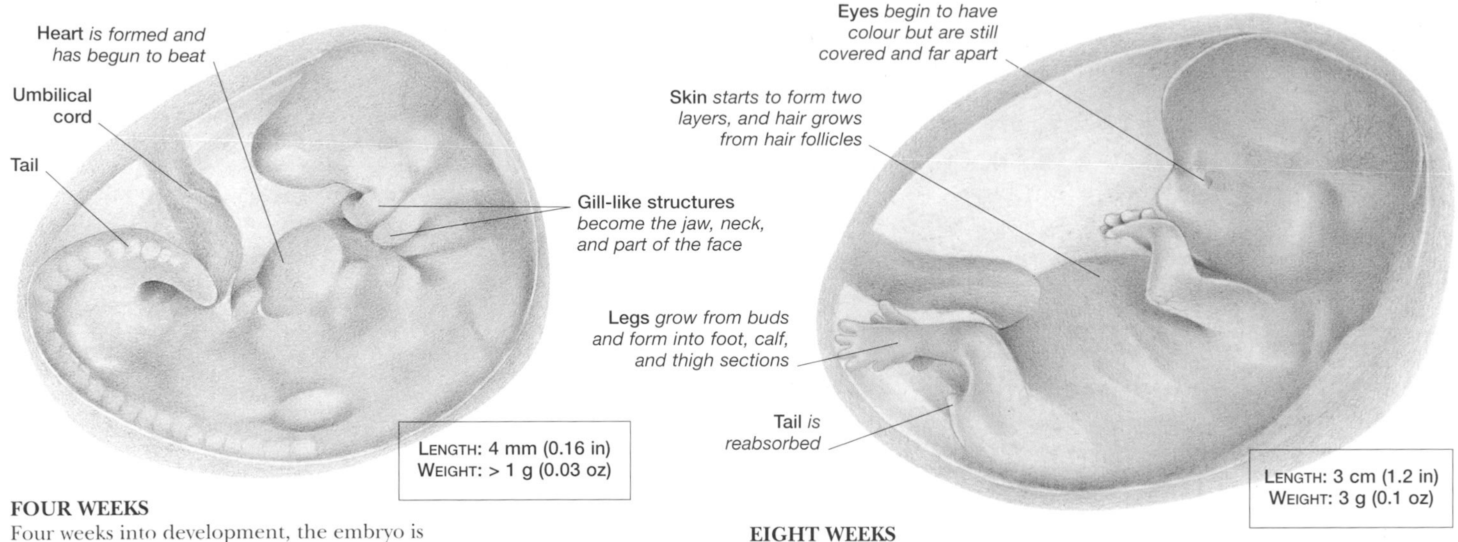

Heart is formed and has begun to beat
Umbilical cord
Tail
Gill-like structures become the jaw, neck, and part of the face
LENGTH: 4 mm (0.16 in)
WEIGHT: > 1 g (0.03 oz)
Eyes begin to have colour but are still covered and far apart
Skin starts to form two layers, and hair grows from hair follicles
Legs grow from buds and form into foot, calf, and thigh sections
Tail is reabsorbed
LENGTH: 3 cm (1.2 in)
WEIGHT: 3 g (0.1 oz)

FOUR WEEKS
Four weeks into development, the embryo is roughly the size of a pea. It resembles a tadpole more than a human, and has a tail-like protrusion instead of legs. Within the embryo, many changes have taken place. Cells have become specialized, so the heart is forming and has been beating for about a week, and the beginnings of a spinal cord is present. Many other vital organs, such as the brain, liver, and intestine, have begun to develop.

EIGHT WEEKS
The developing baby is now about the size of a strawberry and is called a fetus. Most organs are already established, and the head has grown to accommodate the rapidly developing brain, the basic structure of which is in place. The main features of the face are formed, and the tail-like protrusion is gone. The arms and legs have grown from buds, and the fingers and toes – initially paddle-like limbs – are now separated from one another as individual digits. The heart is pumping blood around new blood vessels, as the fetus moves around inside the uterus.

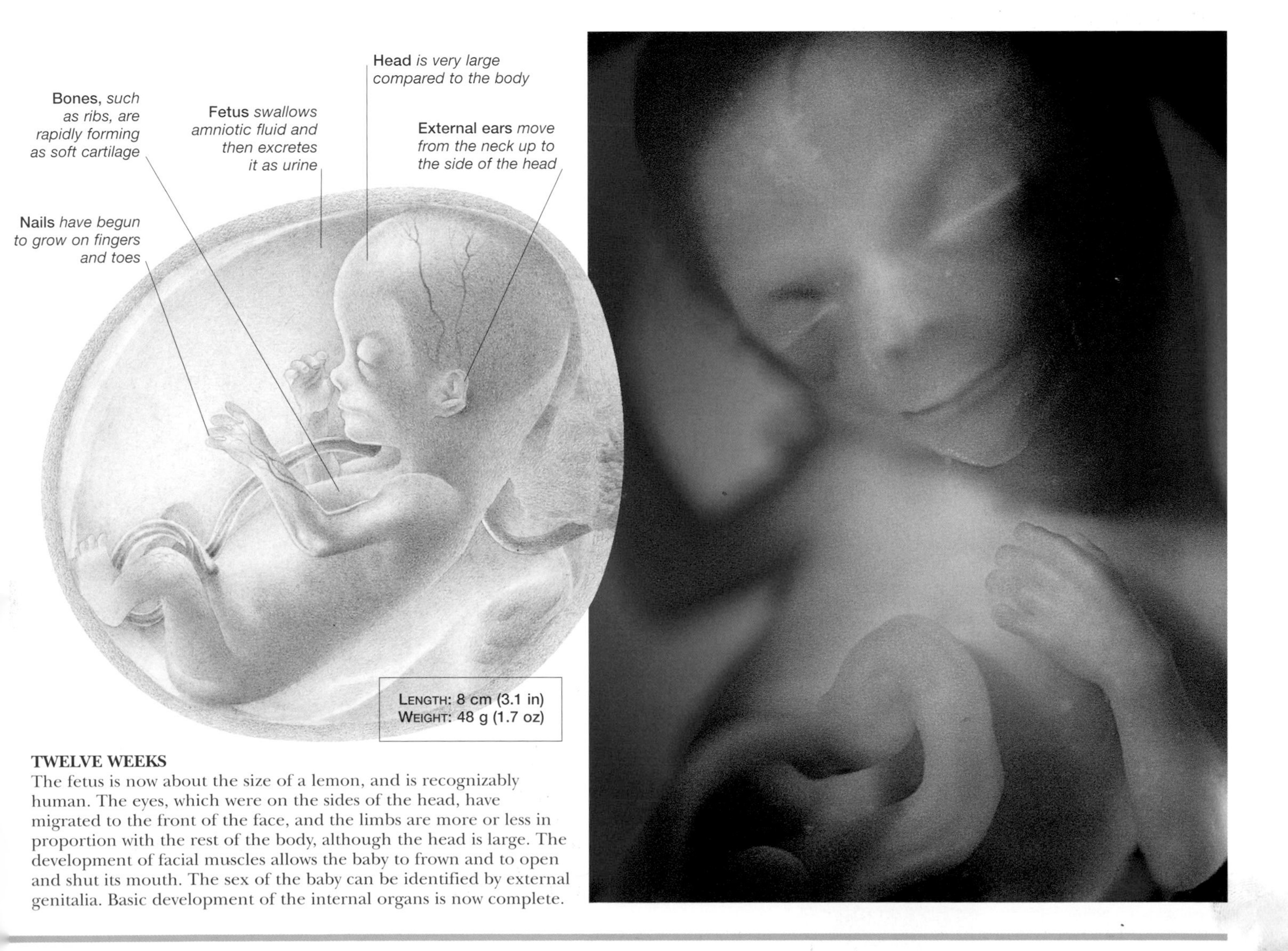

Bones, such as ribs, are rapidly forming as soft cartilage
Fetus swallows amniotic fluid and then excretes it as urine
Head is very large compared to the body
External ears move from the neck up to the side of the head
Nails have begun to grow on fingers and toes
LENGTH: 8 cm (3.1 in)
WEIGHT: 48 g (1.7 oz)

TWELVE WEEKS
The fetus is now about the size of a lemon, and is recognizably human. The eyes, which were on the sides of the head, have migrated to the front of the face, and the limbs are more or less in proportion with the rest of the body, although the head is large. The development of facial muscles allows the baby to frown and to open and shut its mouth. The sex of the baby can be identified by external genitalia. Basic development of the internal organs is now complete.

Growing fetus

BY THE END OF THE FIRST TRIMESTER of pregnancy, the major organs of the fetus are in place. The second and third trimesters (weeks 13–28 and 29–birth) are primarily concerned with fetal growth and final maturation. At first, the fetus grows rapidly in size but gains little weight, but later he or she becomes heavier and plumper as fat accumulates beneath the skin. Growth of the fetus causes the uterus to expand, so that the mother's abdomen swells noticeably. She may also experience effects such as breathlessness and frequency of urination as the uterus begins to press upon nearby organs such as the diaphragm and bladder. Towards the end of pregnancy the fetus undergoes final preparations for life outside the uterus, and usually turns head down, ready to pass through the birth canal head first.

Fetus *is thin but has grown in length, and muscles and bones are maturing*

Skin *is sensitive to touch, so the fetus moves if the mother's abdomen is pressed*

LENGTH: 22 cm (8.7 in)
WEIGHT: 800 g (1 lb 12 oz)

24-WEEK FETUS
Fetal development in the second trimester is mainly concerned with growth. After six months the baby's length has increased rapidly so the head is more in proportion, but the fetus is only about a quarter of its final weight. Brain and nerve cells and an immune system are maturing, and the ears begin to hear sounds inside and outside the uterus. By the end of the second trimester, the hands, fingers, and face are fully developed.

PREGNANCY IN THE SECOND TRIMESTER

During the second trimester, the mother's abdomen begins to swell noticeably as the uterus expands beyond the boundaries of the pelvis in order to accommodate the growing fetus. A pigmented streak called the *linea nigra* may appear down the centre of the abdomen, and other pigmented areas such as the nipples may darken. Changes in the digestive system can cause heartburn and reduced frequency of bowel movements, and the gums may become spongy. In order to supply extra blood to the uterus and kidneys, the heart works twice as hard as normal.

Hair and skin *becomes healthier in some women*

Uterus *enlarges, making the abdomen swell and sometimes causing stretchmarks*

Fetal *movements are first felt by the mother as a "fluttering" sensation between weeks 16 and 22*

Back pain *may be caused by the increased weight of the fetus*

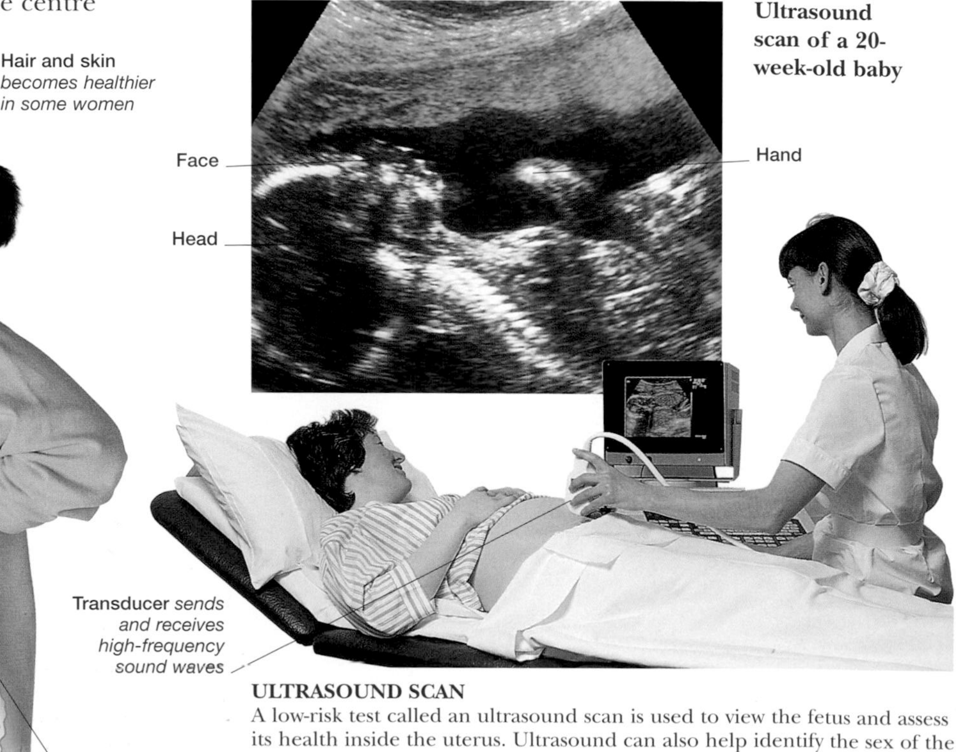

Ultrasound scan of a 20-week-old baby

ULTRASOUND SCAN
A low-risk test called an ultrasound scan is used to view the fetus and assess its health inside the uterus. Ultrasound can also help identify the sex of the baby. High-frequency sound waves are beamed into the abdomen through a hand-held instrument called a transducer, which is placed on the skin. Waves are reflected back from the contents of the abdomen into the transducer, creating an image of the uterus and fetus on a television screen.

Exercising in Pregnancy

Regular, suitable exercise during pregnancy has several benefits – it helps the mother's body adapt to the pregnancy and prepare for labour, and it enables her to recover more quickly following childbirth. Exercise also helps with relaxation, and can promote improved sleep. Swimming can be particularly enjoyable and beneficial. Muscles are toned and stamina improves, while the water supports the mother's increased weight, making injuries very unlikely.

Amniotic sac *is now filled with about 1 litre (1.76 pints) of fluid*

FULL-TERM BABY

The weight of the fetus increases rapidly during the last trimester of pregnancy. Fat is deposited beneath the skin in order to provide the baby with energy and insulation after birth, and as a result the fetus becomes more rounded. The lungs become well-developed and produce a substance called a surfactant, which will play a vital role in breathing. In many ways the fetus behaves as it will after birth – it has periods of sleep and wakefulness, responds to music and voices, and is startled by loud noises.

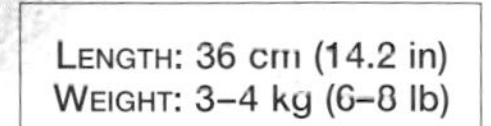

Eyes *are blue but may change colour after birth*

Pregnancy in the Third Trimester

The growing fetus causes the uterus to fill the abdomen, so that it displaces the digestive tract and presses against the diaphragm. This, combined with the oxygen demands of the fetus, causes the mother to breathe more rapidly and deeply. Meanwhile, pressure of the uterus on the bladder often leads to an increased frequency of urination. Aches and pains occur due to stretching of ligaments, especially in the hips and pelvis.

Breasts *may secrete colostrum, a yellowish fluid that provides nourishment for the baby after birth*

Squashed intestine

Fetus *has undergone a growth spurt and takes up much of the abdomen*

Mucus plug *protects the fetus from infection*

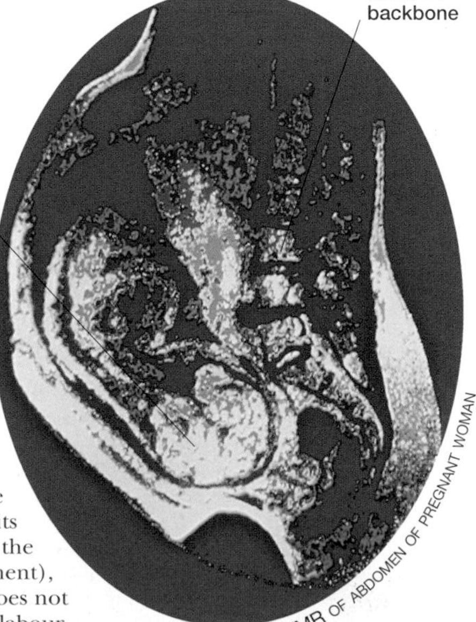

NMR OF ABDOMEN OF PREGNANT WOMAN

Fetus *usually turns head down to fit into the pelvis during engagement*

READY FOR BIRTH

By the end of the pregnancy the fetus is quite plump – it only just fits into the uterus by curling up. If the fetus is not already positioned head down in the uterus, it usually turns now so that it can pass through the birth canal head first. Once it has turned, the fetus may descend slightly so that its head settles into the lowest part of the uterus (a process called engagement), although in many cases this does not happen until the start of labour.

Birth

BIRTH OCCURS WHEN THE FETUS LEAVES the uterus to enter the outside world. This usually happens about 38 weeks after fertilization. In most cases birth begins when the mother goes into labour, which is the process by which powerful contractions of the muscular walls of her uterus gradually expel the fetus through the vagina. There are three distinct stages of labour: the first of these prepares the birth canal for the baby to pass through it, the second involves the birth of the baby, and the third consists of the delivery of the placenta. The length of time that these stages take varies, and is influenced by the number of previous pregnancies the mother has had. As soon as the baby is born it takes its first breaths and begins to adjust to its new environment, allowing it to survive outside its mother's body.

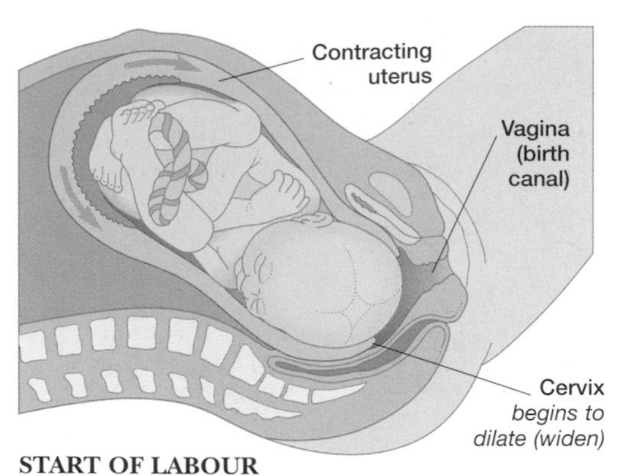

START OF LABOUR
Labour begins with contractions of the uterus that become painful and frequent. Often, the amniotic sac lining the uterus breaks at this stage, and the fluid surrounding the fetus pours out through the vagina. The cervix begins to widen so that the fetus can pass through it. Widening of the cervix is brought about by the softening of its tissues, and by pressure from the head of the fetus as it is pushed downwards with each contraction.

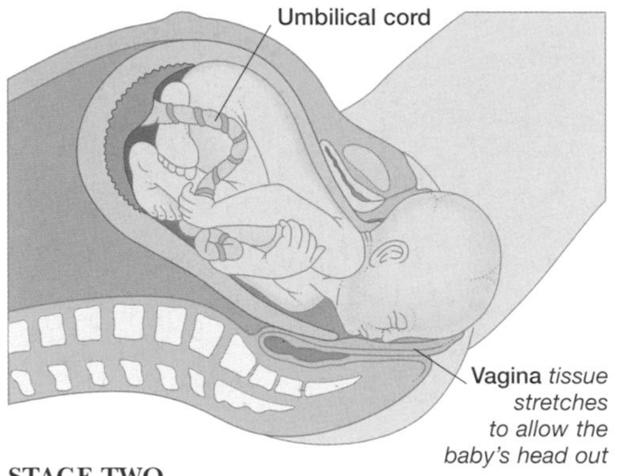

STAGE TWO
During the second stage of labour, which begins once the cervix has widened fully, the baby is born. At first, the fetus faces the mother's side, but as it gradually descends its head rotates so that its face is towards the mother's back. The mother usually now develops an irresistible desire to push with each contraction, adding to the forces that propel the fetus along the birth canal.

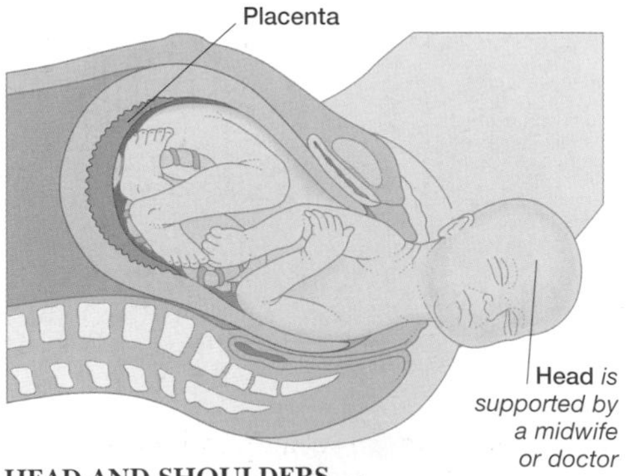

HEAD AND SHOULDERS
As soon as the head of the fetus has emerged from the vagina it rotates back into line with the rest of the body, allowing the shoulders to be delivered with the next contraction. Normally, the shoulder nearest the mother's belly appears first, soon followed by the other. Once the shoulders have been delivered, the rest of the body can pass through the birth canal quickly, with little difficulty.

NEWBORN BABY

Sudden contact with the world outside the uterus normally stimulates the baby to take its first breath, moments after it leaves the birth canal and is lying between its mother's legs. The baby's circulatory system begins to adjust so that blood circulates through the lungs rather than the placenta. Much of the nutritious blood in the placenta flows back along the umbilical cord into the baby's body so that it is not wasted when the cord is cut. As soon as the cord has been cut, the baby can be given to the mother to suckle. On average, a baby is 51 cm (20 in) long and weighs 3.5 kg (7 lb 11 oz) at birth. The baby's skull bones are not completely fused, so its head might be temporarily misshapen by the journey down the birth canal.

FORCEPS SECRET

18th-century forceps

Obstetric forceps consist of two interlocking paddles that are inserted into the vagina during difficult deliveries to help the baby's head move through the birth canal. They were developed in the 17th century by an English family, the Chamberlens, and represented a significant breakthrough, as previous means of assisting in difficult deliveries generally resulted in the death of the baby. However, the Chamberlens kept their invention secret, and it was not until the 18th century that the forceps became more widely used. Numerous versions have evolved since, although many forceps in use today are based on early designs.

Cut umbilical cord

Placenta *peels away from the uterus lining*

FINAL STAGE

The final stage of labour begins as soon as the baby has been born, and lasts until the placenta is delivered. Mild contractions continue after the birth of the baby, and the placenta gently peels away from the inner surface of the uterus and follows the umbilical cord out through the vagina. Blood vessels in the wall of the uterus that previously supplied the placenta clamp shut, preventing excessive bleeding. The placenta, or afterbirth, is usually delivered about 5–30 minutes after the birth of the baby, and consists of a disc of spongy, blood-rich tissue about 20 cm (8 in) in diameter.

APGAR SCORE

At birth, and during the following minutes, the baby's well-being can be assessed using a tool called the Apgar score, developed by an American doctor, Virginia Apgar, in which vital signs are evaluated. A low Apgar score indicates that the baby is unwell.

SIGN	SCORE: 0	SCORE: 1	SCORE: 2
Activity	Limp	Some bending of limbs	Active movements
Pulse	None	Below 100 bpm	Over 100 bpm
Grimace	None	Grimace or whimpering	Cry, sneeze, or cough
Appearance	Pale, blue	Blue extremities	Pink
Respiration	None	Slow or irregular breaths, weak cry	Regular breaths, strong cry

Vernix *is a greasy substance that covers and protects the baby's skin until it is born*

Umbilical cord *is clamped and cut once blood has flowed back to the baby*

BREASTFEEDING

During pregnancy, hormones activate the milk-producing glands of the breasts, so that food is available for the newborn. At first, the breasts produce colostrum, a yellowish fluid that is rich in antibodies, which protect the baby from infection. During the second day after birth, colostrum is replaced by ordinary breast milk, which contains lactose (a type of sugar), protein, and fat. Usually, a mother produces about 1 litre (1.76 pints) of milk a day for as long as she wishes to breastfeed.

SAVING MOTHERS

THE BIRTH OF A BABY SHOULD be a time of happiness for parents. Yet, although the process is natural, it is never risk-free and, tragically, a few mothers die during or following childbirth. The number of fatalities today is tiny when compared with earlier centuries. In the 18th and 19th centuries, for example, many women died within a week or two of giving birth from a disease called puerperal (childbed) fever, particularly in hospitals. In Europe and America, a handful of doctors realized both how puerperal fever might be spread and be prevented. But their findings were ignored, and it took decades before measures were taken to save mothers.

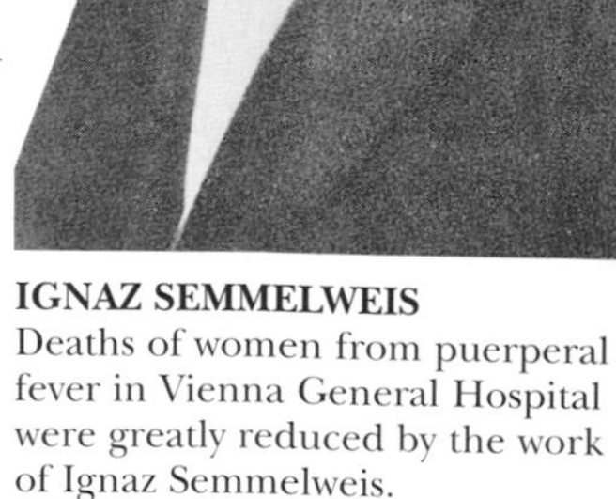

IGNAZ SEMMELWEIS
Deaths of women from puerperal fever in Vienna General Hospital were greatly reduced by the work of Ignaz Semmelweis.

WASHING HANDS
To many 19th-century doctors, the idea that washing their hands would stop the spread of disease seemed ridiculous. Today, hand washing is a vital part of stopping infection passing from doctor or midwife to patient.

THE MENACE OF INFECTION

By the end of the 18th century there had been great advances in the understanding of pregnancy and childbirth, with the training of specialist doctors, called obstetricians, and the opening of maternity hospitals. Although these hospitals provided bed rest for poorer expectant mothers, they also had appalling death rates. Within days of giving birth, many mothers died from puerperal

EARLY MATERNITY WARD
This illustration, taken from the bo *Microcosm of London* published in 18(shows the scene inside a woman's ward in the Middlesex Hospital, London, England. The view was probably idealized by the artist, as conditions were certainly more crowded and dirty than shown here.

fever, an infection that spreads through the vagina and uterus. Why such high death rates occured in these hospitals was unknown.

In 1846, Hungarian obstetrician Ignaz Semmelweis (1818–65) started work in Vienna General Hospital. He was shocked to discover that many more women died from puerperal fever in Maternity Ward 1 than in Ward 2. The difference between the wards was that in Ward 1, women were attended by doctors who also carried out autopsies (dissections of dead patients), while in Ward 2, women were attended by midwives who did not. Semmelweis deduced that the doctors, who rarely washed their hands, were transferring infection from corpses to their women patients. He ordered medical staff to wash their hands carefully before attending to their patients and, as a result, the death rate in Ward 1 dropped to the same level as Ward 2. But, when he reported his findings, Semmelweis was ridiculed by fellow doctors. He had dared to question the accepted medical view that infections were spread not by human contact but by a mysterious miasma, or "bad air".

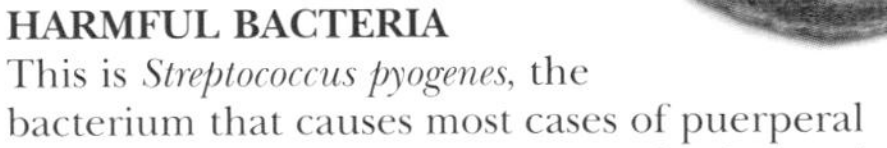

HARMFUL BACTERIA
This is *Streptococcus pyogenes*, the bacterium that causes most cases of puerperal fever. Streptococcal bacteria are spherical and are linked together in a chain. Infection may take place during or after birth when a woman is examined by someone whose fingers are contaminated with *S. pyogenes*.

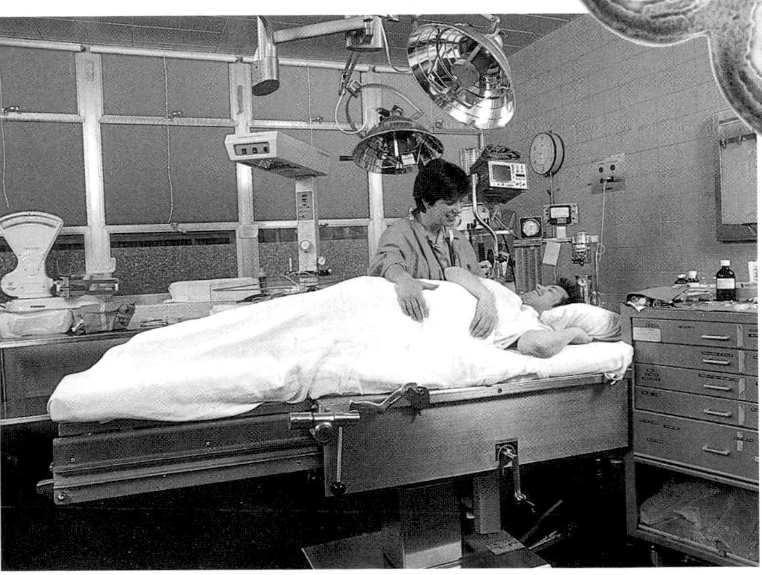

MODERN DELIVERY ROOM
In the photo above, an expectant mother is being attended to by a nurse in the delivery room of a modern hospital. She will give birth here safe in the knowledge that conditions are clean and hygienic, and that medical care is at hand should there be any complications during the birth.

Identifying the cause

Semmelweis was not alone in his views. In America, the physician and writer Oliver Wendell Holmes (1809–94) presented a paper in 1843 entitled *On the Contagiousness of Puerperal Fever* at the Boston Society of Medical Improvement. Like Semmelweis he suggested that infection was spread to women by unhygienic medical staff. Holmes was also ridiculed by other doctors, but he persisted in his campaign to stop the unnecessary deaths of new mothers. Both Semmelweis and Holmes were finally vindicated by Louis Pasteur (1822–95), the great French scientist who, in 1879, proved that puerperal fever was caused by the *Streptococcus* bacterium, thus destroying the miasma theory forever.

Modern practices

As a result of Pasteur's findings, and those of physicians such as Joseph Lister (see p. 270), there was a gradual acceptance of the need for clean conditions in hospital wards. Modern maternity wards provide a safe and sterile environment for childbirth, and obstetricians and midwives are educated to a high standard. If bacterial infection does occur, it can be treated using antibiotics.

Childhood

CHILDHOOD SPANS THE YEARS from infancy to adolescence. During early childhood, the achievements of infancy provide a foundation for further learning and development. The child begins to walk and run, and gradually becomes more skilled and graceful in his or her movement. The secure emotional bond between a child and his or her parents allows and encourages an interaction with others, and learning takes place through active engagement with the world. During the ensuing years, intellectual skills such as reading are acquired, and abstract thinking and complex social behaviour evolve. Rapid physical change also takes place – the child grows steadily, the body shape becomes more mature, and milk teeth emerge, which are later replaced with permanent teeth. By the time adolescence approaches, the child has become a sophisticated being, and has undergone remarkable personal, intellectual, and physical development.

MOVEMENT
Basic movement skills that were acquired during infancy develop further during early childhood. Walking and running are achieved by the age of two, by which time a child is also adept at manipulating objects. With each passing year, new skills are acquired, such as jumping, walking on tiptoe, and riding a tricycle. As the child develops, these skills are refined, so that by the age of ten, movement is smooth and coordinated, and complex tasks such as juggling can be carried out with practice.

Girls and boys *interact well together for the first ten years*

Reading *usually begins by the age of six*

Play *is an important part of development and learning*

LEARNING
Early childhood is largely concerned with physical exploration, for example learning about colours and shapes. With time, concepts such as size, direction, and time are grasped. As new knowledge is gained, it is incorporated into the child's understanding of the world. Specific skills such as reading and counting are acquired. Analytical thinking develops, so that by about the age of 12 the child is capable of complex reasoning.

SOCIAL BEHAVIOUR
The secure emotional bonds a young child shares with his or her parents encourage interaction with others. By the age of three or four he or she enjoys playing with other children, and is upset when others are hurt. Play becomes more imaginative and complex, and an independence from adults develops. Older children begin to work well in groups and enjoy team-based activities. Personal development continues so that by adolescence, children are well developed socially and capable of complex thought.

Speech and vocabulary *becomes complex as childhood progresses*

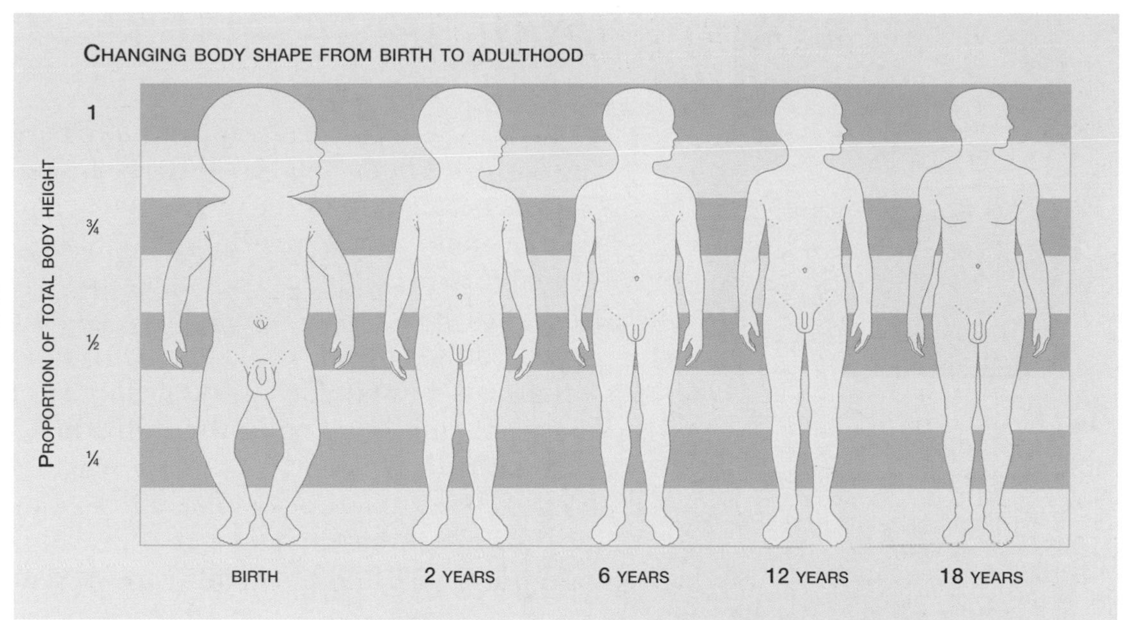

SPEECH

Speech develops at varying rates in young children. Gradually actual words replace the "babble" of infancy. From about the age of two the child begins to combine groups of words to convey more complex ideas, and by four he or she is usually able to engage in simple conversation. During the remaining years of childhood, speech rapidly increases in complexity as vocabulary and grammatical understanding develop.

BODY PROPORTIONS

Children grow rapidly during the first two years of life. Growth then continues at a reduced rate for the remainder of childhood – usually about 7 cm (2.75 in) in height and 2.7 kg (6 lb) in weight each year. Different parts of the body grow at varying rates. In a young child the head accounts for a large part of the total height of the body, but it becomes less prominent as the limbs and trunk grow longer. Over time, the proportions of the body gradually begin to resemble those of an adult.

COLOURED X-RAY OF A CHILD'S SKULL

FACE SHAPE

At birth and throughout infancy, copious reserves of fat beneath the skin give the face a rounded appearance. This persists well into childhood, so that during the first years of life the finer features of the face have yet to become apparent. At about the age of five, this begins to change – the fat reserves of babyhood ebb away, and the face becomes leaner and the features more distinct. Family resemblances that may previously have been vague often become more pronounced at this time. Facial bones become larger and stronger as childhood passes, also changing the face shape.

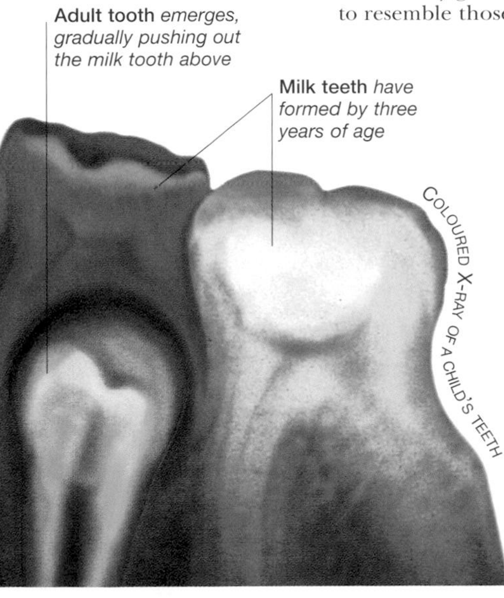

Adult tooth *emerges, gradually pushing out the milk tooth above*

Milk teeth *have formed by three years of age*

COLOURED X-RAY OF A CHILD'S TEETH

TEETH

By the end of infancy, a child usually has two front teeth in the upper jaw and two in the lower. More teeth emerge during the following months, so that the full set of 20 milk teeth (deciduous teeth) is in place by about the age of three. Teething can be painful, so while a tooth is emerging the child may be irritable and tearful. At about the age of six, permanent teeth begin to emerge, and as they do so, each dislodges the overlying milk tooth, which falls out. Permanent teeth have usually replaced milk teeth by adolescence (see pp. 180–1).

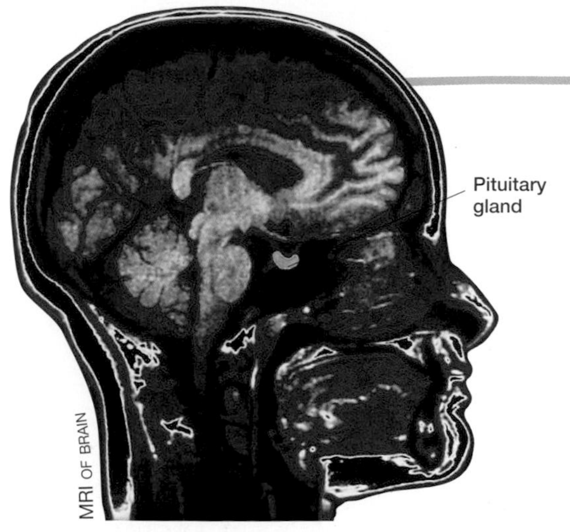
Pituitary gland
MRI OF BRAIN

SWITCHING ON
Puberty is triggered by the hypothalamus (see p. 95) which sends hormones to the pituitary gland, located at the base of the brain. The pituitary gland then releases hormones of its own that stimulate the ovaries and testes to mature. As these organs become active, they too secrete hormones. These cause changes such as female breast growth and the deepening male voice.

Adolescence

ADOLESCENCE IS A TIME OF transition from childhood to adulthood, and involves both physical and psychological changes. Through a process called puberty, the sexual organs become functional. As a result, sex cells (sperm in males and ova in females) become available so that the individual is capable of reproduction. Other physical characteristics, such as facial hair in males and breasts in females, also appear, and a growth spurt takes place so that adult height is achieved. Meanwhile, vital personal development takes place. As childhood ends, the individual becomes more self-aware and sophisticated, and independence and an adult identity are forged through a process that can be emotionally turbulent. By the end of the late teens, the individual has progressed both physically and personally, and is ready for adult life.

PUBERTY FOR BOYS

In boys, puberty usually begins between the ages of 12 and 14, although this varies greatly. The first sign of puberty is enlargement of the testes as they prepare to produce sperm. The penis then begins to grow, and reaches its full adult size after about two years. Pubic, facial, armpit, and chest hair grows. Structures of the larynx (voice box) increase in size, making the voice deeper. As sexual development progresses, the growth rate accelerates, the shoulders broaden, and muscles develop.

FACIAL HAIR
An increase in the hormone testosterone causes facial hair to grow as boys progress through puberty. This photo (left) shows facial hairs that have been shaved and are now growing back. The first hairs are fluffy, but later become thicker and stronger.

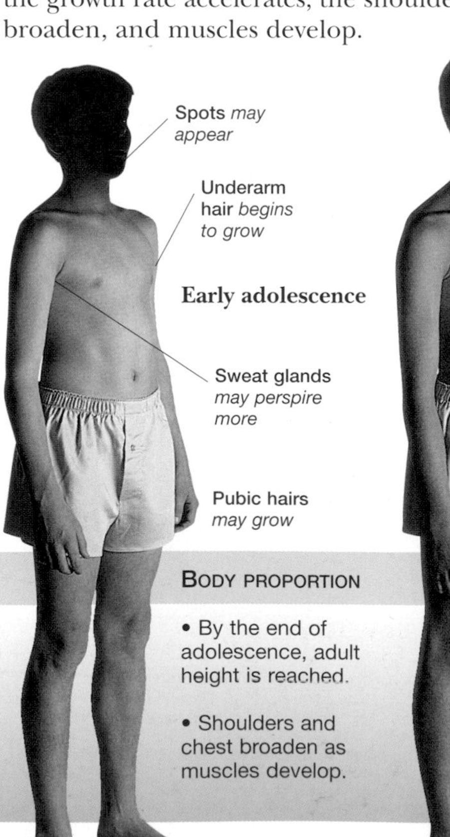
Spots *may appear*
Underarm hair *begins to grow*
Sweat glands *may perspire more*
Pubic hairs *may grow*

Early adolescence

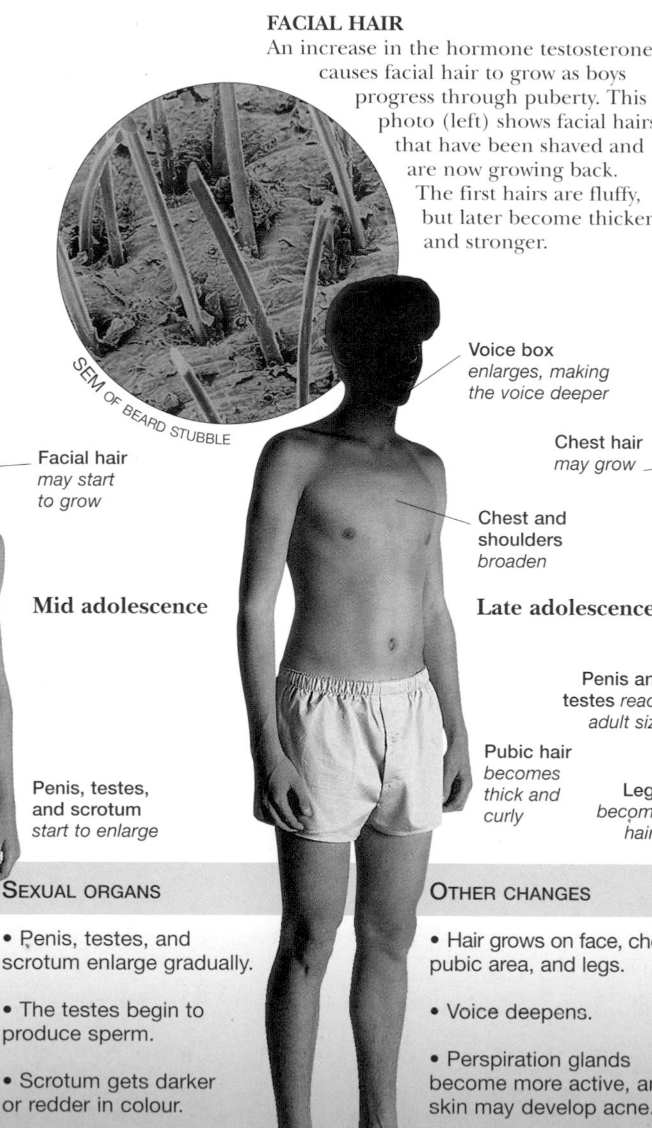
SEM OF BEARD STUBBLE
Facial hair *may start to grow*
Penis, testes, and scrotum *start to enlarge*
Voice box *enlarges, making the voice deeper*
Chest and shoulders *broaden*

Mid adolescence

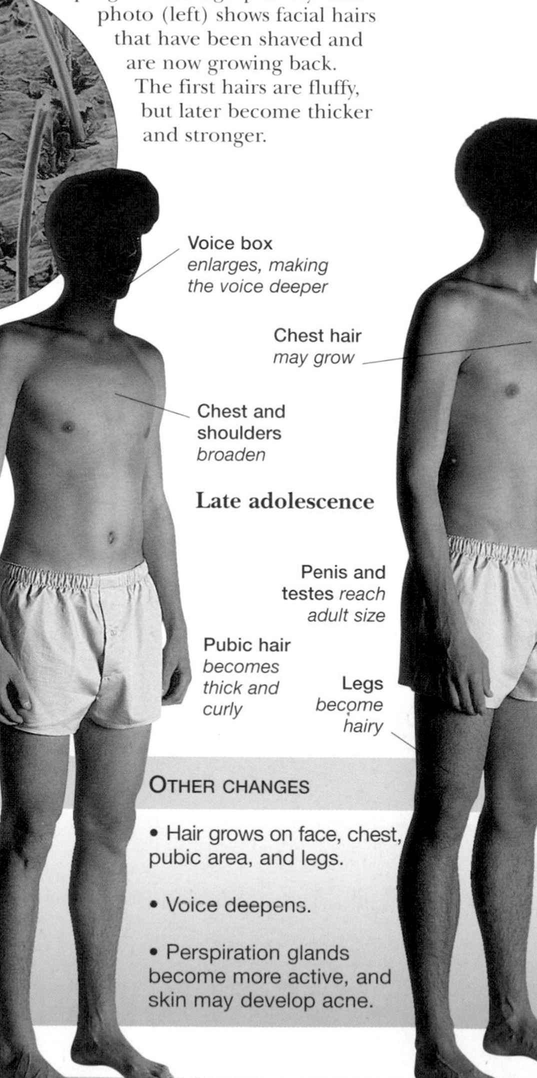
Chest hair *may grow*
Penis and testes *reach adult size*
Pubic hair *becomes thick and curly*
Legs *become hairy*

Late adolescence

BODY PROPORTION

- By the end of adolescence, adult height is reached.
- Shoulders and chest broaden as muscles develop.

SEXUAL ORGANS

- Penis, testes, and scrotum enlarge gradually.
- The testes begin to produce sperm.
- Scrotum gets darker or redder in colour.

OTHER CHANGES

- Hair grows on face, chest, pubic area, and legs.
- Voice deepens.
- Perspiration glands become more active, and skin may develop acne.

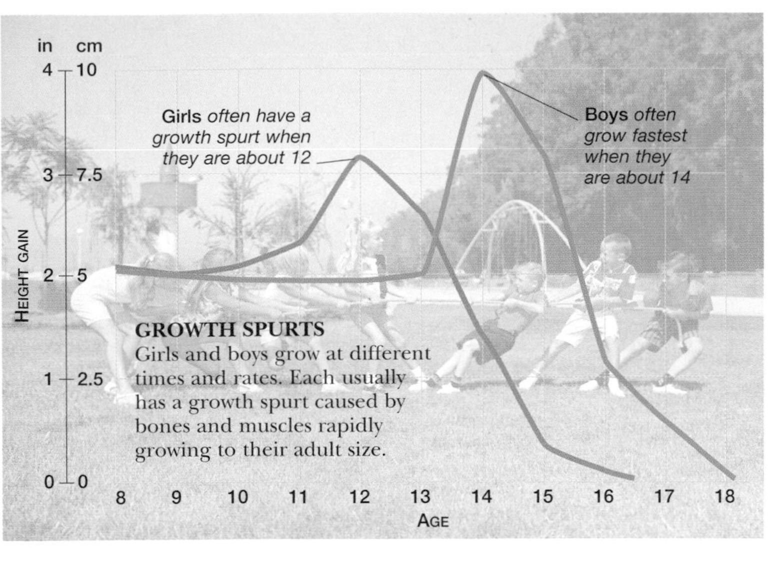
in cm
4 10
3 7.5
2 5
1 2.5
0 0
HEIGHT GAIN
8 9 10 11 12 13 14 15 16 17 18
AGE
Girls *often have a growth spurt when they are about 12*
Boys *often grow fastest when they are about 14*

GROWTH SPURTS
Girls and boys grow at different times and rates. Each usually has a growth spurt caused by bones and muscles rapidly growing to their adult size.

PUBERTY FOR GIRLS

Girls usually begin puberty between the ages of 10 and 12, and the first period (when ova start to be released) occurs about two years later. Puberty in girls is heralded by enlargement of the ovaries, but as this is hard to see, the first sign is usually breast growth. By the time of the first period, pubic and armpit hair are well developed. At first, periods may be irregular, but they gradually become predictable. Meanwhile, rapid growth, a change in the distribution of body fat, and widening of the pelvis take place.

ACNE
During puberty, hormones affecting both boys and girls cause glands in the skin to secrete more oil than previously. If the duct leading from the gland to the skin's surface becomes blocked with dead cells or hardened oil, spots form. The surrounding skin can become infected, making it sore and red.

CHANGING FEELINGS
During adolescence, individuals become more sophisticated and independent. Complex choices about issues such as sexual identity, relationships, and the future are faced. Childhood is abandoned, and the individual faces the daunting search for an adult identity. This can be a challenging, stressful time, and as a result, adolescents may experience feelings of anxiety, depression, and poor self-esteem. However, adolescents are generally less confrontational than is commonly perceived.

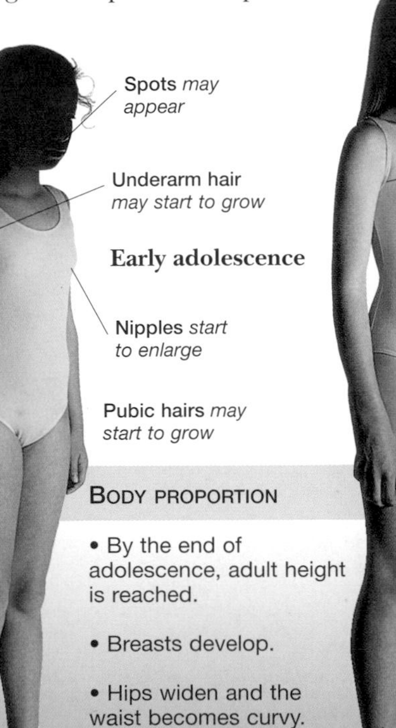
Spots *may appear*
Underarm hair *may start to grow*
Early adolescence
Nipples *start to enlarge*
Pubic hairs *may start to grow*

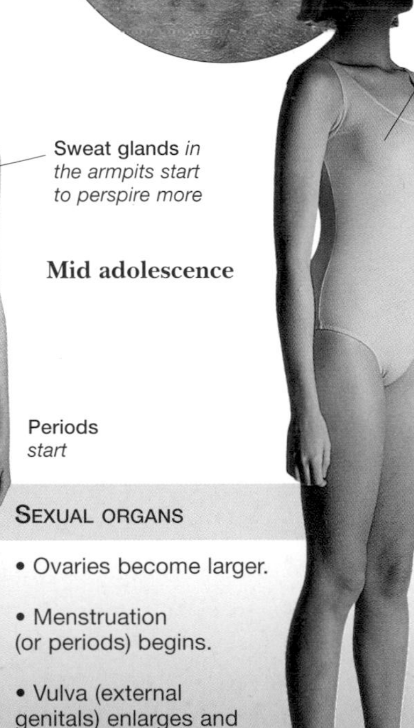
Sweat glands *in the armpits start to perspire more*
Mid adolescence
Periods *start*
Breasts *become rounder*

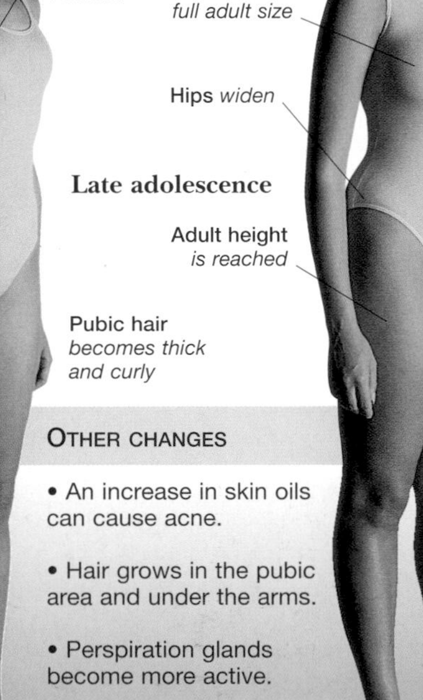
Breasts *reach their full adult size*
Hips *widen*
Late adolescence
Adult height *is reached*
Pubic hair *becomes thick and curly*

BODY PROPORTION

- By the end of adolescence, adult height is reached.
- Breasts develop.
- Hips widen and the waist becomes curvy.

SEXUAL ORGANS

- Ovaries become larger.
- Menstruation (or periods) begins.
- Vulva (external genitals) enlarges and becomes fleshy.

OTHER CHANGES

- An increase in skin oils can cause acne.
- Hair grows in the pubic area and under the arms.
- Perspiration glands become more active.

Adulthood and ageing

AN INDIVIDUAL'S PERSONALITY AND lifestyle evolve as he or she progresses through adulthood. The way in which this takes place is unique to everyone, and is influenced by factors such as personal preference, opportunities and circumstances, and social and cultural expectations. Indeed, the lifestyles of two adults of similar age often differ just as much as those of two people from separate generations. As an individual's life progresses, physical changes also take place. The dramatic development of childhood and adolescence is replaced by a subtler, more gradual transition, and in due course signs of ageing begin to appear in the bones, skin, senses, and other organs. Despite these changes, modern medicine allows many elderly people to have an enjoyable and healthy old age.

COLOURED X-RAY OF A LOWER SPINE WITH OSTEOPOROSIS

ADULTHOOD

The nature of adulthood varies enormously depending on personality, decisions made, circumstances, and social context. However, commonly this is a time of independence and responsibility. Adulthood allows personal freedom but this is often accompanied by duties – adults are generally held accountable for their actions, and are expected to support themselves and their families financially. Meanwhile, choices must be made about careers, relationships, and parenthood, all of which involve taking on responsibilities.

Osteoporosis *has caused this vertebra to collapse*

BONE DEGENERATION
This X-ray shows the spine of a 65-year-old woman who has developed a bone condition called osteoporosis. This disease causes the outer layer of bones to become thinner, while the inner layer becomes more porous, with fewer blood cells and less calcium. Bones are therefore more brittle and likely to be damaged. Here, a vertebra has become compressed and wedge-shaped, distorting the spine. Osteoporosis is common in older women, particularly after the menopause.

MENOPAUSE

The menopause is when a woman stops having periods, usually between the ages of about 45 and 54. The menopause represents the end of a woman's reproductive potential, and this can have emotional and psychological impacts. Also, as the ovaries cease to work, levels of the hormone oestrogen decline. This may cause symptoms such as hot flushes, loss of sexual desire, and wrinkling of the skin, and also increases the risk of disorders such as stroke and osteoporosis. Women may therefore choose to use hormone replacement therapy, in which artificial oestrogen is taken in pills (right) or through patches (top right).

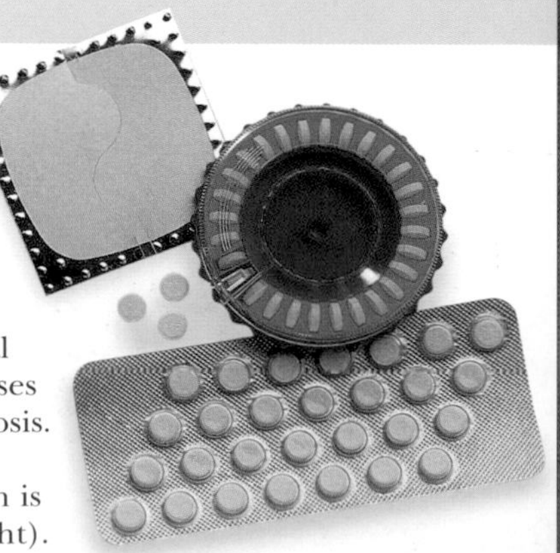

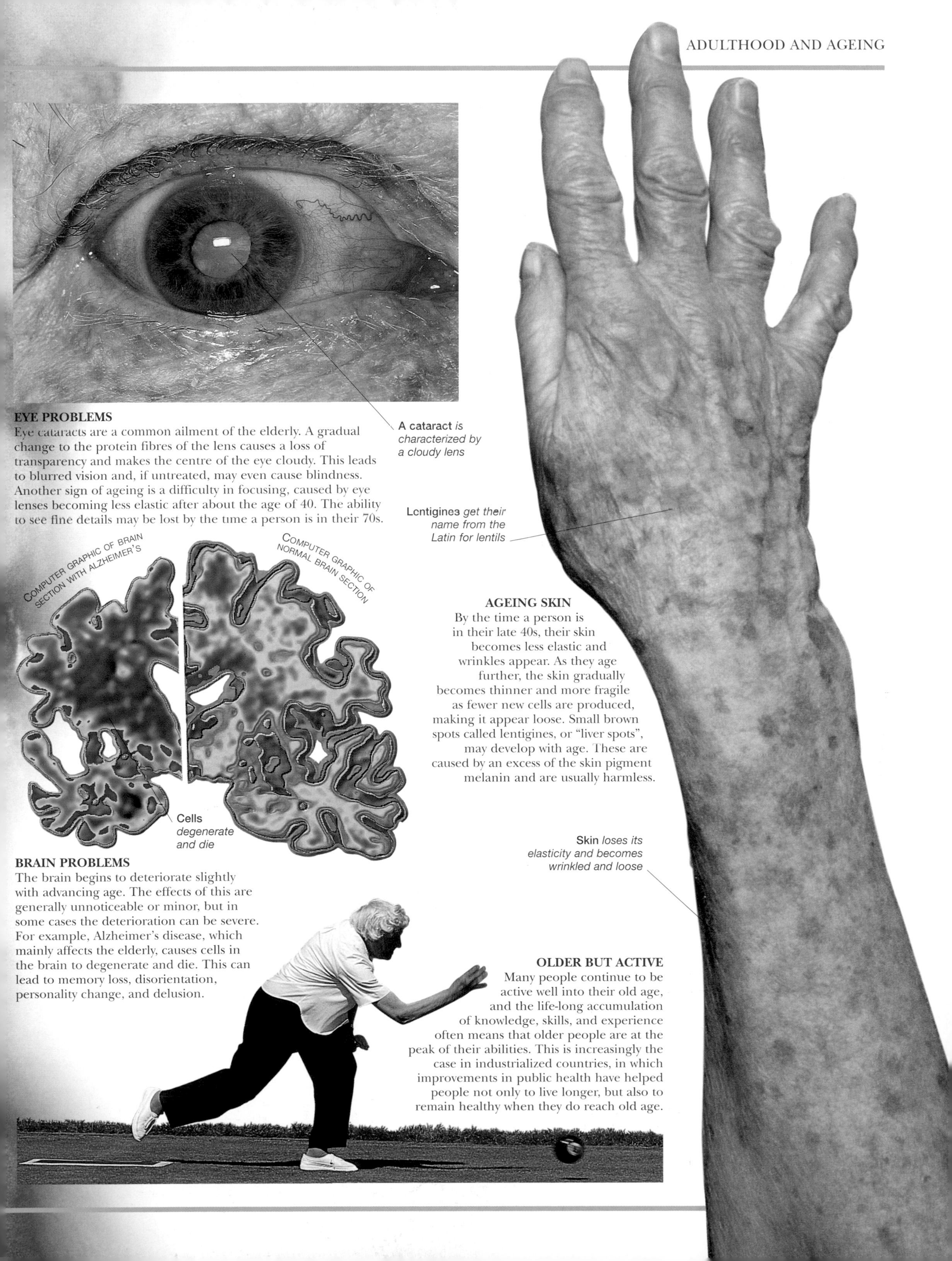

A cataract is characterized by a cloudy lens
Lentigines get their name from the Latin for lentils
COMPUTER GRAPHIC OF BRAIN SECTION WITH ALZHEIMER'S
COMPUTER GRAPHIC OF NORMAL BRAIN SECTION
Cells degenerate and die
Skin loses its elasticity and becomes wrinkled and loose

EYE PROBLEMS
Eye cataracts are a common ailment of the elderly. A gradual change to the protein fibres of the lens causes a loss of transparency and makes the centre of the eye cloudy. This leads to blurred vision and, if untreated, may even cause blindness. Another sign of ageing is a difficulty in focusing, caused by eye lenses becoming less elastic after about the age of 40. The ability to see fine details may be lost by the time a person is in their 70s.

AGEING SKIN
By the time a person is in their late 40s, their skin becomes less elastic and wrinkles appear. As they age further, the skin gradually becomes thinner and more fragile as fewer new cells are produced, making it appear loose. Small brown spots called lentigines, or "liver spots", may develop with age. These are caused by an excess of the skin pigment melanin and are usually harmless.

BRAIN PROBLEMS
The brain begins to deteriorate slightly with advancing age. The effects of this are generally unnoticeable or minor, but in some cases the deterioration can be severe. For example, Alzheimer's disease, which mainly affects the elderly, causes cells in the brain to degenerate and die. This can lead to memory loss, disorientation, personality change, and delusion.

OLDER BUT ACTIVE
Many people continue to be active well into their old age, and the life-long accumulation of knowledge, skills, and experience often means that older people are at the peak of their abilities. This is increasingly the case in industrialized countries, in which improvements in public health have helped people not only to live longer, but also to remain healthy when they do reach old age.

HUMAN INHERITANCE

IT IS OFTEN EASY TO recognize members of a family because of the remarkable similarity between them. Just as striking, though, are the differences. Why individuals inherit some features but not others has been studied over the past century by scientists. The key to inheritance lies inside body cells, where chromosomes store the information needed to construct and run each cell in "units" called genes. Since cells make up the body, they also determine its appearance. When humans reproduce, the chromosomes in their sperm and ova are passed on to the next generation.

FAMILY TREE

This photo shows four generations from the same family. Each individual developed from a zygote, or fertilized ovum, produced when their father's sperm fertilized their mother's ovum. Both ovum and sperm have the same number of chromosomes, but some of the genes (genetic instructions) that they carry differ. Passing on a new combination of genes to an offspring means that, while one generation resembles the previous one, a daughter is not identical to her mother (or father). Everyone, therefore, apart from identical twins, has a unique combination of genes in their chromosomes.

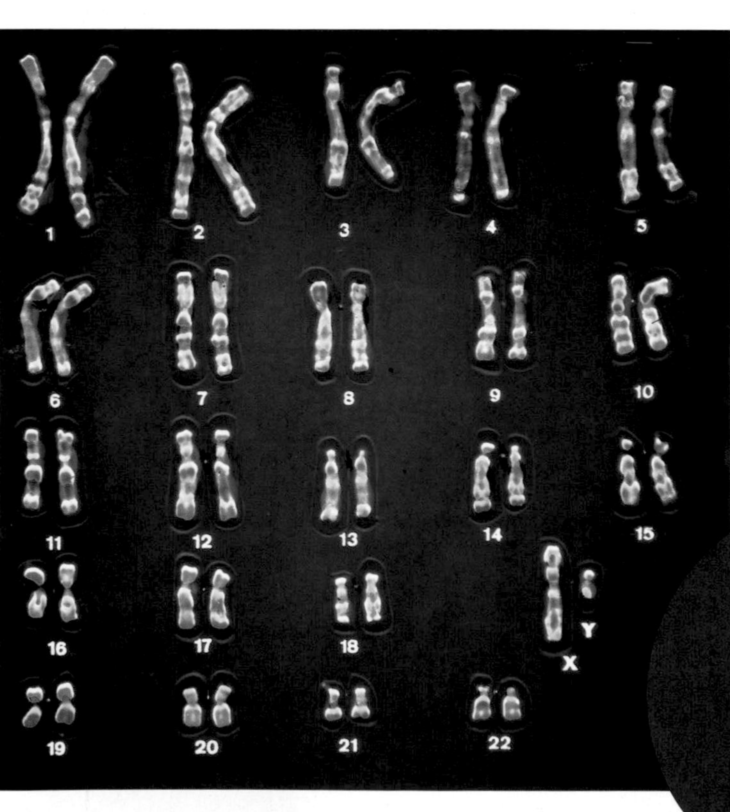

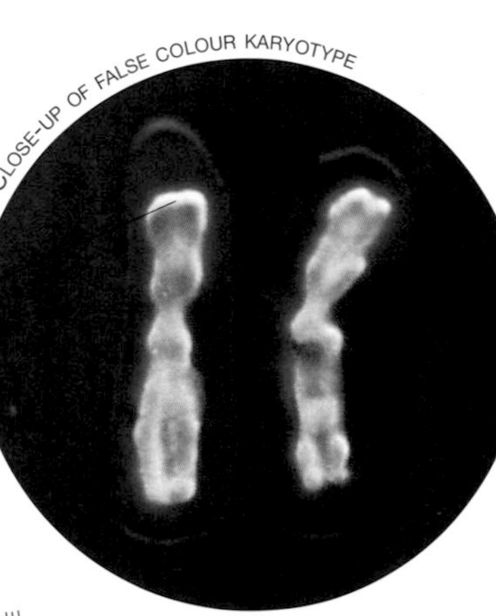

Female sex chromosome

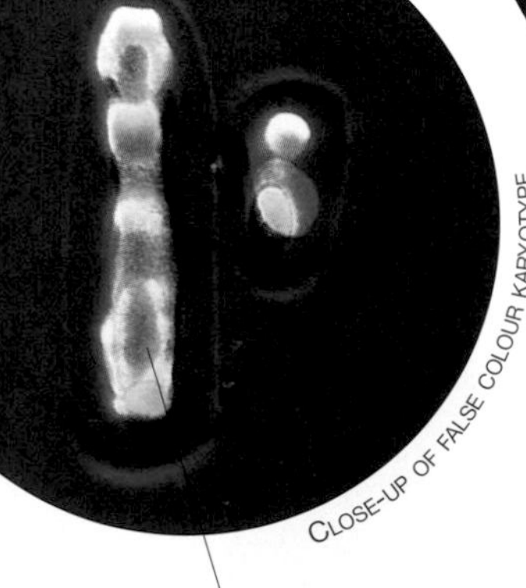

Male sex chromosome

HUMAN CHROMOSOMES

Within the nucleus of almost all body cells are 23 pairs of thread-like chromosomes. Normally, chromosomes are difficult to see under the microscope because they are long, thin, and tangled. But they can be seen when a cell is about to divide because they get shorter and thicker. A photograph of the chromosomes is cut up to produce a karyotype like this one. Pairs of matching chromosomes are arranged in order of size from 1 (largest) to 22 (smallest). These 22 pairs of chromosomes, called autosomes, control most body characteristics. The remaining pair are the sex chromosomes.

SEX CHROMOSOMES

This single pair of chromosomes is responsible for determining a person's sex. Not surprisingly, they differ between females and males. A female's body cells each contain a matching pair of sex chromosomes called X chromosomes. A male's sex chromosomes do not match. His body cells each contain one X chromosome paired with a Y chromosome.

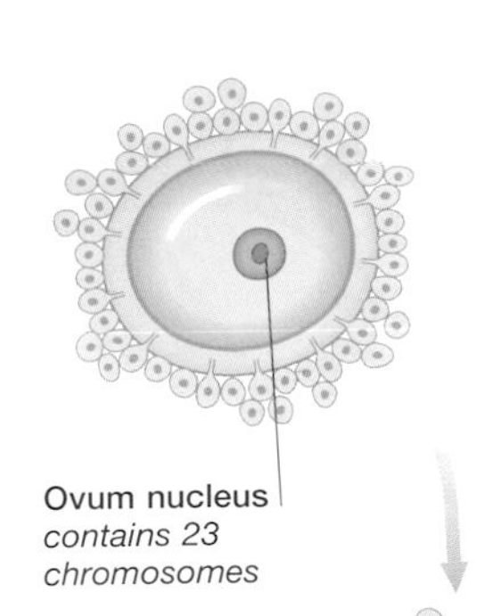

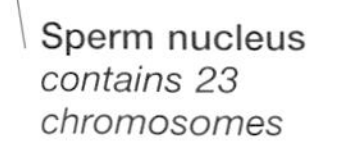

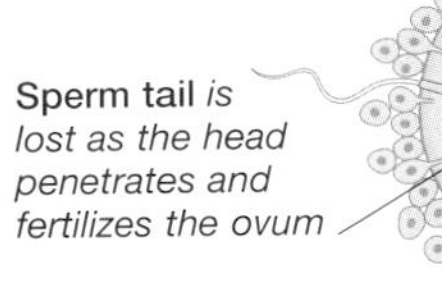

Ovum nucleus *contains 23 chromosomes*

Sperm nucleus *contains 23 chromosomes*

Sperm tail *is lost as the head penetrates and fertilizes the ovum*

Fertilization *brings the two sets of chromosomes together to make a total of 46*

HAPLOID TO DIPLOID
Most body cells are diploid ("double") because they have two sets of chromosomes. But sex cells – sperm and ova – are different. Produced by a special kind of cell division called meiosis (see p. 213), sperm and ova contain just one set of chromosomes and are referred to as haploid ("single"). After fertilization, when a sperm and an ovum combine, the resulting fertilized ovum, or zygote, has a full complement of 46 chromosomes. Two haploid cells have joined to form a diploid one.

Cloning

The process of cloning produces new individuals, called clones, that are genetically identical to their "parent", and does not involve sexual reproduction. In 1997, scientists in Scotland produced Dolly the sheep, the first mammal to be cloned from a cell taken from an adult. The cell's nucleus was injected into an ovum from another sheep that had had its nucleus removed. This "ovum" was then implanted into a sheep's uterus, where it developed into Dolly. Could this happen in humans? Cloning a human has never been achieved, and most countries have made it illegal to even try.

Dolly the sheep

Tongue-rolling *is a characteristic controlled by a person's genes, which means some people can do it while others cannot*

GENETIC VARIATION
Around 99.8 per cent of the genetic information contained within chromosomes is identical in all people. It ensures that bodies are constructed and work in the same way. The remaining 0.2 per cent is variable and produces differences between individuals. Some features show discontinuous variation – for example, people can either roll their tongue or not, there is no in-between stage. Other characteristics show continuous variation – for example, people's height covers a wide range from short to tall.

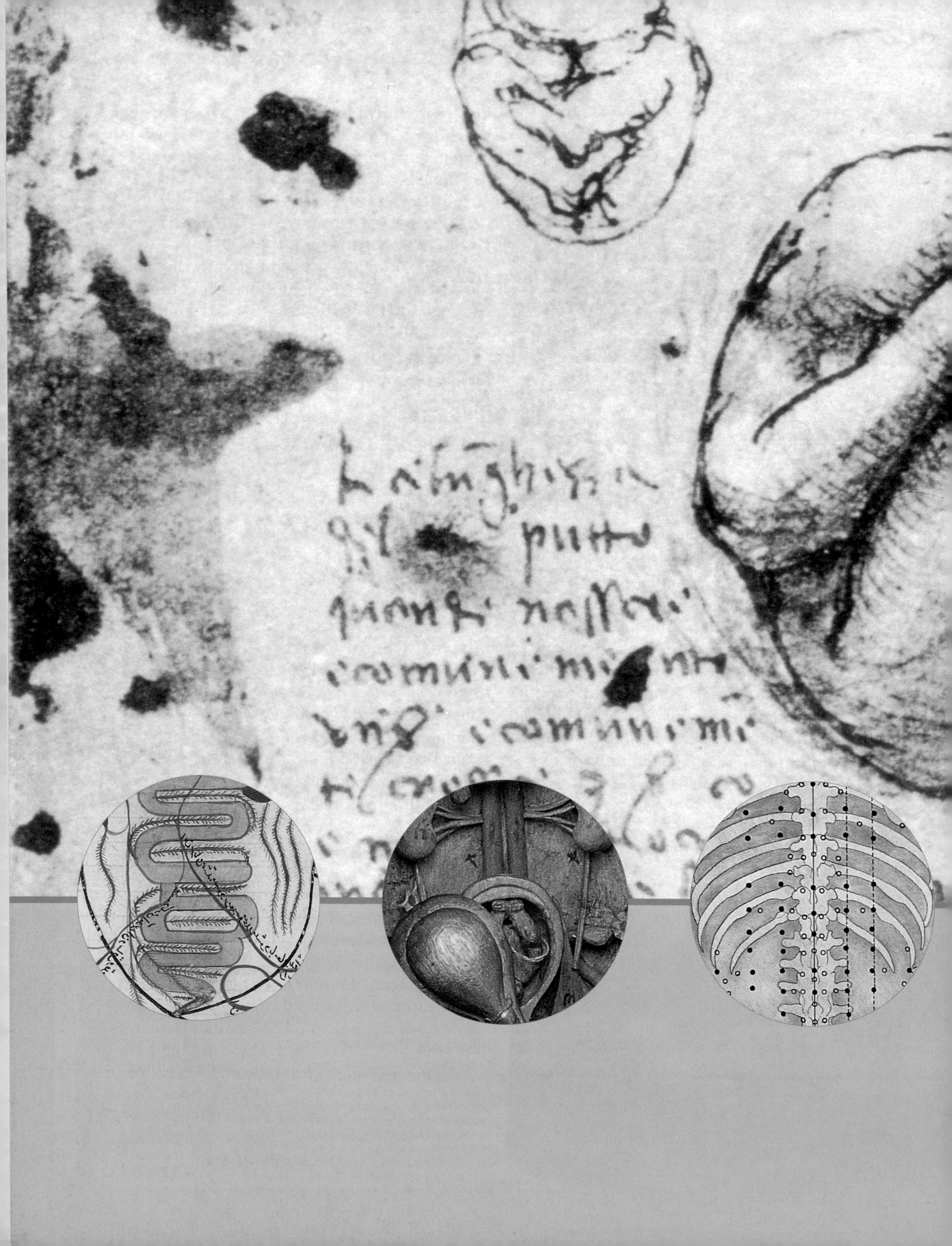

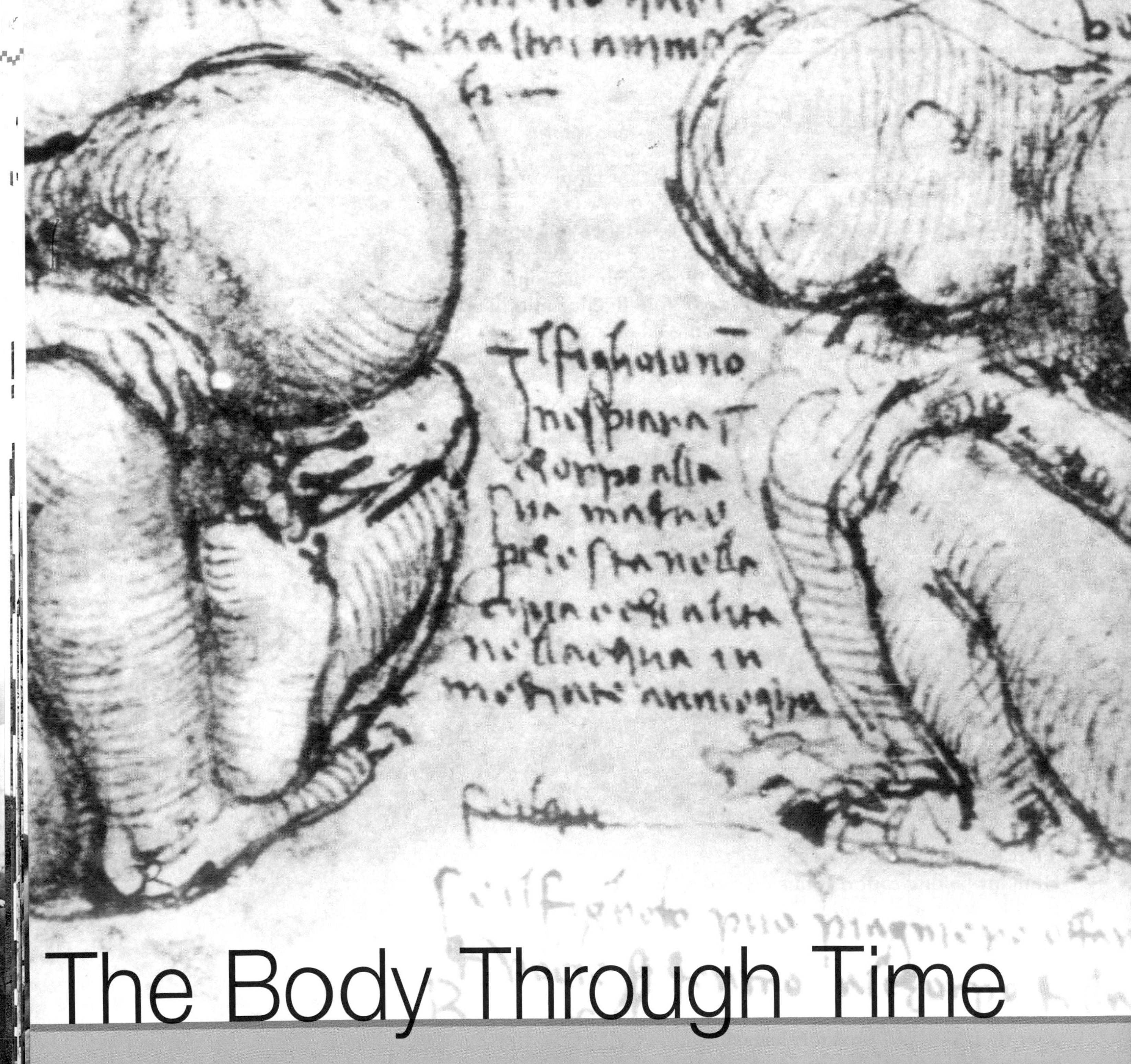

The Body Through Time

THEIR POWERFUL INTELLIGENCE, and ability to analyse, communicate, and record, make human beings unique in the living world. Driven by a natural curiosity, these skills have allowed humans to study themselves in order to understand both how the body works and why it goes wrong. Over the millennia, human biology and medicine have evolved together across many cultures. The resulting accumulation of knowledge enables 21st-century humans to stay healthier and live longer than their ancient relatives.

FOOD GATHERERS
Humans evolved as energetic hunter-gatherers, spending part of every day collecting food. The body has changed very little over the last 10,000 years, since agriculture and settled civilization developed.

The body today

MOST HUMAN SOCIETIES have left their Stone Age hunter-gatherer origins far behind. Our reasoning ability and technological skills have provided large sections of the population in the developed world with work that needs little daily physical effort. Our diet delivers surplus energy from fatty food, and we have a choice of tempting leisure activities that exercise our minds but not our bodies. Lifestyles in the developed world have changed a great deal but our bodies, which evolved to cope with the daily physical stresses of gathering food, have not. Obesity and stress-related illness now rival infectious diseases as major threats to human health. Medical science can provide us with a better understanding of the way our body works, and with better treatments for its illnesses, but regular exercise and a balanced and varied diet are as important for our survival today as they were for our ancestors.

STONE AGE VERSUS MODERN AGE

Imagine, for a moment, that you are a hunter-gatherer. As dawn breaks, your first thought is to collect and catch food for the day. You will need to travel many miles on foot to find it, always alert to dangers posed by predatory animals and ready to run for your life. Physical fitness is vital for your survival. Now imagine daily life as an urban office worker. Your first thought in the morning is to grab a quick breakfast, then you worry about the traffic jams that will make you late for work. All day you sit at a computer, dealing with stressful problems, with just a short break for a cigarette and burger at lunch. When you arrive home, you are so tired that all you can do is microwave some pre-prepared processed food and sit in front of the television until bedtime.

MODERN DIET
Pre-prepared foods that can be cooked quickly may be very convenient for busy modern lifestyles, but a diet based on these alone, combined with lack of physical exercise, is a recipe for obesity and heart disease. Fresh foods, and particularly fruit and vegetables, are essential for a healthy diet.

HEALTHY EXERCISE
Enjoying the full benefits of modern society depends on recognizing the nutritional and physical needs of the body. We no longer chase and trap our next meal, but by taking full advantage of sporting and exercise facilities, we can maintain bodily health.

YOUR BODY

Physical fitness is still vital for your survival today, because your 21st-century body works in exactly the same way as the body of a Stone Age hunter. Our lifestyles may have changed, but our bodies have not evolved to automatically cope with a life that involves sitting down all day, eating a diet with a high fat intake and fewer vegetables and fruits, and living with constant levels of background stress. If we ignore the physical and nutritional requirements of the body, we can expect to suffer from obesity, heart problems, diseases like diabetes, and a shortened lifespan. The key to maximizing our chances for a long and healthy life in our modern society is to take the time and trouble to maintain fitness through a balanced diet and regular exercise.

STRESS
Stresses of modern life are founded on daily frustrations at work or during travel. These stresses operate continuously over long periods of time, undermining mental and physical health.

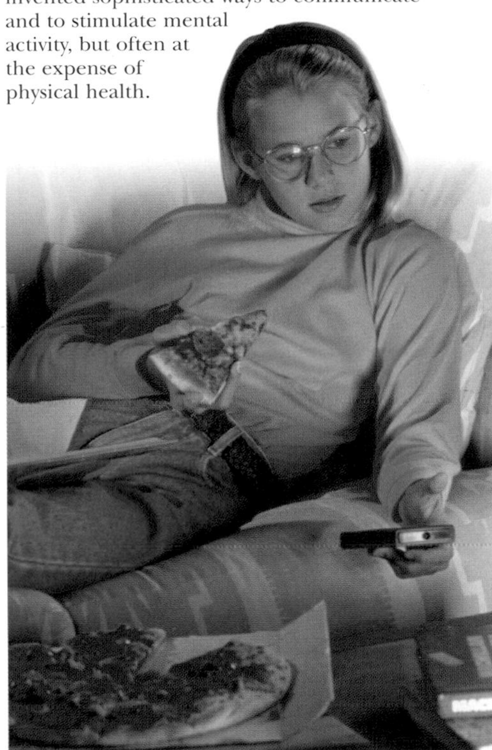

LESS ACTIVITY
Technological advances have removed much of the physical effort from daily life. Travel, work, and leisure entertainment involve more sitting down. Modern society has invented sophisticated ways to communicate and to stimulate mental activity, but often at the expense of physical health.

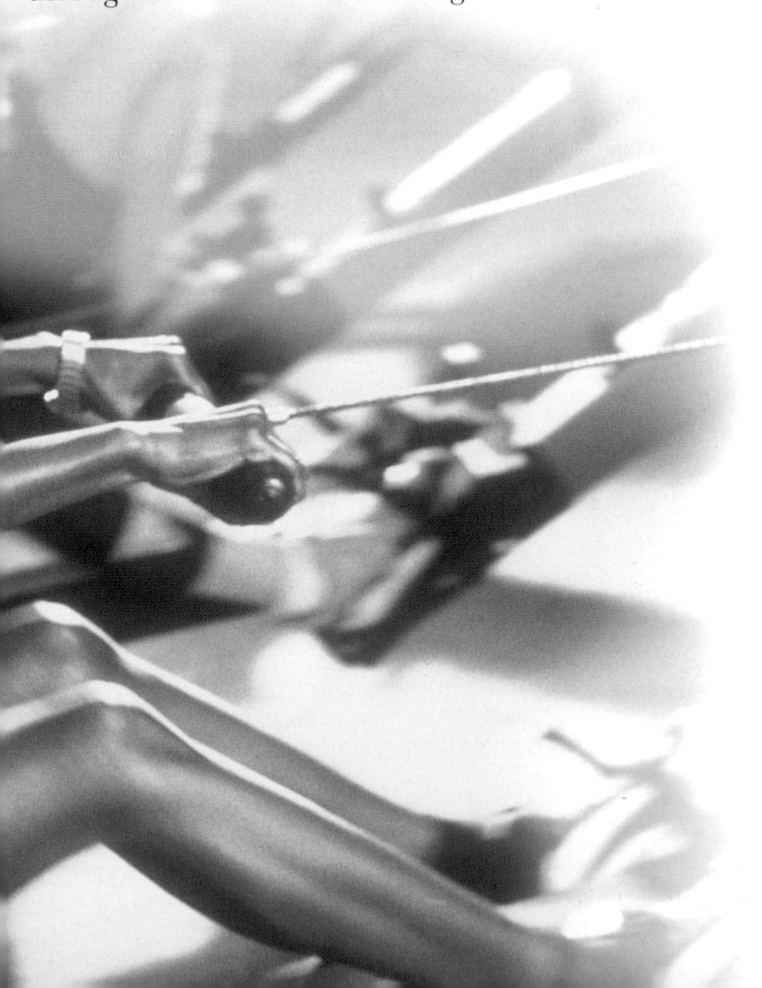

LEVELLING INEQUALITIES

While health problems in the developed world are often rooted in too much food and too little activity, millions of people in the developing world live shortened lives through lack of basic human needs. In many major cities in the developing world, sanitation and clean water, housing, simple health care, and adequate food are still in short supply. Providing these basic essentials for the world's poor is the major challenge.

THE CHALLENGE

Medical science has come a long way since Mesopotamian priests called upon the gods to cure the sick. With the development of modern biomedical technology, we can hope to cure diseases that only a decade ago were considered to be untreatable. But maintaining bodily health requires more than just scientific skill. For millions of the world's poorest people, it depends on providing the basic necessities for healthy living. For people in the rich nations, it depends on realizing that we need to take responsibility for maintaining the fitness of our own bodies.

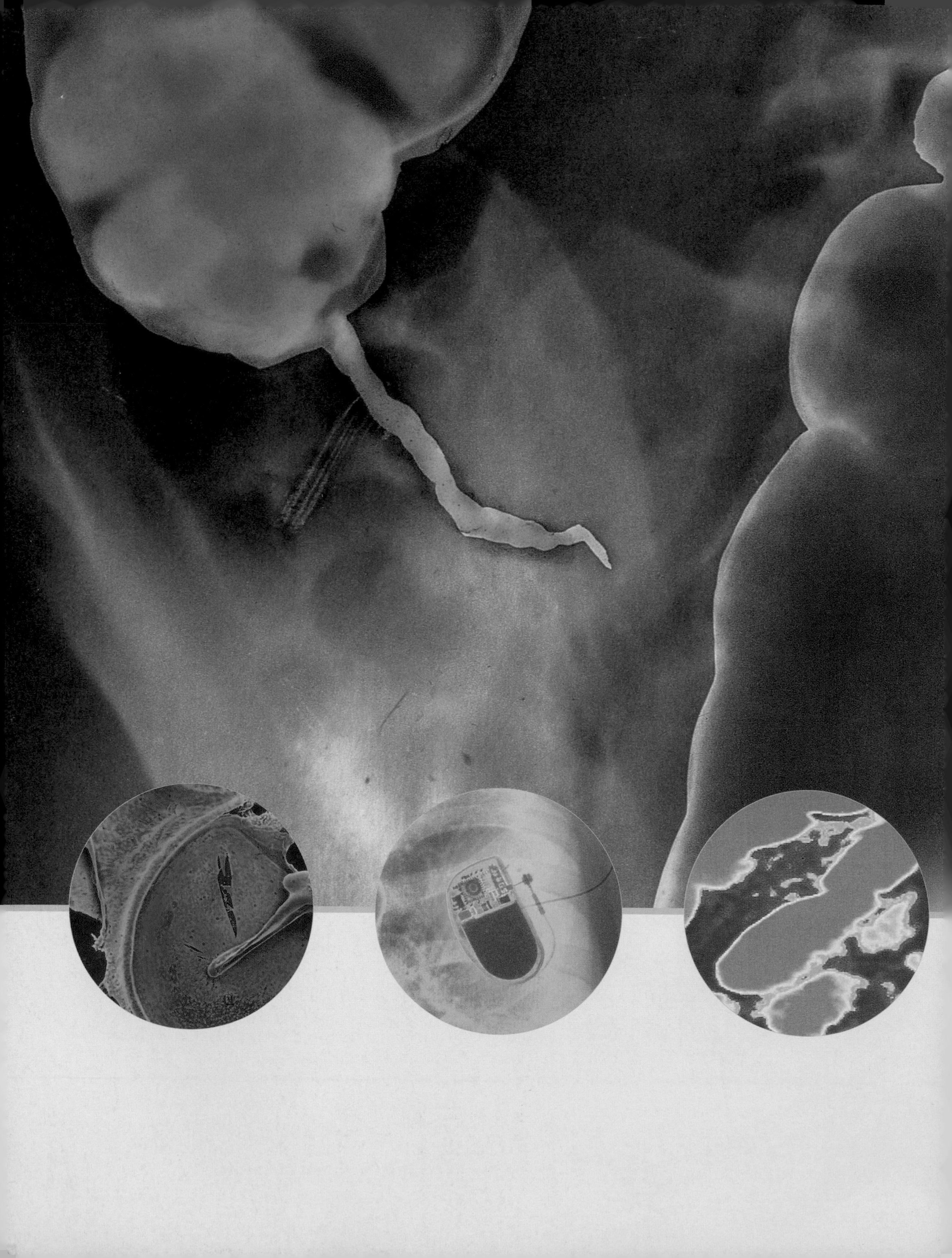

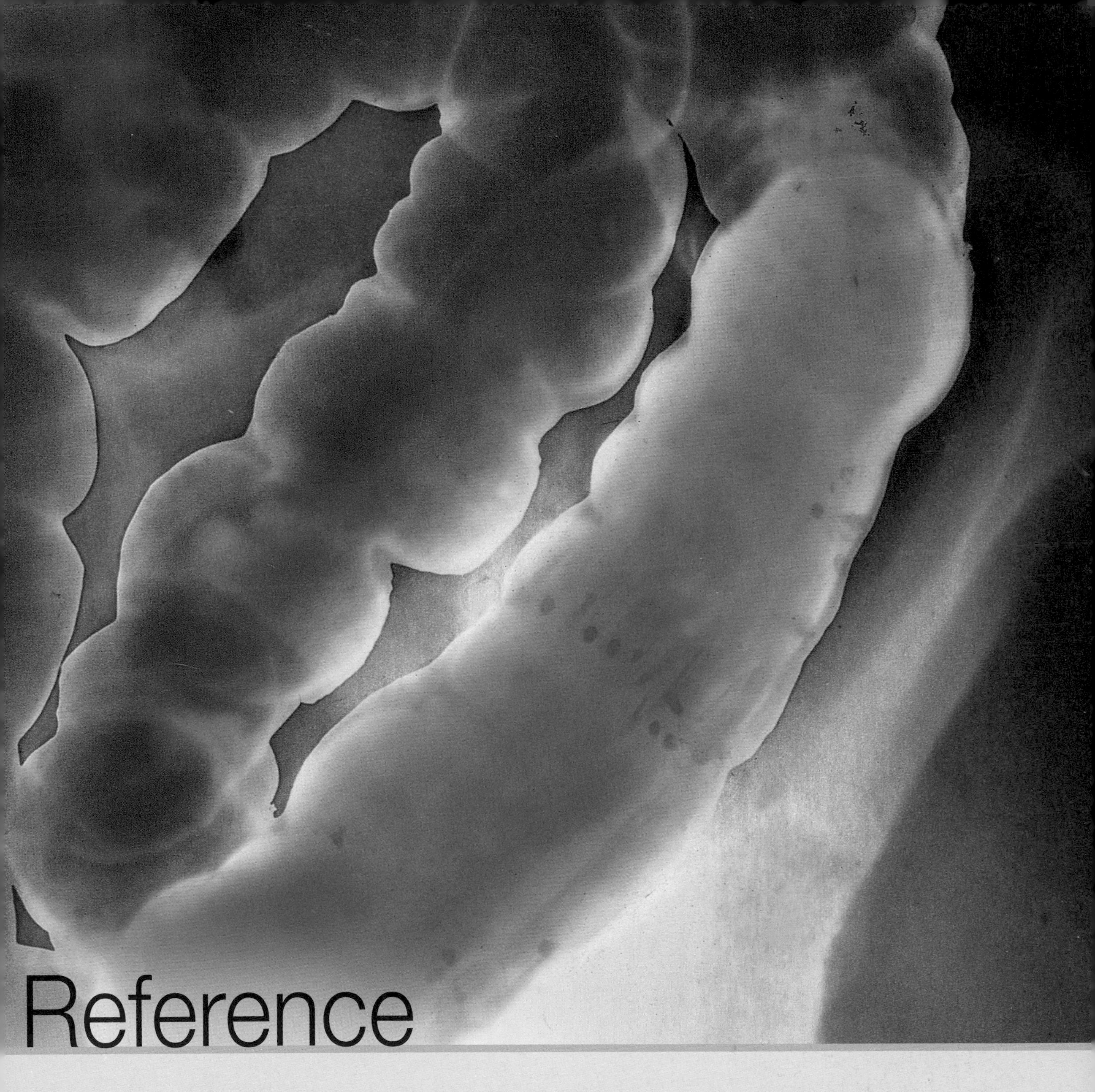

Reference

THIS FINAL SECTION PROVIDES invaluable reference resources in the form of a timeline and glossary. Spanning the millennia from 100,000 BC to the present day, the timeline lists significant milestones in both the study of the human body and the development of medicine. Any confusion over words used in the book can be solved by looking them up in the glossary, which gives easy-to-understand explanations of nearly 300 scientific terms, from "Abdomen" to "Zygote".

TIMELINE

c. 100,000 BC
Modern humans (*Homo sapiens*) first appear in Africa.

c. 70,000 BC
Humans spread from Africa to other continents.

c. 30,000 BC
Humans produce cave paintings and sculptures of themselves and other animals.

c. 10,000 BC
Transition from hunter-gatherer lifestyle to agricultural, settled communities.

c. 2650 BC
Earliest known physician, the Egyptian Imhotep, later given full status as a god.

c. 2600 BC
Chinese Emperor Huang Ti believed to have laid down the basic principles of the *Nei Ching*, a standard manual of Chinese medicine.

c. 1750 BC
King Hammurabi of Babylonia establishes a set of laws, called the Code of Hammurabi. The laws help to regulate the work of physicians.

c. 1500 BC
Date of origin of the Ebers papyrus (discovered in Egypt in 1873), which remains the oldest known medical text.

c. 500 BC
Greek physician and philosopher Alcmaeon of Croton proposes that the brain, and not the heart, is the organ of thinking and feeling.

c. 420 BC
Greek physician Hippocrates teaches the importance of observation and diagnosis over magic and myth in medicine.

c. 350 BC
Greek philosopher Aristotle states that the heart is the organ of feeling and intelligence, a belief held centuries before by the ancient Egyptians.

c. 280 BC
Herophilus of Alexandria reinstates the brain as the organ of thinking. He also describes the cerebrum and cerebellum, and discovers that nerves are channels of communication.

AD 40
Roman philosopher Cornelius Celsus publishes *On Medicine*, a medical handbook based on earlier Greek sources.

c. AD 200
Influential Greek-born Roman doctor Claudius Galen describes, often incorrectly, the workings of the body. With his ideas remaining unchallenged until the 1500s, few advances are made in the understanding of the human body.

AD 890–932
Persian physician Abu Bakr ar-Razi (Rhazes) produces many influential medical texts, and accurately describes measles and smallpox.

1000

c.1000
Publication of medical texts by Arab doctor Ibn Sina (Avicenna), which influences European and Middle-Eastern medicine for the next 500 years.

c.1000
Arab surgeon al-Zahrawi (Albucasis) publishes surgery textbooks that describe complicated operations.

1200

1268
Roger Bacon, an English scientist, records the use of glasses to correct eye defects.

c. 1280
Arab physician Ibn an-Nafis shows that blood flows through the lungs.

1300

1347–1350
Black Death (a bubonic plague pandemic) spreads through Europe, killing more than a quarter of its population.

1500

c. 1500
From his own dissections, Italian artist and scientist Leonardo da Vinci produces accurate anatomical drawings of the human body.

1543
Andreas Vesalius, a Flemish doctor, publishes *The Structure of the Human Body*, the first accurate description of human anatomy.

1545
Ambroise Paré, a French surgeon, publishes *Method of Treating Wounds* in which he describes his less painful, more successful techniques for treating wounds.

1561
Italian anatomist Gabrielle Fallopio (Fallopius) publishes *Anatomical Observations* in which he describes the duct linking the ovary to the uterus.

1562
Italian anatomist Bartolommeo Eustachio is the first to describe the ear in detail in his book *The Examination of the Organ of Hearing*.

1565
Swiss physician Paracelsus (Philipp von Hohenheim) publishes *Opus Chyrurgicum*, in which he attacks the works of Galen and Avicenna.

1590
Dutch instrument maker Zacharias Janssen invents the microscope.

1600

1603
Italian anatomist Hieronymus Fabricius publishes *On the Valves of Veins*, the first detailed description of vein structure.

1614
Italian physician Santorio Santorio (Sanctorius) publishes *The Art of Statistical Medicine*, the results of a 30-year study of his own bodily functions.

1628
William Harvey, an English doctor, publishes *On the Movement of the Heart*

and Blood, describing how blood circulates around the body, pumped by the heart.

1662
René Descartes' book *De homine*, published 12 years after his death, puts forward ideas about the brain and mind, and describes reflexes.

1663
Marcello Malpighi, an Italian physiologist and microscopist, discovers blood capillaries, helping to confirm that blood circulates around the body.

1664
Thomas Willis, an English doctor, describes the blood supply to the brain.

1665
English physicist Robert Hooke publishes *Micrographia*, in which he coins the term "cell".

1667
English physician Richard Lower carries out first blood transfusion to a human, using blood from a sheep.

1672
Regnier de Graaf, a Dutch doctor, describes the structure and workings of the female reproductive system.

1674–1677
Dutch draper and pioneer microscopist Antoni van Leeuwenhoek observes and describes red blood cells, sperm, and skeletal muscle cells using an early microscope.

1691
English doctor Clopton Havers makes the first description of the microscopic structure of bones.

1700

1717
Lady Mary Wortley Montagu, a British writer, brings the Turkish practice of smallpox inoculation to England.

1747
British naval doctor James Lind discovers that citrus fruits prevent the deficiency disease scurvy during long sea voyages.

1763–1793
British doctor John Hunter makes significant advances in knowledge about human anatomy, and elevates surgery from a craft to a science.

1775
French chemist Antoine Lavoisier discovers oxygen, and later shows that cell respiration is, like burning, a chemical process that consumes oxygen.

1780
Italian doctor Luigi Galvani experiments with nerves, muscles, and electricity.

1785
British doctor William Withering shows that the extracts of the foxglove plant could be used to treat heart failure.

1792
Austrian doctor Franz Gall begins his investigations into the link between behaviour and bumps on the skull. The investigations help to form the basis of his "science of phrenology".

1796
First vaccination against smallpox is carried out by British doctor Edward Jenner, when he takes pus from a cowpox blister and introduces it into the arm of an eight-year-old boy.

1800

1800
French doctor Marie François Bichat publishes *Traité des Membranes* in which he shows that organs are made of different groups of cells called "tissues".

1801
Philippe Pinel, a French doctor, suggests that mentally ill people should be treated more humanely.

1811
Scottish anatomist Charles Bell describes spinal nerve roots and shows that nerves are bundles of nerve cells.

1816
The stethoscope is invented by French doctor René Laënnec.

1817
English doctor James Parkinson first describes a brain disorder that affects movement in some older people, later to be called Parkinson's disease.

1818
British doctor James Blundell performs the first successful transfusion of human blood to a human patient.

1833
American surgeon William Beaumont publishes *Experiments and Observations on the Gastric Juice and the Physiology of Digestion*. The book records the results of his researches into the mechanism of digestion made on Alexis St. Martin, a man seriously wounded in a shooting accident.

1837
Czech biologist Johannes Purkinje first observes neurons (nerve cells) in the cerebellum of the brain, later called Purkinje cells.

1838
German scientists Theodor Schwann and Jakob Schleiden put forward their "cell theory", which states that all living things are made from cells.

1840
Jakob Henle, a German anatomist, states in his book *On Miasmas and Contagions* that infectious diseases are caused by micro-organisms.

1840
Charles Laveran, a French doctor, identifies the protist *Plasmodium* as the cause of the disease malaria.

1842
British surgeon William Bowman first describes the microscopic structure and function of the kidney.

1844
German doctor Carl Ludwig shows that nephrons in the kidneys act as filters for the production of urine.

1846
American dentist William Morton uses ether as an anaesthetic to make a patient unconscious and pain-free during an operation in Massachusetts General Hospital, USA.

1848
American railroad worker Phineas Gage survives an accident that drives an iron rod through the front of his brain, but suffers a behaviour change. This indicates to scientists that the frontal lobe of the cerebrum controls personality.

1848
Claude Bernard, a French scientist, demonstrates the function of the liver, and later shows that body cells need stable surroundings, thereby establishing the principles of what will later be called homeostasis.

1848
Hungarian doctor Ignaz Semmelweis demonstrates that hand washing by medical staff dramatically reduces deaths of women from puerperal (childbirth) fever.

1849
English-born American Elizabeth Blackwell becomes the first woman to qualify as a doctor in the USA.

1851
Hermann von Helmholtz, a German physicist, invents the ophthalmoscope (an instrument for looking inside the eye).

1854
British doctor John Snow halts an outbreak of cholera in London by removing the handle pump in Broad Street, suggesting that the disease is spread by contaminated water.

1858
In his book *Cellular Pathology*, German biologist Rudolf Virchow states that all cells are made from existing ones, and that diseases occur when cells stop working normally. This establishes the basis for the branch of medicine called pathology.

1859
Charles Darwin, a British scientist, puts forward the theory of evolution in his ground-breaking book *The Origin of Species*.

1860s
Louis Pasteur, a French scientist, explains how micro-organisms cause infectious diseases.

1861
Pierre Paul Broca, a French doctor, identifies the area of the brain (now Broca's area) that controls speech.

1865
British doctor Joseph Lister first uses carbolic acid as an antiseptic during surgery and dramatically reduces deaths from infection.

1870
Elizabeth Garrett Anderson begins practising as the first woman doctor in Britain.

1872
Italian doctor Camillo Golgi devises a stain that, for the first time, shows the brain's nerve cells clearly under the microscope.

1874
Carl Wernicke, an Austrian doctor, identifies the area on the left side of the brain (to be called Wernicke's area) that controls the understanding of spoken and written words.

1882
German doctor Robert Koch identifies the bacterium (*Mycobacterium tuberculosis*) that causes TB (tuberculosis).

1888
French microbiologists Emile Roux and Alexandre Yersin show that bacteria release toxins (poisons), which cause the symptoms of many diseases.

1889
Spanish physiologist Ramón Santiago y Cajal states that the nervous system is made up of a network of distinct nerve cells (later called neurons) that do not touch.

1895
X-rays are discovered by German physicist Wilhelm Roentgen.

1897
Ronald Ross, a British doctor, shows that the micro-organism (a protist) causing malaria is spread from person to person by *Anopheles* mosquitoes.

1900

1900
Christiaan Eijkmann, a Dutch doctor, shows that the deficiency disease beriberi can be treated by a change in diet. This helps to establish the concept of "essential food factors", later called vitamins.

1900
Austrian doctor Sigmund Freud publishes *The Interpretation of Dreams*, which contains the basic ideas of psychoanalysis.

1900–1901
American army surgeon Walter Reed and his team demonstrate that yellow fever is transmitted by *Aëdes* mosquitoes and is caused by a virus.

1901
Austrian-American doctor Karl Landsteiner demonstrates the existence of blood groups (later classified as A, B, AB, and O), paving the way for safe blood transfusions. He was awarded the Nobel Prize for Medicine in 1930.

1901
Japanese biochemist Jokichi Takamine is the first scientist to isolate crystals of a pure hormone, adrenaline.

1902
British physiologists Ernest Starling and William Bayliss isolate secretin, the first substance to be named a hormone (a term that was later devised by Starling in 1905).

1903
An early version of the ECG (electrocardiograph), a device for monitoring heart activity, is invented by Dutch physiologist Willem Einthoven.

1905
Ernest Starling devises the term "hormone" to describe the newly discovered "chemical messengers" that co-ordinate body processes.

1906
Charles Sherrington, a British physiologist, publishes *The Integrative Action of the Nervous System*, a landmark work describing how the nervous system works.

1906–1912
Frederick Gowland Hopkins, a British biochemist, demonstrates the importance of "accessory food factors" (vitamins) in food.

1907
German neurologist Alois Alzheimer first describes the brain disorder (later to be called Alzheimer's disease) that causes a progressive decline in mental abilities.

1910
German scientist Paul Ehrlich discovers salvarsan, the first drug used to treat a specific disease.

1912
Polish-American biochemist Casimir Funk coins the term "vitamin" to describe essential nutrients needed in small amounts for normal body functioning.

1912
American Harvey Cushing publishes *The Pituitary Gland and its Disorders* in which he describes the functioning of the gland.

1914
American doctor Joseph Goldberger shows that pellagra is not an infectious disease but is caused by poor diet (later shown to be lack of the vitamin niacin).

1916
American birth control pioneer Margaret Sanger opens her first clinic in Brooklyn, USA.

1918
Edward Mellanby, a British scientist, discovers vitamin D, which is essential for normal bone growth.

1921
British birth control pioneer Marie Stopes opens her first clinic in London.

1921
Canadian physiologists Frederick Banting and Charles Best isolate the hormone insulin, enabling the disease diabetes to be controlled.

1921
German-born American scientist Otto Loewi detects chemicals called neurotransmitters involved in carrying signals between neurons.

1926
William Castle, an American doctor, demonstrates the intrinsic factor secreted by the stomach that aids the body's intake of vitamin B_{12}.

1928
Scottish bacteriologist Alexander Fleming discovers penicillin, the first antibiotic, when he notices mould growing on a plate of bacteria.

1928
Hungarian-born Albert von Szent-Györgyi, an American biochemist, isolates vitamin C.

1929
English physiologists Henry Dale and H.W. Dudley demonstrate the chemical transmission of nerve impulses between neurons, and identify acetylcholine as the first neurotransmitter.

1930
American physiologist Walter Cannon devises the term "homeostasis" – from the Greek for "standing still" – to describe the mechanisms whereby the body maintains a stable internal state.

1933
German electrical engineer Ernst Ruska invents the electron microscope.

1937
German-British biochemist Hans Krebs discovers the sequence of reactions called the Krebs cycle (or the citric acid cycle), which breaks down glucose during aerobic respiration to release energy.

1943
Dutch doctor Willem Kolff invents the kidney dialysis machine to treat people with kidney failure.

1948
World Health Organization (WHO) formed within the United Nations.

1952
British scientists Alan Hodgkin and Andrew Huxley describe nerve impulses.

1953
Using research by British physicist Rosalind Franklin, US biologist James Watson and British physicist Francis Crick discover the structure of DNA.

1953
American surgeon John Gibbon develops the heart-lung machine to pump blood during heart surgery.

1954
First use of polio vaccine developed by American doctor Jonas Salk.

1954
First successful kidney transplant carried out in Boston, USA.

1958
Ultrasound first used to check health of fetus in its mother's uterus by British professor Ian Donald.

1965
American biochemist Marshall Nirenberg finishes deciphering the genetic code through which DNA controls production of proteins inside a cell.

1967
South African surgeon Christiaan Barnard carries out first successful heart transplant.

1967
Introduction of mammography, an X-ray technique for detecting breast cancer.

1969
British biochemist Dorothy Hodgkin determines the structure of insulin using X-ray crystallography, having previously described that of penicillin in 1946 and vitamin B_{12} in 1956.

1970s
Discovery of natural painkillers, called endorphins and enkephalins, produced by the body.

1972
CT (computerized tomography) scanning first used to produce images of body organs.

1977
Last recorded case of smallpox; the disease is declared eradicated in 1979.

1978
Successful IVF (in vitro fertilization) by British doctors Patrick Steptoe and Robert Edwards results in first "test tube" baby, Louise Brown.

1980
Introduction of "keyhole" surgery, using an endoscope to look inside the body through small incisions.

1980s
PET scans first used to produce images of brain activity.

1981
AIDS (acquired immune deficiency syndrome) identified as a new disease.

1982
First artificial heart, invented by US scientist Robert Jarvik, implanted into a patient.

1984
Luc Montagnier, a French scientist, discovers the virus called HIV (human immunodeficiency virus), which causes AIDS.

1990
Human Genome Project is launched to identify the genes in human chromosomes.

1999
Chromosome 22 becomes the first human chromosome to have its DNA sequenced.

2000

2000
First "draft" of Human Genome Project completed.

2002
Gene therapy used to cure boys suffering from an inherited immunodeficiency disease which would otherwise leave the body defenceless against infection.

GLOSSARY

Abdomen
Lower part of the trunk (central part of the body) between the thorax (chest) and the hips.

Absorption
Process by which the products of digestion pass through the wall of the small intestine into the bloodstream.

Accommodation
Adjustment made by changing the shape of the lens of the eye so it can focus on near or distant objects.

Acne
Skin disorder causing spots that results from inflamed sebaceous glands and hair follicles.

Adolescence
Transition period between childhood and adulthood that occurs during the teenage years.

Aerobic respiration
Release of energy from glucose that takes place inside cells and requires oxygen.

Alimentary canal
Hollow tube which extends from the mouth to the anus, and includes the pharynx, oesophagus, stomach, and small and large intestines.

Allergy
Illness caused by over-reaction of the body's immune system to a normally harmless substance.

Alveoli (sing. Alveolus)
Microscopic air bags inside the lungs through which oxygen enters, and carbon dioxide leaves, the bloodstream.

Amino acid
One of a group of 20 chemical compounds that are the basic building blocks from which proteins are made.

Amputation
Surgical removal of all or part of an arm or leg.

Anaerobic respiration
Release of energy from glucose that takes place inside cells and does not use oxygen.

Anaesthetic
Drug used to temporarily abolish feelings of pain in a patient during surgery or while giving birth.

Anatomy
Study of the structure of the body, and how its parts relate to one another.

Angiogram
Special type of X-ray that reveals the outline of blood vessels after a dye that absorbs X-rays has been injected into them.

Antibody
Substance released by lymphocytes of the immune system that disables a pathogen and marks it for destruction.

Antigen
Foreign substance, usually found on the surface of pathogens such as bacteria, that triggers the immune system to respond.

Antiseptic
Chemical applied to the skin to destroy bacteria and other micro-organisms before they can cause infection.

Apgar score
System of scoring used to assess the condition of a newborn baby.

Appendicular skeleton
Part of the skeleton made up of the bones of the pectoral and pelvic girdles, and those of the upper and lower limbs.

Arteriole
Very small artery that delivers blood to a capillary.

Artery
Blood vessel that carries blood from the heart towards the tissues.

Association neuron
Neuron (nerve cell) that relays nerve impulses from one neuron to another, and processes information.

Atom
Smallest particle of an element, such as carbon or hydrogen, that can exist, and one of the building blocks from which all matter is made.

ATP (adenosine triphosphate)
Substance that stores, carries, and releases energy.

Atrium (pl. Atria)
Left or right upper chamber of the heart.

Autonomic nervous system (ANS)
Part of the nervous system that controls the involuntary activities of internal organs, such as heart rate and blood pressure.

Axial skeleton
Central part of the skeleton consisting of the skull, backbone, ribs, and sternum.

Axon
Also called a nerve fibre, this is the long "tail" of a neuron that carries nerve impulses away from its cell body.

Bacteria (sing. Bacterium)
Group of single-celled micro-organisms, commonly known as germs, some of which cause diseases such as typhoid and TB.

Base
One of four nitrogen-containing substances – adenine, cytosine, guanine, and thymine – that spell out the instructions written in genetic code in molecules of DNA.

Biopsy
Removal of a small piece of tissue from the body for examination under the microscope to look for signs of disease.

Blind spot
Also called the optic disc, this is the part of the retina of the eye where the optic nerve leaves the eye, and where light cannot be detected.

Blood vessel
Tube, such as an artery or vein, that carries blood through the body.

Body language
Form of non-verbal communication that uses body position, gestures, and facial expressions.

Brain stem
Lowest part of the brain. It is connected to the spinal cord, and controls vital functions such as breathing.

Cancer
One of a number of different diseases caused by body cells dividing out of control and producing growths called tumours.

Capillary
Microscopic blood vessel that carries blood from arterioles to venules and supplies individual cells.

Carbohydrate
One of a group of organic compounds, made up of carbon, hydrogen, and oxygen, that includes glucose and glycogen, and provides the body's main energy supply.

Carbon dioxide
Gas that is a waste product of cell respiration, which is released into the air from the lungs during exhalation (breathing out).

Cardiac muscle
Type of muscle found only in the heart.

Cartilage
Tough, flexible connective tissue that helps support the body and covers the ends of bones where they meet at joints.

Catalyst
Substance that speeds up the rate of a chemical reaction but does not change itself.

Cell body
Part of a neuron (nerve cell) that contains its nucleus.

Cell division
Process by which cells multiply by dividing into two.

Cell membrane
Also called a plasma membrane, this is the thin membrane that surrounds a cell and separates it from its environment.

Cell (Internal) respiration
The release of energy from glucose and other fuels that takes place inside cells.

Central nervous system (CNS)
The part of the nervous system that consists of the brain and spinal cord.

Cerebellum
Part of the brain that controls balance and ensures that movements are smooth and co-ordinated.

Cerebral cortex
Thin surface layer of the cerebrum that processes information relating to thought, memory, the senses, and movement.

Cerebrospinal fluid
Watery fluid that circulates within and around the central nervous system, and helps protect and nurture the brain and spinal cord.

Cerebrum
The largest part of the brain, which is involved in conscious thought, feelings, and movement.

Chemical digestion
The breakdown of food into simple molecules using enzymes.

Chemoreceptor
Receptor, such as those in the nose and tongue, that responds to chemicals dissolved in water.

Chromosome
One of 46 thread-like packages of DNA found inside most body cells that contain genes, the instructions needed to construct and run a body.

Chyme
Creamy, soup-like liquid containing semi-digested food that passes from the stomach into the small intestine during digestion.

Cilia (sing. Cilium)
Microscopic, hair-like projections from certain cells that beat in a rhythmic, wave-like manner to move things, such as mucus, across their surface.

Cochlea
Coiled structure inside each ear that detects sounds.

Collagen
Tough, fibrous protein that helps to strengthen cartilage, tendons, and other types of connective tissue.

Compact bone
Also called cortical bone, this is the very hard material that forms a bone's outer layer.

Computed tomography (CT)
Scanning technique that uses X-rays and computers to produce "slices" through living tissues.

Conception
The period between fertilization and the implantation of an embryo in the lining of the uterus.

Cone
One of two types of light receptors in the retina of the eye, cones provide colour vision and work in bright light.

Connective tissue
Tissue, such as bone or cartilage, that supports the body and holds together its various structures.

Consciousness
Awareness of self and surroundings produced by the cerebrum, which enables a person to make decisions and to know what they are doing.

Contagious disease
Infectious disease, such as the common cold or measles, that is easily passed from person to person.

Contraception
Use of various methods to prevent pregnancy.

Convex lens
Lens, such as the one found in the eye, that curves outwards on both surfaces, and which makes light rays converge (come together).

Cornea
Clear area at the front of the eye that allows light in and refracts (bends) light rays.

Cranial nerve
One of the 12 pairs of nerves that arise from the brain.

Cranium
Upper part of the skull, made from eight interlocking bones, that surrounds the brain.

Cytoplasm
Jelly-like fluid that fills a cell between the cell membrane and nucleus.

Deficiency disease
Disease caused by a lack of a particular nutrient in the diet, especially a vitamin or a mineral.

Dendrite
Short filament that carries nerve impulses to the cell body of a neuron.

Dentine
Hard, bone-like tissue that surrounds the pulp of a tooth, and gives the tooth its basic shape.

Dermis
Lower, thicker layer of the skin that contains blood vessels, sweat glands, and sensory receptors.

Diaphragm
Dome-shaped sheet of muscle that separates the thorax (chest) from the abdomen, and which plays a key role in breathing.

Diastole
Part of the heartbeat cycle when either the atria or ventricles are relaxed.

Diffusion
Random movement of molecules in a gas or liquid from an area of high concentration to one of low concentration until they are evenly distributed.

Digestion
Breakdown of complex molecules in food into simpler substances that can be absorbed into the bloodstream.

Diploid cell
One, like most body cells, that contains two sets of 23 chromosomes.

Dissacharide
Sugar, such as maltose, sucrose, or lactose, that is made up of two monosaccharide units.

DNA (Deoxy-ribonucleic acid)
One of a number of large molecules, each consisting of two intertwined nucleic acid strands, found inside body cells, which carry the genetic instructions needed to build and operate that cell.

Double helix
Name given to the twin strands of nucleic acids that spiral round each other like a twisted ladder in each DNA molecule.

Duct
A tube that leads from a gland and carries its products, such as the tear duct that carries tears from the tear glands.

Eardrum
Thin membrane at the end of the ear canal that vibrates when sounds hit it.

Egestion
Removal from the body in the form of faeces the undigested waste remaining after digestion.

Elastin
Protein whose fibres can stretch and recoil like a rubber band and give elasticity to connective tissues, such as those in the dermis of the skin.

Electrocardiogram (ECG)
Recording of the electrical changes that occur as the heart beats made by an electrocardiograph.

Electroencephalogram (EEG)
Recording of brain waves produced by electrical activity in the brain made by an electroencephalograph.

Electron microscope
Powerful microscope that uses an electron beam instead of light to produce highly magnified views of body cells and tissues.

Embryo
Name given to an unborn child during the first eight weeks of development after fertilization.

Enamel
Hardest material found inside the body, which covers the crown of a tooth.

Endocrine gland
Gland, such as the pituitary gland, that secretes hormones into the bloodstream.

Enzyme
Protein that acts as a biological catalyst to speed up the rate of chemical reactions both inside and outside cells.

Epidemic
Outbreak of an infectious disease that affects many people in the same location at the same time.

Epidermis
Upper, thinner, protective layer of the skin, from which the topmost layer of dead cells is constantly worn away and replaced from below.

Epithelium
Also called epithelial tissue, a sheet of cells, one or more cells thick that covers the body, lines its internal cavities, and forms glands.

Excretion
Elimination from the body of waste products produced by cell metabolism or of substances that have entered the bloodstream and are surplus to requirements.

Exhalation
The movement of air out of the lungs; also called expiration or breathing out.

Exocrine gland
Gland, such as a salivary or sweat gland, that secretes chemicals along a duct onto the body surface or into a body cavity.

Faeces
Solid waste consisting of undigested food, dead cells, and bacteria that remains after digestion and is eliminated from the body through the anus.

Fat
Type of lipid that is solid at room temperature, which is found in many foods.

Fatty acid
Building block, with glycerol, of fats and oils.

Feedback system
Control mechanism that maintains a stable state in the body by correcting unwanted changes and which regulates, for example, body temperature.

Fertility
Ability of a man and a woman to produce children without undue difficulties.

Fertilization
Joining together of an ovum and a sperm to make a new individual.

Fetus
Name given to the unborn child from the ninth week after fertilization until birth.

Fibre (dietary)
Also called roughage, this is indigestible plant material that gives bulk to food and improves the efficiency of intestinal muscles.

Fibre (muscle)
Name given to a muscle cell.

Follicle
Cluster of cells found inside the ovary that contains an ovum (see also Hair follicle).

Fracture
Break in a bone, often caused by a fall.

Fungi (sing. Fungus)
Group of living organisms, including yeasts and mushrooms, some of which are parasitic on humans, causing diseases such as athlete's foot.

Gas exchange
Movement of oxygen from the lungs into the bloodstream, and that of carbon dioxide in the opposite direction.

Gene
Carries the instructions to make a specific protein, and is one of the 30,000–50,000

genes stored in the DNA that makes up chromosomes inside a body cell.

Genetic code
Code used to convert the "message" carried by the sequence of bases in DNA into a sequence of amino acids to make a protein.

Genetic engineering
Artificial alteration made to the genetic make-up of an organism.

Genetics
Study of inheritance and transmission of genes from one generation to the next.

Girdle
Ring of bones that attaches the limbs to the rest of the skeleton.

Gland
Group of cells that produce chemical substances and release them into or onto the body.

Glial cells
Also called neuroglia, these cells protect and nurture neurons (nerve cells).

Glucose
Main sugar found circulating in the bloodstream, and the body's primary energy source.

Glycogen
Polysaccharide made of glucose subunits that forms a carbohydrate energy store in liver cells and muscle fibres.

Grey matter
Surface layer of the cerebrum, and inner part of the spinal cord, which consists mainly of neuron cell bodies.

Haemoglobin
Oxygen-carrying, iron-containing protein found inside red blood cells.

Hair follicle
Deep, hollow space in the skin from which a hair grows.

Haploid cell
A cell, such as a sperm or ovum, that is formed by meiosis and contains only a single set of 23 chromosomes.

Haversian system
Also called an osteon, this is a cylindrical collection of concentric bony tubes that is the basic units of compact bone.

Hepatic
Related to the liver, for example the hepatic artery that supplies blood to the liver.

Heredity
The passing on of characteristics controlled by genes from one generation to the next.

Homeostasis
Maintenance of stable conditions, including body temperature and blood glucose levels, regardless of external conditions.

Hormone
Chemical messenger that is produced and released by an endocrine gland and carried to its "target" by the blood.

Hunter-gatherer
Typical of early human societies but rarer today, person who lives by hunting animals and gathering plants rather than through agriculture.

Hypothalamus
Small but important part of the brain that regulates many body activities, including thirst and body temperature, by way of the pituitary gland and the autonomic nervous system.

Immune system
Collection of cells within the circulatory and lymphatic systems that protect the body from disease-causing micro-organisms.

Immunity
Ability of the immune system to "remember" and provide resistance to specific disease-causing micro-organisms.

Immunization
Provision of immunity against a disease by injecting a vaccine containing a weakened form of the micro-organism that causes that disease.

Infant mortality rate
Number of infants who die during the first year of life per 1,000 live births.

Infection
Establishment of disease-causing micro-organisms, such as bacteria, in the body.

Infectious disease
Disease, such as chickenpox, which is caused by a specific micro-organism.

Infertility
Inability of either a man or a woman, or both partners, to produce a child.

Ingestion
Taking food or drink into the body through the mouth.

Inhalation
The movement of air into the lungs; also called inspiration or breathing in.

Insertion
Attachment point of a muscle, through its tendon, to a bone that moves.

Insoluble
A substance that does not dissolve in water.

Integumentary system
External protective covering of the body provided by the skin, hair, and nails.

IVF (in vitro fertilization)
Technique used to help infertile couples conceive by fertilizing an ovum outside the body, then returning it to the uterus to develop.

Karyotype
Complete set of chromosomes inside a cell, photographed and arranged in pairs in descending size order.

Keratin
Tough, waterproof protein found inside cells making up hair, nails, and the upper epidermis of the skin.

Labour
Contractions of the muscular wall of the uterus before and during birth.

Ligament
Tough strips of fibrous connective tissue that hold bones together where they meet at joints.

Light microscope
Instrument that uses light rays focused by glass lenses to produce a magnified image of an object.

Limbic system
Part of the brain at the base of the cerebrum that controls emotions.

Lipid
One of a group of organic compounds, made up of carbon, hydrogen, and oxygen, that includes fats and oils (made of fatty acids and glycerol), phospholipids, and steroids, such as cholesterol.

Lymph
Fluid that flows through the lymphatic system from the tissues to the blood.

Lymphocyte
Type of white blood cell that plays a key role in the immune system.

Macronutrient
Nutrient – such as carbohydrate, fat, or protein – needed in large amounts by the body.

Macrophage
White blood cell present in a number of tissues that engulfs bacteria and foreign debris and plays a part in the immune system.

Magnetic resonance imaging (MRI)
Scanning technique that uses magnetism, radio waves, and a computer to produce images of the inside of the body.

Magnetoencephalography (MEG)
Scanning technique that produces real-time images of brain activity.

Mammal
Living organism that belongs to a group of animals that are "warm blooded", have a covering of fur, and feed their young with milk.

Marrow
Soft fatty tissue, either red or yellow, found in the spaces within bones.

Mechanical digestion
Breakdown of food into smaller particles through chewing or the churning action of stomach muscles.

Mechanoreceptor
Receptor that can detect pressure produced by touch, sound waves, or the stretching of muscles.

Medieval
Relating to the Middle Ages between the 5th and 15th centuries.

Meiosis
Type of cell division that occurs in the ovaries and testes to produce sex cells – ova and sperm – that contain a single set of chromosomes.

Melanin
Brown-black pigment found in the skin, hair, and the iris of the eye that gives them their colouring.

Membrane
Thin layer made up of epithelial tissue supported by connective tissue that covers or lines an external or internal body surface (see also Cell membrane).

Meninges (sing. Meninx)
Protective membranes that cover the brain and spinal cord.

Menopause
Period of a woman's life, between the ages of 45 and 55, when ovulation and menstrual periods cease.

Menstrual cycle
Sequence of changes, repeated about every 28 days, that prepares the lining of a woman's uterus to receive an ovum should it be fertilized.

Mesopotamia
Ancient region of south-western Asia between the Rivers Tigris and Euphrates, known today as Iraq.

Metabolism
Sum of all the chemical processes that take place within the body, particularly within its cells.

Metabolic rate
Rate at which energy is released by metabolism.

Microbe
General name for a micro-organism that causes disease.

Micrograph
Photograph taken with the aid of a microscope.

Micronutrient
Nutrient – such as a vitamin or mineral – needed in small amounts by the body.

Micro-organism
Tiny organism, such as a bacterium, that can only be seen with a microscope.

Middle Ages
Period of Western European history between the 5th and 15th centuries.

Mineral
One of about 20 chemical elements, including calcium and iron, that must be present in the diet to maintain good health.

Mitochondria (sing. Mitochondrion)
Organelles inside cells that carry out aerobic respiration to release energy.

Mitosis
Type of cell division used for growth and repair that produces two identical cells from each "parent" cell.

Molecule
Chemical unit that is made up of two or more linked atoms, such as the two hydrogen atoms and one oxygen atom in a water molecule.

Monosaccharide
Sugar such as glucose, fructose, or galactose, that is the simplest type of carbohydrate.

Motor neuron
Neuron (nerve cell) that carries nerve impulses from the central nervous system to muscles and glands.

Mucous membrane
Layer that lines the body cavities which open to the exterior, for example the respiratory system, and which secretes mucus.

Mucus
Thick, slimy fluid secreted by mucous membranes, which moistens, protects, and lubricates.

Muscle tone
Partial contraction of a muscle that maintains the body's posture.

Mutation
Change to the DNA in one of a cell's chromosomes that may have harmful effects, and can be passed on to the next generation.

Myelin sheath
Insulating sheath wrapped around most axons (nerve fibres) that increases the speed of conduction of nerve impulses.

Myofibril
One of thousands of tiny rod-like strands inside a muscle fibre (cell).

Nephron
One of a million filtration units inside each kidney that produce urine.

Nerve
Cable-like bundle of neurons (nerve cells) that relays nerve impulses between the body and central nervous system.

Nerve fibre
Also called an axon, this is the long "tail" of a neuron that carries nerve impulses away from its cell body.

Nerve impulse
Tiny electrical signal that passes along a neuron (nerve cell) at high speed.

Neuron
One of the billions of interconnected nerve cells that carries electrical signals at high speed and makes up the nervous system; the brain, spinal cord, and nerves.

Neurotransmitter
Chemical released when a nerve impulse reaches the end of a neuron, which triggers a nerve impulse in a neighbouring neuron.

Nitrogen
Gas, like oxygen, that is found in the air but which normally plays no part in body functions.

Nobel Prize
Prestigious award given annually for outstanding achievement in one of five fields, including physiology or medicine.

Non-infectious disease
Disease, such as cancer or heart disease, that is not caused by a disease-causing micro-organism.

Nucleic acid
Organic compound, such as DNA or RNA, which contains carbon, hydrogen, oxygen, nitrogen, and phosphorus, and is made up of units called nucleotides.

Nucleotide
Basic building block of nucleic acids such as DNA, consisting of a phosphate group, a deoxyribose sugar, and a nitrogenous base (adenine, cytosine, guanine, or thymine).

Nucleus
Control centre of a cell that contains chromosomes.

Nutrient
Substance – such as carbohydrate, protein, fat, vitamin, or mineral – needed in the diet to maintain good health and normal body functioning.

Obstetrician
Doctor who specializes in pregnancy and childbirth.

Ophthalmoscope
Instrument used to view the inside of the eye.

Orbit
Socket in the skull that surrounds, supports, and protects the eyeball.

Organ
Body part, such as the kidney or brain, with a specific role or roles that is made up of two or more different types of tissues.

Organic compound
Substance, such as carbohydrate, protein, lipid, or nucleic acid, that has a carbon "skeleton", and is made only by living systems.

Organelle
Microscopic structure inside a cell, such as a mitochondrion, that has a specific function.

Origin
Attachment point of a muscle, through its tendon, to a bone that is stationary.

Osmoregulation
Maintenance of the correct levels of water and salts in blood and tissue fluids carried out by the kidneys.

Ossicle
One of the three small, sound-transmitting bones found inside the middle ear.

Ossification
Process of bone formation.

Ovarian cycle
Sequence of changes, repeated about every 28 days, that causes an ovum to be released from a woman's ovary.

Ovulation
Release of an ovum from a woman's ovary.

Ovum (pl. Ova)
Also called an egg, this is the female sex cell, which is produced by, and released from, a woman's ovary.

Oxygen
Gas found in the air which is taken into the bloodstream through the lungs and used by body cells in aerobic respiration to release energy from glucose.

Palaeontologist
Scientist who studies fossils.

Papillae (sing. Papilla)
Small bumps projecting from the tongue's surface, some of which house taste receptors called taste buds.

Paraplegic
Weakness or paralysis of both legs and sometimes part of the trunk caused by damage to the spinal cord.

Pathogen
Disease-causing micro-organism such as a bacterium, virus, protist, or fungus.

Pectoral girdle
Girdle formed by the two collar bones and two shoulder blades, which attaches the arms to the skeleton.

Pelvic girdle
Girdle formed by the two hip bones which anchors the legs to the skeleton and, with the sacrum, forms the basin-like pelvis.

Periosteum
Membrane covering the surface of bones, which contains blood vessels.

Peripheral nervous system (PNS)
The part of the nervous system that consists of the nerves that relay nerve impulses between the body and the central nervous system.

Peristalsis
Wave of muscular contraction through a hollow organ that, for example, pushes food down the oesophagus or urine down a ureter.

Phagocyte
General name for white blood cells – including neutrophils and macrophages – that engulf and digest disease-causing micro-organisms.

Phagocytosis
Process by which phagocytes engulf and digest disease-causing micro-organisms.

Phantom pain
Sensation of pain felt in a limb that is no longer present because it has been amputated.

Phospholipid
Type of phosphate-containing lipid that makes up the cell membranes around cells, and the membranes around organelles.

Physiology
Study of how the body works and functions.

Pitch
Quality of a sound – whether high- or low-pitched – that depends on the frequency of sound waves, that is, how quickly one wave is followed by the next.

Placenta
Organ that develops in the uterus during pregnancy that forms an interface between the blood supplies of the mother and fetus, through which the fetus receives food and oxygen.

Plasma
Liquid part of the blood, which is mainly water.

Polysaccharide
Complex carbohydrate, such as glycogen, that does not have a sweet taste and is made up of long chains of monosaccharides, such as glucose.

Portal system
Veins that carry blood from one organ to another rather than towards the heart.

Positron emission tomography (PET)
Scanning technique that uses radioactive substances injected into the body to show parts of the body at work, especially the brain.

Pregnancy
The period from conception to birth, but which is dated from the start of a woman's last menstrual cycle, and so lasts about 40 weeks.

Primates
Group of mammals which includes monkeys, apes, and humans.

Protein
One of a group of organic compounds, made up of carbon, hydrogen, oxygen, nitrogen, and sulphur, that perform many roles inside the body including making enzymes.

Protists
Group of single-celled organisms, some of which cause diseases in humans such as malaria.

Puberty
Period during adolescence when the body grows and develops an adult appearance, and the reproductive system starts working.

Pulmonary circulation
Part of the circulatory system that carries blood from the heart to the lungs and back to the heart.

Pupil
Opening in the centre of the iris through which light enters the eye.

Radionuclide scanning
Scanning technique that uses radioactive substances to reveal the functioning of organs such as bones.

Receptor
Special cells or neurons that detect stimuli, such as light, and trigger sensory neurons.

Reflex
Automatic, unconscious, split-second response to a stimulus that often protects the body from danger.

Reflex arc
Nervous pathway, often through the spinal cord but not the brain, involved in a reflex.

Renaissance
Term meaning "rebirth", which describes the period between the 14th and early 17th centuries in Europe when there was a creative revolution in the arts, sciences, and medicine.

Renal
Related to the kidney, as in the renal artery that supplies blood to the kidney.

Retina
Inner lining of the eyeball that is packed with light receptors.

Rod
One of two types of light receptors in the retina of the eye, rods provide black and white vision, and work best in dim light.

Saliva
Fluid released into the mouth, especially during chewing, by the salivary glands.

Scanning electron micrograph (SEM)
Photograph produced using a scanning electron microscope.

Sebaceous gland
Gland connected to a hair follicle that produces an oily liquid called sebum.

Sebum
Oily liquid that keeps hair and skin soft, flexible, and waterproof.

Secretion
Chemical substance made and released by a gland.

Semen
Fluid produced by male reproductive glands, which activates and nurtures sperm, and in which they swim.

Semicircular canal
Part of the inner ear that is involved in balance.

Sensory neuron
Neuron (nerve cell) that carries nerve impulses from sensory receptors to the central nervous system.

Septum
Dividing wall within body parts, such as in the nose.

Skeletal muscle
Type of muscle that is attached to the skeleton and moves the body.

Smooth muscle
Type of muscle found inside the walls of organs that, for example, pushes food along the small intestine.

Soluble
Describes a substance that dissolves in water.

Solution
Mixture of one substance (called a solute) dissolved in another (the solvent); glucose and carbon dioxide are both solutes that dissolve in the water (solvent) in blood.

Species
A group of living things in the natural world that can breed with each other.

Sperm
Also called spermatozoa, these are the male sex cells, which are made in and released from a man's testes.

Spermatogenesis
Process of sperm production inside the testes.

Sphincter
Ring of muscle around a passage or opening that opens or closes to control the flow of, for example, food or urine along it.

Spinal cord
Column of nervous tissue that runs down the back and relays nerve signals between the brain and body.

Spinal nerve
One of the 31 pairs of nerves that arise from the spinal cord.

Spongy bone
Also called cancellous bone, this is the tough but lightweight honeycomb of struts and cavities that forms the inner part of a bone.

Stain
Dye used to colour cells and tissues so they can be seen under the light microscope.

Sterilization
Surgical procedure that stops a person being able to reproduce. Also the process used to destroy any micro-organisms on surgical instruments and other materials used in hospitals to minimize the risk of infection to patients.

Steroid
One of a group of lipids that includes cholesterol and some hormones and vitamins.

Stethoscope
Instrument used to listen to heart sounds and breathing.

Stimuli (sing. Stimulus)
Any change in external conditions that causes a change in body activities, such as the smell of food causing the release of saliva.

Sugar
A simple, sweet-tasting carbohydrate such as glucose or sucrose.

Surgery
Treatment of disease or injury by direct intervention, often using instruments to open the body.

Sweat
Watery liquid produced by sweat glands that cools the body when it evaporates from the skin's surface.

Symptom
Indication of a disease or disorder that is noticed by a patient.

Synapse
Junction between two neurons in which they do not touch but come very close to each other.

Synovial joint
Most common type of joint in the body, a movable joint with a fluid-filled space between the bones.

System
Group of linked organs that work together to perform a particular task or tasks.

Systemic circulation
Part of the circulatory system that carries blood from the heart around the body and back to the heart.

Systole
Part of the heartbeat cycle when either the atria or ventricles are contracted.

Taste bud
Taste receptor found in the surface of the tongue.

Tendon
Cord or sheet of strong connective tissue that connects muscle to bone.

Terminal hair
Thick hair which grows on the scalp, and forms the eyebrows and eyelashes.

Thermogram
Colour-coded image produced by a special camera that shows the amount of heat released by different parts of the body.

Thorax
Also called the chest, the upper part of the trunk (central part of the body) between the neck and abdomen.

Tissue
Group of one type of cell, or similar types of cells, that work together to perform a particular function, such as epithelial cells forming a protective lining to the mouth.

Toxin
Poisonous substance released into the body by a disease-causing bacterium.

Transmission electron micrograph (TEM)
Photograph produced using a transmission electron microscope.

Transplant
Replacement of a diseased organ or tissue with a healthy living organ or tissue provided by a donor, who has usually just died.

Trepanning
Ancient practice of cutting or drilling holes into the skull, probably to release "evil spirits".

Trunk
Also called the torso, the central part of the body, consisting of the thorax and abdomen, to which the head and limbs are attached.

Tumour
Abnormal growth of tissue produced when tissue cells multiply at an increased rate.

Ultrasound scanning
Scanning technique that uses high-frequency sound waves to produce images of the inside of the body, including those of the developing fetus.

Ultraviolet (UV) radiation
Radiation that occurs naturally in sunlight, but can be harmful if the skin is over-exposed to it.

Umbilical cord
Rope-like structure that connects the fetus to the placenta.

Urination
Release of urine from the bladder to the outside of the body.

Urine
Liquid produced by the kidneys, which contains wastes and surplus water and salts removed from the blood.

Vaccination
Also called immunization, this is the process whereby a vaccine is injected into the bloodstream to stimulate the body to produce antibodies against a disease.

Vaccine
Medication containing a weakened form of a disease-causing micro-organism.

Vein
Blood vessel that carries blood from the tissues towards the heart.

Vellus hair
One of the millions of fine, soft hairs that grow all over the body.

Ventricle
One of the two (left and right) lower chambers of the heart.

Venule
Very small vein that collects blood from a capillary.

Villi (sing. Villus)
Tiny, finger-like projections from the lining of the small intestine that greatly increase its surface area for absorption.

Viruses (sing. Virus)
Infective non-living agents, much smaller than bacteria, that invade cells and cause diseases such as the common cold and measles.

Vitamin
One of over 13 organic compounds, including vitamin A and niacin, that are needed in small amounts in the diet for normal body functioning.

White matter
The inner part of the brain, and outer part of the spinal cord, which is made up mainly of axons (nerve fibres).

X-ray
Form of radiation that reveals bones when projected through the body onto a photographic film.

Zygote
Cell produced when a sperm fertilizes an ovum.

INDEX

Page numbers in bold indicate the principal reference to a subject. Page numbers with the suffix g indicate a glossary entry.

A

B

C

D

E

F

I

J

K

L

M

N

O

P

Q

R

S

ACKNOWLEDGEMENTS

The team at Dorling Kindersley would like to thank:
Carole Oliver, Clair Watson and Willie Wood for design help; Caryn Jenner and Brad Round for editorial help; Sophie Young for DTP help; Alyson Lacewing for proofreading; and Marie Lorimer for the index.

The publisher would like to thank the following for their kind permission to reproduce their photographs:
Key: a=above; b=below; c=centre; l=left; r=right; t=top; bi=background image; mi=main image

10 Telegraph Colour Library/Getty Images: Alistair Berg bl; **10–11** Science Photo Library: mi, l; Simon Fraser c; r Professors P. M. Motta, S. Makabe & T. Naguro r; **12–13** Harry Cutting mi; **14–15** Liam Bailey bl; source unknown cl; **16–17** Science Photo Library: Professor P. Motta, mi (circular), bl, tr (circular), Astrid & Hanns-Frieder Michler tr (cutout), Pascal Goetgheluck cb; Science Museum ct (cutout); Ann Ronan Picture Library: ct (circular); **18–19** Science Photo Library: mi, c (circular), br, Eric Grave tl, Quest bl, Professor P. Motta cl (circular); Mary Evans: tr; **20–21** Denis Models mi; Science Photo Library: Alfred Paskia l; Professors P. M. Motta, S. Makabe & T. Naguro tr, CNRI cr (circular), Dr Kari Lounatmaa br; **22–23** Science Photo Library: bi, Dr Arther Tucker cl, Professor K. Seddon & Dr T. Evans cr, Ken Eward br; **20–21** Denis Models mi; Science Photo Library: Alfred Paskia l, Professors P. M. Motta, S. Makabe & T. Naguro tr, CNRI cr (circular); Dr Kari Lounatmaa br; **22–23** Science Photo Library: bi, DR Arther Tucker cl, Professor K. Seddon & Dr T. Evans cr, Ken Eward br; **24–25** Science Photo Library: bl; A. Barrington Brown ct, Dr Gopal Murti tr; **26–27** Science Photo Library: Quest mi and ct, David M. Martin cl (circular), BSIP Laurent/H. Amercain cb (box), Professor P. Motta rt, Manfred Kage rc, cr; **28–29** Science Photo Library: Simon Fraser mi; **30–31** Science Photo Library: Volker Steger mi, SNRI tl, GJLP c, Dr John Mazziotta et al/Neurology rt, Alfred Paskia tr (cutout), cl, Simon Fraser/Royal Victoria Infirmary, Newcastle-upon-Tyne br; **32–33** Science Photo Library: Professor P. Motta mi, all other images Science Photo Library; **34–35** Geoff Brightling l and tr (cutout); Corbis Images ct; Science Photo Library: Martin Dohrn c (box), Andrew Syred tr (circular); Andy Crawford cb (cutout and circular); Dave King br (cutout) and (circular); **36–37** Science Photo Library: Dr Jeremy Burgess mi, Christina Pedrazzini c, Dr P. Marazzi cr, D. Phillips tr; Richard Wehr/Custom Medical Stock Photo/SPL cl; **38–39** Science Photo Library: bl, Dr P. Marazzi ct, Quest c, Geoff Brightling cb; Corbis Images: tr; Anthony Duke, digital artwork br; **40–41** Science Photo Library: D. Phillips mi; Andrew Syred bl and cr, Eye of Science r and br; Andy Crawford ct photograph (digitally amended by Peter Bull); **42–43** Science Photo Library: b (box), Professor P. Motta bi, Biophoto Associates tr, CNRI bl, GJLP/CNRI bl; Mary Evans c (box); Andy Crawford mi; **44–45** Philip Dowell tl; Science Photo Library: Alfred Paskieka c (oval), Dave Roberts tr, Dave King br; **46–47** Science Photo Library: r, Allsport c, GJLP/CNRI tr (circular); **48–49** Science Photo Library: Department of Clinical Radiology, Salibury District Hospital bi, CNRI tl; Mehau Kulyk cb, tr; Ann Ronan Picture Library c (box); Philip Dowell r; **50–51** Science Photo Library: Hugh Turvey tl and ct, Mehau Kulyk bl; Andy Crawford c (x2); **52–53** Science Photo Library: bi; sources unknown bl and tr; Corbis Images: br; **54–55** Science Photo Library: Andrew Syred l, c and r, Professor P. Motta ct, Robert Becker/Custom Medical Stock Photo br; **56–57** Science Photo Library: Department of Clinical Radiology, Salibury District Hospital bi, bl, and tr, Quest cl, Scott Camazine ct; Mary Evans: bc (box); **58–59** Ray Moller cb; Dave King tr (cutout); Science Photo Library: Dr P. Marazzi and Department of Clinical Radiology, Salibury District Hospital (digitally amended by Anthony Duke) tr, Mike Devlin br (circular), Brad Nelson/Medical Stock Photo br; **60–61** Andy Crawford: mi photography, tc (digital artwork Peter Bull), and r; 6 artworks surrounding skeleton by Colin Salmon; **62–63** Science Photo Library: cl and c, Dr R. Clark & M. Goff br; M.I. Walker ct and br (box); **64–65** Andy Crawford c circular images (x2); Raymond Evans/Unit of Art in Medicine, University of Manchester, London r; **66–67** Science Photo Library: bl and rc; Andy Crawford tr; **68–69** The Royal Collection Picture Library bl (and bi); Wellcome Library, London: tl and cb; Science Photo Library: c; **70–71** Corbis c ; Getty Images cb; **72–73** Getty Images: mi; Corbis Images: bl; Quest c (circular); Andy Crawford and Steve Gordon tr; Science Photo Library: Will McIntyre br; **74–75** Science Photo Library: GJLP/CNRI r; all other images Science Photo Library; **76–77** Getty Images: mi; Science Photo Library: BSIP VEM tl, bl (box) and c (circular); Andy Crawford br photography (head and hand); brain artwork, eye and heart DK; **78–79** Science Photo Library: Quest bl, Nancy Kedersha ct, BSIP, Sercomi cb (box); Corbis Images tr; **80–81** Peter Bull bl; Science Photo Library: Eye of Science ct (circular), tr and br; Frank Greenaway cr; **82–83** Science Photo Library: Quest bl, Sue Ford tr; **84–85** Andy Crawford mi photography; Getty Imags br; **86–87** Andy Crawford l photography (superimposed image DK); Corbis: ct; Andy Crawford cb; Science Photo Library: CC Studio tr; Mary Evans: br; **88–89** Science Photo Library: bl, clt, and crt (box); Andy Crawford r photography (superimposed image DK); **90–91** Imperial War Museum: crb (and bi); Department of Neuorology and Image Analysis Facility, University of Iowa cl; Andy Crawford ct; Science Photo Library: bl; Hank Morgon tr, BSIP Astier br; **92–93** Science Photo Library: GJLP/CNRI bi, ct, Getty Images: tr; **94–95** Steve Gorton bl; Corbis Images: c; Science Photo Library: tr; Dave King br (x3); **96–97** Andy Crawford bl photography (superimposed image DK); Science Photo Library: Hank Morgan ct, BSIP Buntschucr cr; **98–99** Andy Crawford mi photography (superimposed artwork DK); Science Photo Library: Jean-Loup Charmet cr; **100–101**Geoff Dann and Andy Crawford l; Andy Crawford ct, c, and cb; Science Photo Library: Montreal Neuro Institute, McGill University/CNRI cr; **102–103** Science Photo Library: Mehau Kulyk, bi, US National Library of Medicine cl, Simon Fraser (RNC Newcastle-upon-Tyne) cr; Mary Evans: cb; Science Museum: tr; **104–105** Andy Crawford mi photography (digitally amended by Anthony Duke); Corbis: bl; The Bridgeman Art Library: tr (box); Corbis Images: cr; Science Photo Library: BSIP Leca br; **106–107** Andy Crawford bi and l photography (superimposed artwork DK); Getty Images: tr; Science Photo Library: cb and br; **108–109** Andy Crawford tl photography (superimposed artwork DK); Science Photo Library: Quest cl, James King-Holmes cr (circular) and tr; Corbis: br; **110–111** Getty Images: mi; Andy Crawford tl photography (superimposed artwork DK); Science Photo Library: Mark Clarke br; **112–113** Andy Crawford mi photography (superimposed artwork Peter Bull) and br; Science Photo Library: cl and cr; **114–115** Mary Evans bl; Science Photo Library: PIR-CNRI tl, Bill Longcore tr, cbr Mark Burnett cbr; Custom Medical Stock Photo br; **116–117** Science Photo Library: Mehau Kulyk br (plus superimposed artwork DK); Ronald James cl; **118–119** Getty Images: mi; Science Photo Library: bl and tr; **120–21** Andy Crawford cl (superimposed artwork DK); Getty Images: tr; Science Photo Library: br Pascal Goetgheluk; **122–23** Andy Crawford mi photography (superimposed artwork Anthony Duke); Science Photo Library: bl; **124–25** Science Photo Library: Astrid & Hanns-Freider Michler bi, tl, Saturn Stills c, J.C. Revy cr, Volker Steger bl; Banting House National Historic Site tr; **126–27** Science Photo Library: Quest mi, Professor P. Motta l, c, CNRI r; **128–29** Science Photo Library: National Cancer Institute bi; Simon Fraser tr, Quest bl; **130–31** Science Photo Library: Quest bl, tl, St Bartholomew's Hospital ct, Robert Becker, Custom Medical Photo Stock br; Andy Crawford cbr; **132–33** Science Photo Library: bi, CNRI br (circular), Francis Leroy, Bicosmos tr, Eye of Sciencebr; **134–35** Science Photo Library: Jurgen Berger mi, Max Planc Institute, Nigel Dennis bl; CNRI tr (circular) and br, Dr Gopal Murti tr; Quest cr, NIBSC br (circular), Secchi-Lecaque bcr; **136–37** Science Photo Library: bl and tr, NIBSC bcl, tl Volker Steger; Mary Evans: bl; **138–39** Corbis: br; Mary Evans: ct and cb; Science Photo Library: David Scharf bi; Jerry Mason br; **140–41** Science Photo Library: bl and tr, BSIP, Carvallini James ct, Quest cb, Phillipe Plailly br; **142–43** Science Photo Library: Phillipe Plailly ctr, bc; Getty Images: br; **144–45** Science Photo Library: CNRI bi and bl, Will & Denis

Mcintyre tr, D. Ouellette c, Ouellete & Theroux bl; Corbis: ctl ; **146–47** Science Photo Library: CNRI bi; **148–49** Science Photo Library: cl, ct, and tr (digitally manipulated by Anthony Duke); Mary Evans: br; **150–51** Science Photo Library: Saturn Stills tr; Corbis: cbr and br.; **152–53** Science Photo Library: tl John Griem; **154–55** Science Photo Library: Secchi Lecaque bi, Alferd Pasieka ctl, tl, Dr P. Marazzi cr; **156–57** Science Photo Library: Dr Linda Stannard tl; Eye of Science bl and ctr, Dr P. Marazzi bcl, Dr Gary Gaugler c, Quest tr; Mary Evans br; **158–59** Andy Crawford cl phography (superimposed image DK); Science Photo Library: circular images, clockwise from top: Manfred Kage, Professor P. Motta, Science Photo Library, Biophoto Associates, Professors P.M. Motta, K.R. Porter, and P.M. Andrews, Science Photo Library, Philip A. Harington tr, Juergen Bergercr and bi, Oscar Burriel r; **160–61** Science Photo Library: cl, Dr Andrejs Liepins tr, Jerrican br; **162–63** Science Photo Library: tl, br, c, cb, and bi; **164–65** Science Photo Library: Eye of Science br, James King-Holmes tl, cbr, ctr, and br, J. F. Wilson cr; Corbis: tr; **166–67** Science Photo Library: BSP Vem tl; bl, cl, and cr; **168–69** Science Photo Library: Clinique Ste Catherine, CNRI tl, tr, Alain Dex ctr ; Matt Meadows br, H. Schleichkorn br; **170–71** Science Photo Library: CNRI bi; SPL tr; Getty Images: br; **172–73** Andy Crawford bl photography (superimposed images); Science Photo Library: br; **174–75** Mary Evans: tl; Ann RonanPicture Library: bl; Science Photo Library: National Library of Medicine ct, Bill Longcore br; Corbis: bi; Mary Evans; **176–77** Science Photo Library: Mehau Kulyk l, Professor C. Ferland bcl and bcr; Labat tr; Andy Crawford c (x2) photography (superimposed images (x2) Peter Bull); cr (x2) Andy Crawford; Getty Images: br; **178–79** Andy Crawford c photography (superimposed images DK); Science Photo Library: Eye of Science br; **180–81** Andy Crawford mi; Science Photo Library: BSIP Vem tcr, CNRI cr, Dr Tony Brain br; **182–83** Science Photo Library: Biophoto Associates tr; Andy Crawford: bi, l photography (superimposed imges DK); Dr Klaus Schiller cl; **184–85** Science Photo Library: bl and bcr, Eye of Science bcl, Dr K. R. Schiller tr; Corbis: br; **186–87** Science Photo Library: Manfred Cage cl, bl and br, Professor P. Motta tcr; Andy Crawford mi photography (superimposed images DK); **188–89** Science Photo Library: Eye of Science cl, David M. Martin cl, Professor P. Motta tr; Andy Crawford mi and ctr photography (superimposed images DK); **190–91** Science Photo Library: cr, Professors P. Motta & F.M. Magliocca br; **192–93** Andy Crawford bi; Science Photo Library: Adam Hart-Davis far bl, Dr Jeremy Burgess bl, c; Corbis: tcl and tr; Getty Images: br; **194–95** All images DK; **196–97** Science Photo Library: Alfred Pasieka tl, (superimposed image DK), Dr Jeremy Burgess ctr **198–99** Andy Crawford tl photography (superimposed image DK); bl, c, tr and br DK; Professor P. Motta cr, bl; **200–01** Corbis: mi; Andy Crawford bl photography (superimposed image DK); **202–03** Science Photo Library: Brian Yarvin cr, br; **204–5** Science Photo Library: Quest mi, Brad Nelson bl, Biophoto Associates tl, Professors P. Motta and M. Castellucci br (circular); Corbis: bl **206–07** Janos Marffy l; Science Photo Library: BSIP Vem cl, Mauro Fermariello tr, Andrew Syred br (circular); r DK; Andy Crawford bl photography, (superimposed image DK); br Corbis; **208–09** Science Photo Library: Quest bl, Ed Reschke, Peter Arnold Inc tl, tr, CNRI b; James Stevenson br; **210** Science Photo Library: BSIP VEM br; Professor P. Motta, G Macchiarelli, SA Nottola bl; Petit Format– Nestle bc; **210–211** Science Photo Library: TEK Image –211; **212** The Image Bank–Getty Images: Larry Dale Gordon br; **213** Science Photo Library: D. Phillips tr; Science Pictures Ltd br; **215** Corbis: Bohemian Nomad Picturemakers tr; Science Photo Library: Ed Reschke, Peter Arnold Inc. tl; National Library of Medicine br; The Wellcome Institute Library, London: Dr Joyce Harper bc; **216** Science Photo Library: CNRI bc; Astrid & Hanns-Frieder Michler ac; **216-217** Science Photo Library: Science Source tl; **217** Corbis: tr; Science Photo Library: Peter Cull tc; Custom Medical Stock Photo br; **218** Science Photo Library: br; Professor P. Motta, G. Macchiarelli, S. A. Nottola cl; **219** Science Photo Library: CNRI br;Custom Medical Stock Photo tr; Dr Yorgos Nikas tl; **221** Science Photo Library: Professor P. M. Motta & E. Vizza tr; Professors P. M. Motta & J. Van Blerkom tl; **223** Science Photo Library: Dr Yorgos Nikas tl, cla, clb; The Wellcome Institute Library, London: Yorgas Nikas br; **224** Corbis: Bettmann tl; **224–225** Science Photo Library: Pascal Goetgheluck c; **225** Corbis: Bettmann tr; Science Photo Library: Pascal Goetgheluck br; **226** Science Photo Library: Petit Format– Nestle crb; **227** Science Photo Library: Custom Medical Stock Photo br; **228** Science Photo Library: Dept of Clinical Radiology, Salisbury District Hospital crb; **229** Science Photo Library: Petit Format– Prof E. Symonds br; **231** The Wellcome Institute Library, London: l; **232** Corbis: Dick Clintsman cla; Science Photo Library: tr; Jean–Loup Charmet b; **233** Science Photo Library: Lawrence Migdale cr; Alfred Pasieka tr; **234** Corbis: Reed Kaestner br; Barnabas Kindersley: tr; Guy Ryecart: cr; **235** Barnabas Kindersley: c, cr, bc; Guy Ryecart: c; **236** Science Photo Library: Christopher Briscoe tr; Ron Sutherland cl; **237** Telegraph Colour Library–Getty Images: Stephanie Rausser br; **238** Telegraph Colour Library–Getty Images: Jim Cummins b; **239** Corbis: O'Brien Productions tl; Science Photo Library: BSIP VEM clb; Mehau Kulyk br; **240** Science Photo Library: Scott Camazine tl; Quest c; **241** The Image Bank–Getty Images: Ghislain & Marie David de Lossy tr; Science Photo Library: Dr P. Marazzi c; Telegraph Colour Library–Getty Images: tl; **242** Science Photo Library: Dr P. Marazzi r; Telegraph Colour Library–Getty Images: Larry Bray cl; **243** Science Photo Library: Custom Medical Stock Photo r; Dr P. Marazzi tl; Alfred Pasieka cl; **244** Science Photo Library: CNRI cl, crb, bc; Mike Bluestone tr; **245** Pa Photos: bl; **246** Science Photo Library: Peter Menzel tl; **246–247** Science Photo Library: Biophoto Associates; **247** Science Photo Library: tr; **249** Science Photo Library: CNRI bc; Lauren Shear br; **250** Science Photo Library: Andrew Syred bl; **250–251** Science Photo Library: TEK Image; **251** Science Photo Library: BSIP, PIKO tl; BSIP VEM cr; Klaus Guldbrandsen bl; **252** Bridgeman Art Library, London – New York: Bibliotheque des Arts Decoratifs, Paris, France; Science Museum: bc; Science Photo Library: National Library of Medicine bl; **254–255** American Museum Of Natural History: Denis Finnin & Craig Chesek b; **255** Corbis: Kevin Schafer br; Science Museum: tr; Werner Forman Archive: Tanzania National Museum, Dar es Salaam cr; **256** AKG London: Erich Lessing tr; © Michael Holford: cla; Science Photo Library: Christian Jegou, Publiphoto Diffusion br; **257** Science Museum: br; Charles Walker Collection: tl; **258** AKG London: tr; Erich Lessing bc; Charles Walker Collection: c; **258–259** Scala Group S.p.A.: **259** Mary Evans Picture Library: tr; World Health Organisation: cra; **260** Mary Evans Picture Library: tl; Science Museum: tr; Science Photo Library: National Library of Medicine bl; **260–261** AKG London: b; **261** Museum of the Royal Pharmaceutical Society: cla; Hulton Getty Archive: cr; **262** Science Photo Library: Jean-Loup Charmet tl; **262–263** AKG London: b; **263** Corbis: © Bettmann br; Science Photo Library: John Burbidge tl; CAMR–A B Dowsett tr; **264** AKG London: tl; Erich Lessing bc; Bridgeman Art Library, London – New York: Bibliotheque des Arts Decoratifs, Paris, France cr; **265** Science Museum: tl; Mary Evans Picture Library: br **266** Corbis: Bettmann tr; Science Photo Library: cl; **266–267** Science Museum: **267** AKG London: Erich Lessing br; Science Museum: tr; **268** Hulton Getty Archive: bc; Science Photo Library: Sheila Terry tl; **268–269** Science Museum: **269** AKG London: tr; Science Museum: crb; Hulton Getty Archive: bc; **270** AKG London: Musee d'Orsay br; The Image Bank–Getty Images: Roine Magnusson bl; Science Photo Library: Dr Gopal Murti tl; **271** AKG London: br; Mary Evans Picture Library: tl, ac; **272** AKG London: tl; Hulton Getty Archive: br; **273** Science Museum: br; Hulton Getty Archive: c; Sean Hunter: tc; Science Photo Library: tl; **274** Mary Evans Picture Library: tl; **274–275** Science Photo Library: Antonia Reeve b; **275** Science Photo Library: Martin Dohrn tr; James King-Holmes br; **276** The Image Bank–Getty Images: Wayne H. Chasan c; **277** Corbis: Lindsay Hebberd tc; **278** Science Photo Library: Simon Fraser tl; **278–279** Telegraph Colour Library–Getty Images: VCL–Paul Viant b; **279** Science Photo Library: CC Studio tc; Telegraph Colour Library–Getty Images: Bill Losh tr; **280** Science Photo Library: Custom Medical Stock Photo br; **282–83** Science Photo Library: CNRI mi, Quest l, cl; J. Croyle/Custom Medical Stockc; **288–289** Science Photo Library: Nancy Kedershami; **294–95** Science Photo Library: Professor P. Motta mi; **290–291** Science Photo Library: Astrid & Hanns-Frieder Michler; **292–293** Science Photo Library: Dr Gopal Murti; **Endpapers** Science Photo Library: Juergen Berger, Max-Planck Institute.
Jacket: The Image Bank/Getty Images: Fer front c; Science Photo Library: back tl, back tcl; CNRI front tl; Mehau Kulyk front tr, back tr; Dr. Yorgos Nikas front tll; Alfred Pasieka front tc; D. Phillips back tcr.

All other images © Dorling Kindersley.
For further information see: www.dkimages.com